THE HANDBOOK OF AVICULTURE

Frank Woolham

BLANDFORD PRESS
POOLE · NEW YORK · SYDNEY

THE HANDBOOK OF AVICULTURE

For Benson

*First published in the U.K. 1987 by Blandford Press,
Link House, West Street, Poole, Dorset, BH15 1LL.*

*Copyright © 1987 Frank Woolham
© 1987 Disease and Medicine chapter A.G. Greenwood*

*Distributed in the United States by
Sterling Publishing Co., Inc.,
2 Park Avenue, New York, N.Y. 10016.*

*Distributed in Australia by
Capricorn Link (Australia) Pty Ltd
PO Box 665, Lane Cove, NSW 2066*

British Library Cataloguing in Publication Data

Woolham, Frank
   Handbook of aviculture.
   1. Birds
   I. Title
   598   QL673

ISBN 0 7137 1428 X

*All rights reserved. No part of this book may be reproduced or
transmitted in any form or by any means, electronic or
mechanical, including photocopying, recording or any
information storage and retrieval system, without permission in
writing from the Publisher.*

*Typeset in 10/12pt Lasercomp Plantin by
Asco Trade Typesetting Ltd., Hong Kong*

Printed in Yugoslavia

CONTENTS

Foreword by Gerald Durrell 7
Introduction: The Aviculturist's Role in
Conservation 9

PART I: GENERAL CARE

Diets 15
Housing 32
Breeding 48
Disease and Medicine (by A.G. Greenwood MA VetMB
MIBiol MRCVS) 53
The Law and the Ethics 62

PART II: THE BIRDS

Rheiformes 67
Rheas

Tinamiformes 69
Tinamous

Ciconiiformes 70
Ibises, Spoonbills, Flamingos

Anseriformes 74
Ducks, Geese, Swans

Falconiformes 100
Falcons

Galliformes 101
Pheasants, Quail, Partridge, Guineafowl

Gruiformes 120
Button Quails, Cranes, Trumpeters, Rails, Sun Bitterns, Seriemas

Charadriiformes 129
Jacanas, Pratincoles, Plovers, Seedsnipe, Terns

Columbiformes 135
Sandgrouse, Doves, Pigeons

Psittaciformes 148
Lories, Lorikeets, Cockatoos, Macaws, Parrots

Cuculiformes 190
Touracos, Cuckoos, Coucals

Strigiformes 193
Barn Owls, Owls

Apodiformes 196
Hummingbirds

Coliiformes 200
Mousebirds

Trogoniformes 202
Trogons

Coraciiformes 204
Kingfishers, Motmots, Rollers, Hoopoes, Hornbills

Piciformes 212
Barbets, Toucans, Woodpeckers

Passeriformes 222
Broadbills, Cotingas, Manakins, Pittas, Larks, Minivets, Bulbuls, Chloropsis, Waxwings, Thrushes, Chats, Parrotbills, Warblers, Flycatchers, Flowerpeckers, Sunbirds, Spiderhunters, Zosterops, Buntings, Tanagers, Blackbirds, Finches, Waxbills, Weavers, Sparrows, Starlings, Orioles, Birds of Paradise, Crows, Jays

Appendix 360
Bibliography 362
Acknowledgements 363
Index of Scientific Names 365
Index of Common Names 367

FOREWORD BY GERALD DURRELL

I was delighted to be asked to write a foreword to this excellent and comprehensive book. It is just the sort of book one desperately needed when starting out as an amateur aviculturist, and just the sort of book one could never find.

Aviculture over the years has earned itself a somewhat shady reputation and for two major reasons. Firstly, there was the so-called aviculturist who was a mere stamp collector, only interested in adding a new species to his cages, never bothering to breed them. The second were the unscrupulous dealers who fed the avicultural trade and for years had no controls over their activities and whose behaviour in their dealings with literally billions of birds would have put any self-respecting ghoul off its food.

I remember once going to the emporium of a Dutch bird dealer where thousands of birds, both well and sick, stood huddled over three inches of excrement on the cage bottoms, which smelt much as Belsen must have done. In one room, I watched fascinated as three nuns argued over which bird they were going to buy, while around their black robes a lout of a boy was sweeping up piles of seed, dead birds and many that were still fluttering in their death throes. If any recent legislation has put a stop to such diabolic trading, then as a conservationist I am delighted, for this sort of thing only gave aviculture and the true aviculturist a bad name.

Aviculture, as practised by responsible aviculturists has everything to commend it and when practised with skill and dedication can be commended and not denigrated by conservationists. But, by and large, conservationists tend to be very narrow-minded and slow to change their minds. Not many years ago, whenever I suggested that captive breeding be used as an aid to the conservation of species, I was looked at as though I had suggested that infanticide was the best method of keeping the human population down. Yet now, captive breeding for threatened species of birds, mammals and reptiles is being used successfully world wide.

If aviculture as a science had been practised properly through the ages, who knows what wonderful bird species we would still have with us—the Passenger Pigeon, the Carolina Parakeet, an array of Moas, the Elephant Bird from Madagascar, the Solitaire and the great Ground Parrots from Rodrigues and Mauritius and one might even have flourishing colonies of that bird that epitomises extinction, the Dodo.

This handbook is well written, surprisingly detailed, informative and concise and will become an invaluable tool for the aviculturist. I hope to see it included in any true aviculturist's library, every zoo library and indeed in every conservationist's library for there are many books that have been written about the art of aviculture but very few good ones. This one is a good one.

GERALD DURRELL, OBE Hon. Director, Jersey Wildlife Preservation Trust, Les Augres Manor, Trinity, Jersey.

INTRODUCTION: THE AVICULTURIST'S ROLE IN CONSERVATION

An increasing number of exotic bird species may be lost forever to aviculture following legislation in countries around the world to prohibit the export of indigenous flora and fauna. Although such measures are regarded with dismay by aviculturists, they are naturally applauded by conservationists who see the supply of birds to amateur hobbyists as an entirely unwarranted drain on dwindling wild populations.

Many protectionists, of course, maintain a touching, but often misplaced, belief that the pious utterances of politicians will result in long-term protection for wildlife and its associated environments. But one has only to consider the track-record of many of those in government, of no matter what race, colour, creed or political persuasion, to realise that wildlife preservation becomes a priority issue only if it looks likely to be a vote-winner.

On those occasions when protection of a species can be, and is, enforced, a further threat may arise through the rapid destruction of its habitat. There is not, after all, a great deal of point in placing, say, an endangered woodpecker on the protected list if you do not also take steps to prevent destruction of the forests that are the bird's sole habitat.

For myself, I have to acknowledge that although an ardent aviculturist for more than 40 years I am not prepared to dismiss all of the arguments put forward by the conservation lobby which are critical of aviculture. Let us be honest: so far as many exotic species are concerned, ready availability of wild-caught birds has helped create a situation where little or no attempt has been made to establish a reservoir of cage and aviary-bred stock which might, by now, have helped reduce the scale of imports.

With the honourable exception of people whose main interests lie with waterfowl, pheasants, various grassfinches and parrot finches, certain parrakeets, lovebirds, quail and some individual species such as Java Sparrows, Cockatiels and Diamond Doves, most of the hobbyists who keep other kinds of birds from overseas have been far too complacent about the future. Small finches have been shipped out of Africa and many parts of the Far East in their tens of millions during the past 50 or 60 years. Exports of the more popular softbilled species from around the world might be counted in tens, possibly hundreds, of thousands.

Aviculturists have mainly failed to make use of the riches that were once available to them. No more than a handful of concerned and dedicated people have applied themselves to the task of persuading a few species of foreign birds to reproduce fairly consistently in controlled conditions. In some instances, though, species have now been bred to several generations—the sort of effort that is vital if some semblance of what we are still able to enjoy is to be passed on into the twenty-first century.

Now there is increasing recognition of what needs to be done. For aviculturists it is likely to be essentially a matter of self-help. Nobody is going to come to our aid; in fact the reverse will probably apply as some of the hard-liners in various protectionist organisations bring pressure to bear to put an end to aviculture as we know it. Yet, there should be a crucial role for aviculturists to play towards the conservation of many species of birds which, for various reasons, are going to be more and more at risk in the wild.

Habitat loss will continue to be the greatest threat to wildlife around the world. Despite the magnificent work of front-line conservationists (those actively involved in field work as opposed to the armchair-based brigade) the medium to long-term future for several species looks bleak because of this problem. It is all very well to sit back and pontificate about the rapid destruction of the world's rainforests and other irreplaceable parts of our global environment just so long as we recognise the part many of us play towards this sort of carnage.

Concessions for activities ranging from logging to cattle ranching seem to have been tossed around in some countries like confetti at a wedding. In a recent conversation with a South American-based European I was told of a project, already under way, to clear an area of forest covering tens of thousands of acres. The aim was to open it up for cattle ranching. For the indigenous human population there was a prospect of some short-term financial gain whilst the cattle-raising scheme was able to prosper. That would be only for a very limited period, however, before the impoverished land ceased to be capable of sustaining even grass.

'What happens to the wildlife when the forest is bulldozed?' I asked.

The locals, I was informed, are a practical lot; so most of what they could catch would be eaten. Certain of the younger animals and birds would be kept alive, ostensibly as pets, in reality being fattened for the pot if they were of an edible kind.

Nevertheless, to be indignant about this wholesale loss of both habitat and wildlife may be somewhat hypocritical. For the entire project is being undertaken to contribute cattle carcases to the mushrooming 'fast-food' industry in more 'civilised' countries. In other words the forests and their inhabitants have been displaced to make way for beefburgers ... on the hoof.

What, you may ask, has all of this to do with aviculture? Well, if present trends continue and the world's remaining wild places continue to be torn apart by such irresponsible but commercially powerful and overtly concerned interests, quite a lot, I suspect. But first of all aviculture must put its own house in order. For only then is an increasingly vociferous and influential protectionist movement likely to consider *any* kind of role for people they presently view with suspicion. That there are still conflicts between conservationists and aviculturists is sad; and it is a situation that *must* be changed. For it is a fact that responsible and experienced birdkeepers have a great deal to offer towards the preservation of endangered and vulnerable species; many of which will almost certainly have to be the subject of planned captive-breeding programmes at some future date if they are to survive.

Aviculturists, more than most, have the experience, knowledge and, I am certain, despite some of the shortcomings of the past, the will to work towards the establishment of aviary-bred stocks of many exotic birds. It is the sort of commitment that could elevate what is at present no more than an absorbing hobby into a vital link in a concerted effort towards conservation.

The future status of many small (mainly, but not exclusively, passerine) species is unlikely to get much of a boost through the efforts of zoos and the bigger public bird gardens. Aside from the fact that few such establishments are geared for anything other than general mixed exhibits of small birds, they are likely to have their hands full with larger species also facing the prospect of oblivion without a little help from their human friends.

One facet of bird-keeping that makes it difficult for conservationists to take seriously the idea of co-operation between the two factions is exhibiting. Although the idea will be regarded as heresy by most die-hard showmen, I believe it is in the best long-term interests of aviculture to eliminate *all* wild birds from the showbench. Such a move would still permit competition between birds bred in cage or aviary, but it would effectively eliminate most of the rare, and frequently more delicate, imports which, in many instances, are too valuable to expose to the stresses imposed by regular showing.

Exhibitors, I know, will counter such suggestions with the old chestnut that shows are 'the shop window of the fancy'. I doubt that this is the case at all. In fact many non-bird-keepers I have spoken to over the years regard cage bird exhibitions with some suspicion. Some of them even believe small show cages represent the exhibits' permanent accommodation despite the efforts

Introduction

made by many show organisers to emphasise that this is not the case. Far from being a good advertisement for aviculture in general, shows may create doubts in the minds of an uncommitted public. They certainly provoke a hostile reaction from the bird-protection lobby when European and certain exotic species are involved.

Another potentially contentious area is the bird trade itself, or rather the less reputable elements that exist in it. Fortunately unscrupulous operators at all levels (import/wholesale and retail) seem well on the way to extinction themselves. Nevertheless, those who make their living from buying and selling birds, and more especially if they are concerned with any that might be regarded as rare or endangered, must not only maintain the highest standards themselves but also remain vigilant to the activities of others who may, from time to time, join their ranks but resort to less desirable codes of practice in the pursuit of profit.

Nonetheless, there is still plenty of cause for concern about unacceptably high levels of mortality in many quarantine stations where the profit motive clearly overrides commonsense attitudes to proper care of imported birds.

Although at the present time many countries around the world have closed their doors to exports of their native birds, there remains the possibility, no matter how remote it may appear, that circumstances will change and that some of the species presently unavailable will find their way (legitimately) back into aviculture.

In such an event we must be sure we do not repeat the wastages of yesteryear.

F.W. 1987

PART I
GENERAL CARE

DIETS

Diets contribute enormously to the successful care and management of any collection of birds; but more so if the species concerned have diverse tastes which are not satisfied with a basic granivorous menu. Exotic rarities, housed in near-Utopian accommodation and with time lavished on their management, will not do well unless they are properly fed. Conversely, a person with limited space, and a shallow pocket, often achieves spectacular success because he or she has taken the time and trouble to identify particular needs of birds with specialised feeding habits.

A commonsense approach to avian husbandry coupled with some knowledge of the nutritional requirements of groups, families and, in some cases, individual species, will invariably pay handsome dividends. Most of the birds likely to be of interest to amateur aviculturists can be placed into different groups with *basically* similar nutritional requirements.

Thus finches, parrots, pheasants, quail and waterfowl are, with one or two notable exceptions, essentially grain-eaters. Flycatchers, pittas and the like are insect-eaters; but they adapt to eat inanimate insectivorous mixtures when housed in aviaries.

Tanagers are one of a number of groups that feed largely (but not entirely) on soft fruit of various kinds. These are among the easiest non-seedeating species to switch to substitute diets under controlled conditions.

Hummingbirds, sunbirds, honeycreepers and others rely for much of their food intake on nectar which they collect from a variety of flowers. Nowadays modern artificial nectar mixtures provide a means of keeping these tiny feathered gems without difficulty.

Several species, for example jays and magpies, are omnivorous. They need varied diets which include not only fruit and a portion of insectivorous mixture but, in some cases, raw meat, and even the occasional dead mouse or sparrow.

Although it is possible to pigeon-hole many families under these various dietary headings, individual species within many families are often highly specialised in their habits and selective in their choice of food.

Recommended *basic* diets are best regarded simply as just that. Trial and error will ultimately help create a sound and suitable diet for most birds; but in experimenting or making changes do prepare yourself by doing as much homework as possible. Never subject your birds to dramatic changes of diet. Do not try to persuade them to eat items which are clearly unsuitable.

That said, it also needs to be emphasised that the task of feeding even a modest-sized, mixed collection of birds can be greatly simplified by the pre-preparation of many mixtures which can form the main part of diets: various items can be added if need be, according to the individual species or families you are catering for.

Finches and Seedeaters

In terms of basic diets finches and other small seedeating species are among the simplest to cater for. Most thrive on a mixture of seeds. Green food is an obvious necessary addition for many of them and certain species need regular supplies of small live food. There are a number of excellent commercially-blended seed mixtures on the market. They are usually made up of plain canary seed and various kinds of millet. Most provide a good nutritional intake for a wide spectrum of seedeating species.

Few aviculturists bother to make up their own seed mixtures. Unless the collection is very extensive, there is little point, for home-blends rarely produce any cost-saving and are a bit of a chore. On the other hand certain species tend to be very choosy about individual seeds in a mixture and a certain amount of wastage occurs through rejection. It is virtually impossible to eliminate *all* waste; birds are as individualistic in their tastes as human beings. For those who want to prepare mixtures which will go some way to meeting the needs of *some* large and *some* small seedeaters, here are a couple of blends that have served me well over a number of years.

Diet A

This is suitable for most medium to large seedeaters (say mannikin up to Java Sparrow size).

3 parts plain canary seed
3 parts yellow millet
2 parts white millet
2 parts panicum millet

Diet B

This is more suitable for smaller species (waxbills, Bronze Mannikins and the like).

3 parts plain canary seed
3 parts yellow millet
1 part white millet
3 parts panicum millet

To each of these mixtures you can usefully add a small quantity of niger: about a cupful to 10 lb (4.5 kg) of seed mixture. Niger has a very high fat content and is best not used if birds are closely confined in cages. For stock able to take plenty of exercise it can be a valuable additive, especially during winter. Some aviculturists add hemp to a millet/canary mixture; usually if it is to be offered to hawfinches, grosbeaks and the like. It is another seed with a substantial fat content (not as high as niger) and should be used with caution.

Other seeds which can be used, sparingly, in the diets of many foreign seedeaters are maw and linseed. Both are rich in minerals but the latter also has a high fat content. The need to increase input of amino-acids and certain vitamins known to be deficient in basic plain canary and millet mixtures is a factor to be borne in mind in assessing the value of seeds such as niger, maw and linseed. Sprouted seeds are also valuable and are often taken in preference to dry seed as soon as birds are accustomed to them. Plain canary seed and millets should be soaked until germination commences (usually between 48 and 72 hours depending on temperature). Seeds should be rinsed each day whilst soaking, and thoroughly rinsed before being fed to birds.

The heads of various grasses are produced in great profusion in summer and autumn. Such fresh, ripe seeds are an invaluable source of food and should be gathered and used as often as possible. Bundles of them can be tied together and suspended from the aviary top. Take care when gathering these and other natural foods that they have not been sprayed with herbicides or other toxic chemicals. Items collected from the verges of busy carriageways (even in rural areas) may also be suspect if they have been subjected to heavy pollution by petrol and diesel fumes.

Most European and North American seedeaters need a staple seed diet of entirely different composition to that offered to birds from tropical countries. Many seed merchants offer excellent mixtures with a content varied enough to appeal to Fringillid finches and members of the Emberizidae.

Diet C

This is a self-blend mixture suitable for such birds.

2 parts plain canary seed
2 parts rape (equal parts black/red)
3 parts wild seeds (or screenings)
1 part teazle
1 part niger
½ part hemp (cracked for small species)
½ part linseed

There are a number of possible variations of this basic mixture. For example, hawfinches and grosbeaks, housed in aviaries, could have more hemp together with some sunflower seed. Goldfinches, siskins and similar small species would be unlikely to appreciate these two seeds, but would welcome more teazle and the addition of some 'gold of pleasure'.

Millet sprays are greatly enjoyed by practically all exotic seedeaters. They can be offered either dry or soaked: many birds prefer them in the latter state. A bunch of sprays suspended in the aviary will keep its occupants, especially waxbills, occupied for days.

Remember that newly-imported foreign seedeaters are rarely fed a balanced seed diet. Panicum millet is a staple food used by many shippers and dealers and it may take some little time to wean such birds onto more suitable fare. Pin-tailed nonpareils imported from the Far East are invariably fed on paddy rice as a staple diet. It is important not to switch them to a mixed seed mixture immediately. They should be gradually accustomed to change and it may be useful to offer them hemp, niger and other seeds of the kind recommended for Fringillid finches during the changeover.

If you want to know precisely which seeds your birds are eating it is wise to offer the various component items of an appropriate diet in separate dishes. It is then a simple matter to work out which are ignored.

Quail can be loosely divided into two groups for the purpose of sorting out suitable diets for them. The first can include all the smaller species such as Chinese Painted Quail, Jungle Bush Quail, Harlequin Quail and so on. Larger species such as Californian, Scaled and Bobwhite all fit conveniently into a second category.

Most of the smaller quails will do very well on Diet A with the addition of 3 parts of chick starter crumbs. During non-breeding months I prefer to reduce the chick crumb content to one part only, although other people offer more without problem.

For the bigger quail species the following mixture has proved successful.

Diet D

For Californian, Scaled, Bobwhite Quail (also Seedsnipe).

 4 parts plain canary seed/mixed millets (50/50)
 4 parts chick starter crumbs
 3 parts wheat
 3 parts split maize
 1 part groats

Groats and buckwheat are also useful for this type and size of quail.

Green Food

Green food is an important item of diet for many birds—finches, quail, pheasants and waterfowl, for example. A surprising variety can be gathered from patches of wasteland, hedgerows, and the edges of fields.

Inexperienced gatherers of wild green food may find it valuable to invest in one of the many excellent, and generally inexpensive, field guides which will help identify the huge variety of wild plants that abound throughout the countryside. Waste ground may prove to be a safer collecting place for wild plants than cultivated areas. The main problem with the latter is that diligent farmers, growers or even enthusiastic amateur gardeners may well have been using toxic chemicals to improve the yield of a particular piece of ground. Wasteland invariably carries a good crop of 'weeds' which have not been contaminated.

Cultivated green food can also be used, especially in winter when supplies of natural food may no longer be available. The most useful items are probably cress, spinach leaves and lettuce. Little else is of real value to the aviculturist, for birds seem uninterested in the green leaves or tops of other cultivated vegetables.

Cress is probably the most valuable of these foods. It is appreciated by practically all grain or seedeating birds, costs little and is readily available throughout the year. Spinach leaves are another valuable food. Lettuce is also useful but I prefer to use it only occasionally and sparingly. Given in any quantity or each day over a lengthy period it leads to digestive upsets.

While dealing with foods for seedeating species it is appropriate to emphasise the importance of supplying grit at all times. Various types are available: mineralised, oystershell and limestone, for example. Always provide grit in a separate vessel rather than scattering it over the cage or aviary floor.

It is sound practice to have mineralised grit available throughout the year and to supply limestone/oystershell before and during the breeding season. Although the former is regarded mainly as a valuable digestive aid the other two provide essential additional calcium for egg-laying females. Another useful additive for seedeating birds is cuttlefish bone, which is another source of calcium and is valuable for helping to keep mandibles trim.

One of the many multi-vitamin preparations on the market should also be used regularly. Various brands are available in powder form (for dusting onto food) or as a water-soluble liquid. They supply many vital trace elements and vitamins missing from even the most carefully formulated diet.

Live food is essential for many seedeaters. Some such as the twinspots, Violet-eared and Purple Grenadier Waxbills and others need it as a daily part of their diet. Other species become almost exclusively insectivorous when nesting. The various types of live food available to aviculturists are discussed later in this chapter since many of them are valuable for insectivorous, carnivorous and omnivorous birds.

Softbills

Feeding a mixed collection of 'softbilled' birds (fruit, insect and nectar feeding species) is a good deal more complicated than dealing with most other groups. Dietary needs are varied, and things are made more difficult by the wide range of opinions that exists regarding suitable diets.

Fruit is one of the most important items. Frugivorous and omnivorous birds including tanagers, barbets, fruitsuckers, bulbuls, starlings, mousebirds, touracos and toucans need it in varying proportions in their daily diets. Fortunately most of them can be given a mixed fruit diet of similar composition.

Opinions differ as to whether it is best to provide fruit already diced into small cubes or in bigger pieces (say halves) so the birds can break it up themselves. My own experiences over a number of years suggest there is merit in both methods, depending entirely on the species one is dealing with. Many barbets, for example, have bills capable of rending most fruits asunder in seconds; but a Toco Toucan, despite its huge mandibles, would find such a method beyond it.

There is no real rule of thumb. Much depends on individual species and their methods of feeding in the wild; but in broad terms I have tended mainly towards supplying most of my fruit-eaters with very small cubes of things like apples and pears. Grapes and tomatoes I offer halved or sliced, and on the rare occasions I provide orange that too is halved. Many birds are extremely efficient at cleaning out a halved fruit. Zosterops are one species that can remove every vestige of flesh from a pear, orange or tomato, leaving just a thin skin with no waste.

Chopping up a quantity of fruit for a large collection of fruit-eating softbills is unquestionably an onerous and time-consuming task. It is worth the effort in that this method of feeding leads to better visual presentation of diets, and in itself will often persuade a fussy eater to take a renewed interest in the food dish. One other advantage of the dicing method is that some birds lack good eating manners. They can cover themselves in juice when attacking a piece of fruit too big to be swallowed. If they do not then take an immediate bath the effect on their plumage can be disastrous.

Never be tempted to purchase fruit, albeit at a lower price, that has passed its best. Bruised fruit is one thing, because damaged areas can be cut out; but over-ripe fruit of any description is an extremely dangerous, potentially lethal, item of diet for birds. Apart from the fact that damaged or bruised fruit is acceptable, it should otherwise be in perfectly ripe condition for eating. If not, allow it to reach that condition before feeding it to your stock. Under-ripe fruit can also cause problems if fed to delicate softbills.

Among the most frequently used items of fruit are apples, pears, grapes, oranges and tomatoes. Many birds prefer apple varieties with soft, rather than crisp, flesh. Pears should be just ripe when fed. Grapes are mainly useful to persuade diffident feeders to explore the food dish. I use oranges with great circumspection, which means that only the very sweetest are offered to my birds and even these only as an occasional treat.

Tomatoes, on the other hand, are fed daily. Like oranges they are a source of vitamin C, but in my experience less likely to cause digestive upsets than the former which have a not insignificant acid content. Bananas are not safe to feed at all. This particular fruit has probably accounted for the unexplained deaths of more delicate softbills than any other regularly offered item of diet.

There are, of course, other valuable kinds of fruit; many of them more seasonal in their appearance on the market and, at times, expensive. Mango, paw-paw and other exotica can all find their way onto menus for fruit-eating birds, however. Naturally the same criteria should apply as with more readily available items: everything should be ready for eating and any damage confined to bruising or knocks.

Wild berries are much enjoyed by many species and it is well worth the trouble gathering them during the appropriate season. Brambles and elder are common. Various other kinds can be used, but it is wise not to collect known poisonous types—even though wild birds may feed on them.

Many berries are ready for picking in late summer and autumn for a few short weeks, but there is no reason why these valuable items of diet should not be available over a much longer period; right through the year, in fact. For the diligent softbill-keeper will not pass up the opportunity to gather whatever is going in garden or countryside, and then make use of a deep freeze so the harvest can be preserved for use over several months.

Cultivated berries are yet another useful food for many birds. Raspberries, blackcurrants and redcurrants are three kinds that are widely grown. They can be gathered fresh, and then frozen. Tinned berries in syrup can also be purchased from grocers.

Dried fruits such as currants, raisins and sultanas are often used to feed softbills. They need to be soaked in water for a few hours before use. Occasionally one hears of them being used as a near-staple diet for certain species, waxwings among them. They are high in carbohydrates and need to be used with the utmost caution as a food for all species. Waxwings, notoriously lethargic and invariably loathe to take regular exercise, are among the very last birds to be offered soaked raisins ad lib.

Do remember also that fruit alone will not provide a balanced diet for many species. A percentage of their daily intake should consist of a good quality insectile mixture. Some species will also need live food; others may have to have a daily ration of raw meat. So far as the fruit content of their food is concerned, pears and apples usually provide the greatest part of a suitable mixture.

Diet E

This is a basic mix for species being fed diced or chopped fruit.

4 parts diced pear
4 parts diced apple
3 parts sliced tomato
1 part sliced grape

The proportions of tomato and grape used are not critical but should not represent more than about one-third of total bulk.

A smaller but important point is that you must remember that the object of feeding diced or cubed fruit is to provide pieces big enough for birds to pick up and swallow. It follows, therefore, that there will be a difference in size between fruit prepared for small tanagers and that for, say, a Toco Toucan. Halved portions of fruit can be offered in suitable-sized receptacles or impaled on a conveniently-placed nail. Remember that the latter may represent a hazard for birds if left exposed, especially after dark when a night-fright might cause panic.

Birds with entirely different food requirements are those softbills which, in the wild, consume a diet of live insects. Included are such species as pittas, minivets, bee-eaters, tits, woodpeckers and many more. Most of them quickly become accustomed to a mainly inanimate diet, although it is usual to provide some living items of food as well. Some species are more difficult than others in this respect. Bee-eaters and drongos are among those that insist on a high proportion of live insects (even flying insects) in their daily diets.

Minivets and pittas, although less difficult, often prove troublesome when first imported. Some of the flycatchers will resist most attempts to wean them onto substitute foods, but eventually succumb to a dish of insectile mixture with wriggling larvae of some kind giving it the appearance of life.

Nowadays, of course, quarantine regulations mean that most of these problems are faced by importers rather than aviculturists buying from a retail source. By the time insectivorous softbills have undergone a statutory period of isolation they will *usually* have been weaned to suitable diets. There are still occasions when an importer, in a desperate effort to maintain an imported rarity in prime condition during quarantine, will feed an unsuitable staple diet of live insects.

This sort of treatment is likely to succeed only in the short term for a majority of species. Substitute live food is not likely to be more varied than either mealworms or gentles (maggots). So the purchaser may be faced with the task of effecting a switch to a more healthy, balanced diet. Patience is an essential attribute during this frequently difficult process which the old-time bird-catchers referred to as 'meating-off'.

Basic insectivorous mixtures are made up of a variety of items including biscuit meal, meat and fish meal, ground shrimp, soya flour etc. An excellent range of prepared foods is on the market and they provide good basic diets for most insectivorous birds and are certainly recommended for general use. The majority of zoos, bird gardens and the bigger private collections use these prepared diets in preference to making up their own mixtures fresh each day, and it is suggested that the hobbyist follows a similar course.

It would be difficult to improve on the best of these commercial insectile mixtures, and the task of preparing one's own food from scratch is messy and time-consuming. On the other hand there are many items that can usefully be added to a proprietary mix; the additional ingredients depending to a great extent on the type of birds being fed.

Two suggested mixtures are given below. They can be made up fresh each day for a big collection of birds. Alternatively, enough can be prepared to feed a smaller number of birds over a period of about a week and stored in a refrigerator.

Most softbill enthusiasts have their own ideas, often very firmly held, about what additives they will include in their mixtures. I have found the following to be among the most useful: cottage cheese, dried grated cheese, egg yolk (hard boiled), fresh or frozen prawns/shrimps, carrot (grated), apple (grated). Carrot and apple are finely grated and mixed with the insectile food. Cottage cheese can also be added. Scraps of waste cheese from the kitchen should be put aside and allowed to dry (reducing the high fat content) before being grated for the mixture. Egg yolks are pounded before being added. Shrimps and prawns are put through the sort of shredder used for herbs. Whenever possible I use whole unpeeled prawns rather than shelled prawns or shrimps.

Other useful additives include steamed fish roe and steamed beef heart. Both can be used in diets for some highly insectivorous species such as flycatchers, minivets etc. Both items should be thoroughly cooked and allowed to cool. It is vital that all fat is drained from the heart before use. Grated heart is a useful item of food for many delicate species and it is worth taking a little trouble to obtain the best possible since it is likely to be fed mainly to rare and valuable birds. It is often better to obtain hearts from small privately-owned abattoirs (of the kind that deal to a great extent with casualty livestock) rather than the local butcher or a large commercial slaughterhouse. The reason is simply that stock handled by the smaller abattoirs is often much less subject to stress than animals being killed in the bigger, more intensive establishments. It is now recognised that the heart of an animal killed whilst experiencing great stress will be literally drenched with adrenalin; and may prove harmful as a regular diet item for small birds.

Both roe and heart, when cool, should be sectioned into small pieces, wrapped in foil or film, and stored in a deep freeze. Both items grate more easily when used straight from the freezer, and the tiny particles are thawed almost instantly during shredding.

Diet F

This is for flycatchers, redstarts and other small insectivorous species.

1 kg proprietary insectivorous food
5 g yolk of hard-boiled egg

5 g cottage cheese or grated dry cheese
10 g shredded whole prawns (or 15 g shelled prawns/shrimps)
5 g grated beef heart
5 g grated carrot

Diet G
For thrushes, Pekin Robins and other medium to large species.
1 kg proprietary insectivorous food
5 g yolk of hard-boiled egg
5 g grated dry cheese
5 g grated beef heart
5 g grated fish roe
5 g grated carrot

Some species of softbills will remain in good health when fed on basic mixtures without additives. Most of these birds, however, are likely to be frugivorous or omnivorous and their total diet will be balanced with items ranging from fruit to meat. The latter is an important item for many softbills including some of the starlings, jays and other corvids, motmots, kingfishers and shrikes. Fresh minced raw beef is usually offered and it is best to mix the appropriate amount with a quantity of insectile food. This will help prevent the birds taking just the meat and ignoring other nutritious items.

Regular use of a multi-vitamin preparation in softbill diets is strongly recommended. One of the powder-type can be dusted onto chopped fruit and meat or mixed with insectile food.

Parrotlike Species

Parrotlike species are mainly seedeating with green food and fruit the main additions to a staple diet. Some are more specialised in their feeding habits and need artificial nectar as a vital part of their daily intake of food. These include lories and lorikeets, fig parrots and hanging parrots. Sunflower seed, peanuts, hemp and pine nuts form the basis of most parrot diets. For the smaller species (lovebirds and most of the smaller parrakeets) canary seed and white millet are also used.

Many parrots are highly selective in their choice of seeds to form a staple diet. So although there are several ready-blended parrot mixtures on the market there may be a case not only for buying-in individual ingredients but also offering them to the birds in separate containers. The advantage of such an arrangement is that it is immediately possible to see what is eaten and what is ignored; so wastage can be largely eliminated. The disadvantage is that, given ad-lib quantities of black, white and striped sunflower seed, pine nuts and peanuts, it is possible to predict with reasonable accuracy what will be eaten and what will be left; and the birds' choice is unlikely to make economic sense or provide a correct balance of seeds. Pine nuts are likely to disappear first and at the other end of the popularity scale black sunflower seed may be totally ignored.

It is really up to the owner to try and impose a suitable system whereby some kind of variety is being consumed. It is probably expecting too much for a majority of parrotlike birds to relish black sunflower seeds; by and large they seem willing to go hungry rather than eat this particular type of seed.

The smaller species of seedeating parrots pose less of a problem. Plain canary seed and white millet are the two main components of diet, together with some of the seeds offered to their bigger relatives.

Diet H
This diet, which is for larger parrot species, is not a recommended mixture but individual items (in the likely order of preference) which should go to make up a diet.
Pine nuts
White/striped sunflower
Peanuts
Hemp

It is also worth trying groats, oats, safflower, plain canary and maize.

If the parrot enthusiast lives in or near the countryside it is worth trying to obtain some fresh stems of wheat or oats just before harvest time. Parrotlike species enjoy these as much (or even more) than finches and waxbills appreciate bunches of seeding grasses.

Diet J

This diet is recommended for lovebirds, grass parrakeets and similar sized birds.

3 parts plain canary seed
3 parts white millet
1 part striped sunflower
1 part hemp
1 part white sunflower

In the wild parrots enjoy a varied diet of fresh seeds, nuts, buds, shoots and fruit. Sadly it is the lot of many pets to have to endure a never-ending menu of dried peanuts and sunflower seeds as it comes out of the packet.

Some of them are also given welcome variety in the form of fruit, vegetables and other green food. Much less sensible are tit-bits of household scraps that many owners mistakenly believe are of benefit to their pets. Chocolate, sweets and cake (even bacon and egg) are among 'treats' given to some unfortunate birds. Far from being a bonus all such items are likely to do is shorten the bird's life.

Many of these birds appreciate apple and other cultivated fruits, greenstuff (spinach is a favourite) and berries. Sprouted seeds are greatly enjoyed and many of the kinds available from health food shops can be easily sprouted and will be relished by most parrotlike species.

Artificial nectar is an important part of the diet of lories, lorikeets and, to a lesser extent, hanging parrots. All of these attractive and interesting birds need fruit as a regular part of their daily diets. Boiled rice is eaten by some; others appreciate a seed mixture to supplement the nectar and fruit. A number of proprietary nectar mixtures are currently on the market and the best of them can be used for birds ranging from honeycreepers to lories.

The make-up of home-made nectar varies considerably, but white sugar, honey or glucose provide ingredients common to all. Baby cereals, invalid food, high protein foods (such as Gevral or Hydramin), malt extract, yeast and condensed milk all take their place in individual mixtures. Whatever formula the aviculturist eventually settles on for his birds, if they thrive on it, you can be sure it will quickly take its place alongside hundreds of other secret, and allegedly superior, recipes that exist around the world. The following (not claimed to be the best!) gives good results with lories, lorikeets and hanging parrots.

Diet K

113 g honey
113 g condensed milk
2 tablespoons Complan
3 tablespoons Horlicks
5 drops multi-vitamins
3.5 litres water

Honey, condensed milk and Horlicks mix better in a small amount of hot water but Complan is better in cold water; then combine all ingredients and make up to 6 pints (3.4 litres) with warm water. Remember always the importance of keeping vessels used either to mix or to feed liquid mixtures in a scrupulously clean condition at all times. The use of one of the preparations recommended for cleaning utensils for babies is advised. It is also vital to ensure such mixtures are not allowed to go 'off' during hot weather. Many aviculturists prefer to replace milk mixtures with plain honey or sugar and water from late afternoon until the following morning.

Nectar mixtures also form the main part of diets for such birds as honeycreepers, sunbirds and hummingbirds. The first-named also need a variety of fruits together with some small live food. Many sunbirds feed more or less exclusively on nectar and small insects. Hummingbirds, too, thrive on this unvarying regime, but some authorities believe milk-based mixtures may be harmful.

Diet L

This is an artificial nectar mixture suitable for sunbirds, hummingbirds and honeycreepers.

50 g white sugar
1 × 5 ml teaspoon torula yeast
1 × 5 ml teaspoon pure pollen granules
2 × 5 ml teaspoons Minamino compound
600 ml water

The pollen granules should be mixed with a small quantity of water and left for 12–24 hours before being used in the main mixture.

Various other mixtures are in vogue, especially

for feeding to hummingbirds. All have their devotees and most are successful. Whereas hummingbirds were once regarded as difficult in the extreme, they now live for years and many species are breeding in aviaries.

Nectar should be renewed frequently and hummingbirds especially must not be allowed to be without food for more than short periods while bottles are replaced. Hummingbird feeding bottles should have their spouts coloured red, to attract the birds to their feeding stations.

Two other preparations widely used in the composition of diets for hummingbirds are Gevral Protein and Super Hydramin. Both are manufactured in the USA and are also readily available in Europe. They are high in protein (60–65 per cent). With a fat content of less than 2 per cent they are clearly of considerable value in a sugar-based liquid mixture.

Fruit flies or *Drosophila* are an essential part of diets for these birds. Some species are more insectivorous than others but all will benefit from these tiny flies which provide valuable additional protein. They are easily cultured, usually in jars containing decaying soft fruit.

Pheasants and Related Species

Suitable foods for pheasants are given in Diet M and many of the recommended items can also be used to feed guineafowl, turkeys, curassows and guans.

Diet M
For pheasants, guineafowl and turkeys.
  Grain (including wheat, maize, corn and barley), usually fed late in day
  Game bird pellets
  Greenstuff (spinach, cabbage, diced carrots; also dandelion)
  Leaves (chickweed, comfrey)
  Berries as available

Do not overfeed: fat birds do not breed well.

Waterfowl

Wheat is probably the most widely used item of food for waterfowl. Modern pellet foods, however, developed for the poultry industry, also play an important part in the diets of ducks, geese and swans. Other grain such as maize and barley is used but wheat is first choice. Its main use is as a winter feed. Early in the year (much depends on the prevailing climatic conditions) it is desirable to start to reduce the amount of wheat and introduce poultry breeder's pellets to the diet.

By the time waterfowl are ready to lay they should be getting a 100 per cent pellet diet. Remember that pellet foods rapidly disintegrate in water; so they should be fed on dry ground or in covered hoppers.

Sea ducks and stifftails need different treatment. The former can be given a pellet diet throughout the year, but it should include some higher protein items such as trout pellets or flamingo concentrate. Stifftails appreciate plenty of floating aquatic plants and also enjoy small grain such as plain canary seed and mixed millets.

Geese and swans can be given stale wholemeal bread in addition to their staple food; and remember that for many of these birds a 'staple' diet means a good deal of grass and other plants taken as they graze. A large area for foraging in this manner is essential for many of these birds and if they cannot be provided with a good area of grass they are unlikely to do well. Lettuce and other cultivated green food can be provided as a substitute food from time to time. It is not recommended as a long-term alternative to natural grazing.

Many waterfowl of different species can be housed together in large communities. Do ensure that food, especially the last feed of the day in winter which provides 'fuel' to take the birds through a cold night, is well scattered so that the strongest birds are not feeding to the exclusion of smaller or weaker companions.

It is essential that the pellets used are those specifically for breeders. Other kinds are formulated for young stock (growers) and layers. Neither will have the desired effect of helping the waterfowl lay fertile eggs at the right time.

Diet N
For most species of waterfowl.

Grain (preferably wheat—but maize, barley and corn for variety)
Poultry breeder pellets
Grass and green food
Stale wholemeal bread
Duckweed and aquatic plants

Doves and Pigeons

With the exception of fruit pigeons, most doves and pigeons are very easy to feed in aviaries. The smaller species such as Diamond, Zebra and Cape Doves do well on a mixture of small seeds similar to Diet A. For the larger species with non-specialised feeding habits the following mixture will prove suitable.

Diet O

For Barbary and other medium to large doves and pigeons.
   3 parts plain canary seed
   2 parts yellow millet
   2 parts white millet
   1 part split maize
   1 part buckwheat
   ½ part hemp
   ½ part groats

For Nicobar, Bronze-winged and other large species, wheat and other grain should replace smaller millets. Spinach, lettuce, cress and other green food should be offered to all these birds. Many enjoy small amounts of wholemeal bread. In addition to multi-vitamin preparations, salt blocks and mineral mixtures are important for the long-term health of these birds. The latter are widely sold for racing pigeons and are a valuable addition to the diets of exotic species.

The colourful fruit pigeons take no seed or grain. Given free choice they will consume fruit to the exclusion of everything else but will be healthier for a more varied diet. Some species enjoy mealworms and other small live food; but fruit is the main item of diet for these birds. Efforts should be made to persuade them to take some insectile mixture. Diet F without the addition of beef heart can be sprinkled over cubed fruit.

Diet P

For fruit pigeons.
   4 parts diced pear
   2 parts diced apple
   1 part sliced tomato
   1 part sliced grape
   1 part insectile mixture

As well as this fruit mixture Diet P should include a mixture of boiled rice with some soaked trout pellets and chopped egg added.
   30 g boiled rice
   10 g trout pellets
   1 hard-boiled egg (chopped)

Some aviculturists use food items ranging from soaked chick crumbs to mashed banana for these birds. Often they feed avidly on these and equally unlikely things. Care must be exercised, however, in adding foods high in fats or carbohydrates. Fruit pigeons are generally lethargic and unwilling to take strenuous exercise. An excess of the wrong foods will lead to obesity and death.

Cranes

Perhaps because aviculturists have discovered that cranes are not especially difficult to feed or manage, they are increasing in popularity, although surplus birds are rarely on the market. They need a varied, but fairly straightforward, staple diet. If properly fed and housed several species are long-lived and breed well in a controlled environment.

Some need a reasonable amount of animal matter each day (dead mice or small amphibians such as frogs). Others will get by with a few locusts, snails or even earthworms.

Diet R recommends the main components of a suitable diet for these elegant birds.

Diet R

For cranes.
   grain (wheat, maize)
   trout (or other high protein) pellets
   game bird pellets
   chopped or minced beef
   locusts, small mice etc

A small quantity of meat (around 30 g) can be offered to the bigger species each day; smaller species (Demoiselle and Crowned) should have locusts and earthworms. Some species, such as Wattled, may take small strips of raw fish as well as an occasional mouse.

Flamingos

Flamingos are also found in an increasing number of private collections as aviculturists discover the relative ease with which they, too, can be kept, provided that accommodation to suit their rather specialised needs is available. Although a few zoos and bird gardens continue to make up their own flamingo foods on a daily basis, there is increasing use of specially formulated complete diets, in pellet form, marketed by a number of companies. They contain everything necessary to maintain these slender water-lovers in excellent health and have the advantage of helping retain the beautiful pink or red colouring of some species.

Use of these modern diets is recommended. Making up a home-prepared menu on a daily basis will prove costly and extremely time-consuming, for such concoctions need ingredients ranging from cooked rice to ground shrimp, as well as colour-enhancing agents.

If flamingos have access to an area of natural water they will supplement their diet with small molluscs and crustaceans; during the summer months some algae growths are also taken.

Ibis, Spoonbills and Related Species

Ibis, spoonbills and related species generally do well, given spacious quarters and a good diet. They are essentially meat and fish eaters, so their aviaries, if not kept scrupulously clean, quickly begin to smell like a neglected fish market. In a large, natural setting these birds will probably capture a good deal of natural food: frogs, newts, earthworms, water beetles and the like have great appeal. Small mice are also consumed.

Diet S
For ibis, spoonbill and related species.
  150 g minced or chopped raw beef
  50 g raw fish (feed in small strips)
  20 g whole prawns—roughly chopped

Rheas

Rheas need a varied diet; but a mainstay can be a combined mix of equal parts of ratite and game bird pellets.

Diet T
For rheas.
  100 g ratite pellets
  100 g game bird pellets
  100 g chopped carrot/apple
  50 g cubed wholemeal bread
  50 g wheat/maize
  100 g chopped green food (spinach, lettuce, or comfrey)

With the exception of softbilled species, grit is an essential addition to the diets of practically all the birds referred to in this chapter. It will vary in size from that suitable for small finches to flinty mixtures of a kind marketed for poultry breeders. Cuttlefish is an important item for many seed-eaters and some small parrotlike species, as it provides calcium and it helps keep mandibles healthy.

Commercial Foods and Supplements

On the following pages are details, including composition where the information is available, of commercial foods and supplements.

Claus Foods
Honey Food—type I (red)
For chlorpsis, troupials, orioles etc. Also as a supplement for breeding pairs of seedeaters.

Water	11%
Protein	29%
Ash	13%
Fat	6%
Fibre	5%
Sugar and starch	36%
Vitamins per kg	vitamin A: 0.140 Mio; vitamin D_3: 0.014 Mio; vitamin E: 0.42 g

Honey Food—type Ia (black)
For zosterops, tanagers etc. Also as a supplementary rearing food for Gouldian Finches etc.

Water	10%
Protein	30%
Ash	12%
Fat	9%
Fibre	5%
Sugar and starch	34%
Vitamins per kg	vitamin A: 0.140 Mio; vitamin D_3: 0.014 Mio; vitamin E: 0.42 g

Honey Food—type III (brown)
For mynahs, starlings, bulbuls, toucans etc.

Water	13%
Protein	15%
Ash	7%
Fat	6%
Fibre	3%
Sugar and starch	56%
Vitamins per kg	vitamin A: 0.140 Mio; vitamin D_3: 0.014 Mio; vitamin E: 0.42 g

Fat Food—type I (red)
For Pekin Robins, quail, tits, buntings, finches etc.

Water	6%
Protein	24%
Ash	9%
Fat	31%
Fibre	4%
Sugar and starch	26%
Vitamins per kg	vitamin A: 0.140 Mio; vitamin D_3: 0.014 Mio; vitamin E: 0.42 g

Fat Food—type II (green)
For Shamas, European Nightingales, warblers, Wrynecks etc.

Water	6%
Protein	30%
Ash	14%
Fat	31%
Fibre	5%
Sugar and starch	14%
Vitamins per kg	vitamin A: 0.140 Mio; vitamin D_3: 0.014 Mio; vitamin E: 0.42 g

Fat Food—type III (brown)
For thrushes, starlings, woodpeckers, cuckoos, Jackdaws etc.

Water	7%
Protein	17%
Ash	8%
Fat	18%
Fibre	3%
Sugar and starch	47%
Vitamins per kg	vitamin A: 0.140 Mio; vitamin D_3: 0.014 Mio; vitamin E: 0.42 g

Fat Food—type IV (blue)
For rubythroats, flycatchers, niltavas etc.

Water	6%
Protein	31%
Ash	17%
Fat	33%
Fibre	5%
Sugar and starch	8%
Vitamins per kg	vitamin A: 0.140 Mio; vitamin D_3: 0.014 Mio; vitamin E: 0.42 g

Fat foods (types I, red, and III, brown) are prepared with soya oil and are ready to feed. Types II, green, and IV, blue, are also supplied ready for use and contain animal fat. Honey foods (types I, Ia and III) are all ready for use and contain a high proportion of honey.

*Among ingredients of Claus Foods are the following: waffles, wafers, juniper berries, chopped peanuts, sultanas, rolled oats, white bread, mountain ash berries, desiccated coconut, cashew nuts, soya flakes.

*Different proportions of insects are used according to the type of mixture. The following are some of the main kinds used: ant pupae, prawns, white worm, daphnia, crustacea, insect larvae, tiny prawns, dried flies.

*Information supplied by the manufacturers.

Mazuri Zoo Foods

Zoo Diet A
(crushed form suitable for waterfowl)
Ingredients include the following: maize, meat, soya bean, wheat, wheatfeed, vegetable oil, whey powder, yeast calcium carbonate, dicalcium phosphate, methionine, lysine hydrochloride, sodium chloride, ferrous sulphate, copper sulphate, manganese oxide, zinc oxide, cobalt sulphate, calcium iodate, vitamin A and tocopherol acetate, vitamin D_3, thiamine hydrochloride, riboflavin, pyridoxine hydrochloride, vitamin B_{12}, menadione sodium bisulphite, folic acid, nicotinic acid, pantothenic acid, choline and biotin.

Flamingo Diet
Ingredients include the following: maize, meat, meat greaves, soya bean, wheat, wheatfeed, dried blood, fish, vegetable oil, yeast, methionine, calcium carbonate, dicalcium phosphate, ferrous sulphate, copper sulphate, manganese oxide, zinc oxide, cobalt sulphate, calcium iodate, vitamin A and tocopherol acetate, vitamin D_3, thiamine hydrochloride, riboflavin, pyridoxine hydrochloride, vitamin B_{12}, ascorbic acid, menadione sodium bisulphite, folic acid, nicotinic acid, pantothenic acid, choline, biotin and Carophyll Red.

Purina Foods

Game Bird Chow

Protein	19%
Fat	2.5%
Fibre	12%
*Nitrogen Free Extract	40%
Ash	9%

Ingredients include the following: ground yellow corn, ground milo and grain sorghums, soybean meal, dried whey, dehydrated alfalfa meal (preserved with ethoxyquin), wheat midlings, vitamin A supplement, D activated animal sterol, vitamin E supplement, riboflavin, vitamin B_{12} supplement, calcium pantothenate, niacin, calcium carbonate, defluorinated phosphate, iodised salt, manganous oxide, choline chloride, zinc oxide, menadione sodium bisulphate (source of vitamin K activity).

*Nitrogen Free Extract (NFE) is the difference between 100% and the sum of the percentages of moisture, protein, fat, fibre and ash. It is considered to represent carbohydrates, other than fibre.

Trout Chow

Crude protein	not less than	40%
Crude fat	not less than	4%
Crude fibre	not more than	5.5%
Ash	not more than	13%
Minerals	not more than	3%

Ingredients include the following: fish meal, soybean meal, ground wheat, brewers' dried yeast, ground yellow corn, wheat midlings, dried whey, animal fat preserved with BHA (chemical preservatives, propylene glycolipropyl gallate and citric acid), dicalcium phosphate, iodised salt, vitamin A supplement, D activated animal sterol (D_3), menadione dimethylpyrimidinol bisulphate (K), methionine hydroxy analogue calcium, vitamin E supplement, vitamin B_{12} supplement, ascorbic acid, biotin, choline chloride, folic acid, pyridoxine hydrochloride, thiamin, niacin, calcium pantothenate, riboflavin supplement, copper oxide, manganous oxide, iron oxide, zinc oxide, calcium carbonate, cobalt carbonate.

Pigeon Chow

Protein	15%
Fat	2.5%
Fibre	6%
NFE	56%
Ash	2.5%

Ingredients include the following: ground yellow corn or grain sorghums, soybean meal, dried whey, dehydrated alfalfa meal, wheat midlings, vitamin A supplement, D activated animal sterol, vitamin E supplement, riboflavin supplement, vitamin B_{12} supplement, calcium pantothenate, niacin, calcium carbonate, defluorinated phosphate, iodised salt, manganous oxide, choline chloride, zinc oxide, menadione sodium bisulphate.

Sluis Foods
Sluis Bekfin
For small and delicate insectivorous species including tanagers, shamas, nightingales, robins, tits, chloropsis, flycatchers and other small softbills.

Moisture	11.5%
Crude protein	18.5%
Crude fat	12.5%
Minerals	6.0%
Crude fibre	5.0%
Vitamin A	9000 IU/kg
Vitamin D_3	600 IU/kg
Vitamin E	50 mg/kg

Ingredients include the following: cereals, fruit, berries, dried insects, honey, animal fats. Ready for use.

Sluis Universal Food
For Pekin Robins, bulbuls and similar softbilled birds.

Moisture	11.5%
Crude protein	16.5%
Crude fat	12.5%
Crude fibre	5.0%
Minerals	6.5%
Vitamin A	4000 IU/kg
Vitamin D_3	300 IU/kg
Vitamin E	25 mg/kg
Ethoxyquin	
Potassium sorbat	

Note: although Universal Food is ready for use, the manufacturers suggest the occasional addition of cut-up pieces of meat or cheese, fruit, insects, raisins, hard-boiled eggs (in small pieces), rosehips and rowan berries.

Nekton Supplements
Nekton S Vitamin Supplement
(Minimum content per 1000 g—guaranteed analysis.)

Vitamin A	6.56 ml IU
Vitamin D_3	10,000 IU
Vitamin E	6666.60 mg
Vitamin B_1	666.60 mg
Vitamin B_2	1666.65 mg
Calcium D-Pantothenate	3333 mg
Nicotinamide	10,000 mg
Vitamin B_6	666.66 mg
Folic Acid	166.66 mg
Vitamin B_{12}	3333.30 mcg
Vitamin C	16,666.50 mg
Vitamin K_3	1333 mg
Biotin	30 mg

Calcium, iron, phosphorus, L-methionine, L-lysine, zinc, manganese, copper, L-gluten, L-aspartic acid, glycine, L-leucine, L-alanine, L-arginine, iodine, L-valine, L-isoleucine, L-threonine, L-histidine, L-phenylalanine, L-tyrosine, L-proline, L-cystine, L-serine, cobalt, L-tryptophane, glucose.

In water-soluble powder form.

Biotropic
BIO-Nectar
Sucrose, glucose, fructose, Vit. B_1, B_2, B_6, B_{12}, Vit. H, K_3, calcium-D-pantothenate, niacinamide, pteroylglutaminacid, Vit. A, D_3, tri-calciumphosphate, natriumchloride, magnesiumphosphate, di-kaliumhydrogenphosphate, iron II-sulphate, copper II-sulphate 5-hydrate, L-alanine, L-valine, L-leucine, L-isoleucine, L-threonine, L-cysteine, L-cystine, L-methionine, L-arginine, L-lysine, L-phenylalanine, L-histidine, L-tryptophane (all products with an 'L' indicate essential amino-acids), pollen (UV-light treated, pollinin is split), essential fat-acids.

Lory-Nectar (new)
Fructose, glucose, 9 different kinds of wheat concentrate, plant protein, vitamins (see BIO-Nectar), Vit. K_3 is highly concentrated, electrolytes and other supplements, all essential amino-acids (see above), pollen (UV-light treated, pollinin is split), vanillin, essential fat-acids.

Lory-Nectar extra
Sucrose, glucose, fructose, high proportion of pollen (19 kinds), UV-light treated, pollinin is split (for proper digestion), plant protein, all essential amino acids, all vitamins (K_3 is highly con-

centrated), electrolytes, supplements, essential fat-acids, Faex medicinalis.

Lory-Nectar
Sucrose, glucose, pollen, UV-light treated, pollinin is split, all essential amino-acids, fat-acids, all vitamins, minerals, supplements, Faex medicinalis.

BIO-E
Vit. E concentrate of DL-alpha-tocopherol-acetate.

BIO-A-D₃-E
Concentrate of all three vitamins (water soluble), Vit. A acetate, Vit. D₃-80, DL-alpha-tocopherol-acetate.

BIO-MINERAL
Tri-calciumphosphate, magnesiumhydrogenphosphate, di-kaliumhydrogenphosphate, Fe II-sulphate, copper II-sulphate, natriumchloride.

BIO-B-Complex
Thiamin-hydrochloride (Vit. B₁), riboflavin (Vit. B₂), pyridoxin-hydrochloride (Vit. B₆), cyanocobalamin (Vit. B₁₂).

BIO-K
Tri-calciumphosphate, K₃-menadione.

PROTAMIN
All essential amino-acids (see BIO-Nectar), Faec med., tri-calciumphosphate, beta-carotin, canthaxanthin, plant protein.

PROTAVIT
All vitamins, electrolytes and supplements.

SUPRAMIN
Highly concentrated product of all essential amino-acids, vitamins, minerals and supplements, and beta-carotin, canthaxanthin (both water-soluble).

BIO-Moult
All electrolytes and supplements, Faex med., Vit. H, beta-carotin and canthaxanthin, glucose.

BIO CC
Beta-carotin and canthaxanthin, highly concentrated, water-soluble.

PROTAVIT-Emulsion
Emulgated vitamins, highly concentrated for quick resorption (all vitamins included).

BIO-K
High concentrate of Vit. K₃ and tri-calciumphosphate, glucose.

Vionate
Vionate Vitamin Supplement
Degermed corn meal, dibasic calcium phosphate, calcium carbonate, sodium chloride, choline chloride, ascorbic acid (as sodium ascorbate), ferrous carbonate, magnesium oxide, niacin, calcium pantothenate, riboflavin, BHT as a preservative, DL-alpha-tocopherol acetate, vitamin A palmitate, thiamin mononitrate, manganous oxide, cupric sulphate, calcium iodate, pyridoxine hydrochloride, cobalt carbonate, folic acid, D activated animal sterol, vitamin B₁₂ (as cyanocobalamin).

ABIDEC
ABIDEC Multi-vitamin Supplement
(Content of 0.66 cc—approximately 15 drops.)
Vitamin A (palmitate)   (5,000 units) 1.5 mg
Vitamin B₁ (Thiamin hydrochloride)   1.0 mg
Vitamin B₂ (Riboflavin)   1.2 mg
Vitamin B₆ (Pyridoxine hydrochloride) 1.0 mg
Vitamin C (Ascorbic acid)   50.0 mg
Vitamin D (Ergocalciferol)   (400 units) 10.0 mg
Nicotinamide (Niacinamide)   10.0 mg
Pantothenic acid (as sodium salt)   5.0 mg

Minamino Compound
Each 100 ml contains the following.
Biologicals
Liver, extract from: 70 g fresh liver.
Spleen, extract from: 15 g fresh spleen.
Gastric mucosa, extract from: 7 g fresh gastric mucosa.

Vitamins
Vitamin B₁: 300 mg
Vitamin B₂: 40 mg

Vitamin B$_6$: 35 mg
Vitamin PP: 400 mg
Vitamin B$_{12}$: 100 mcg

Amino-acids: 2,000 mg, of which the following are present (typical analysis):
Histidine 175 mg; arginine 140 mg; methionine 73 mg; tryptophane 84 mg; threonine 186 mg; tyrosine 150 mg; glutamic acid 207 mg; aspartic acid 46 mg; proline 14 mg; glycine 53 mg; alanine 85 mg; valine 102 mg; phenylalanine 153 mg; isoleucine 74 mg; leucine 306 mg; lysine 194 mg.

Minerals
Iron citrate 410 mg; manganese sulphate 1.7 mg; copper sulphate 2.8 mg.

Flavoured excipient to 100 ml.

Live Food

Several types of live food are easily propagated and if more than a small number of insect-eating birds are kept it will be found worthwhile to establish cultures. Methods of propagation vary but none is too complicated to be a burden even to those with limited time.

Mealworms, wax moth larvae, crickets, locusts and *Drosophila* are among the most widely used live foods. Mealworms are easily bred and a culture can be established quickly in a suitable plastic or glass container which neither larvae or beetles can crawl out of. A lid with an insert of fine metal gauze or similar wire screen will provide essential ventilation.

Provide a 2 or 3 inch (5–7 cm) depth of dry bran in the vessel and place a couple of pieces of sacking, cut to size, on top. Supply a piece of apple or carrot regularly; but remove and replace when its condition deteriorates. Many recipes are in vogue to improve output of these easily-produced worms: some aviculturists advocate adding items including beer, molasses and yeast to the culture.

There seems little doubt that, used in moderation, they *do* improve the bran-based medium. It is essential, however, to ensure that the latter is at all times friable and wholesome; any suggestion of excess moisture will lead to the rapid formation of mould growth.

Disturb the culture as little as possible after introducing the initial breeding stock; beetles will produce quicker results than mealworms, of course. At a temperature of around 21–24°C (70–75°F) the insects will complete their life cycle in about 4–5 months.

Locusts and crickets are extremely valuable food items for many birds, especially for breeding pairs of insectivorous species. They are not difficult to propagate if the necessary high temperature in their breeding units can be maintained. A minimum of 27°C (80°F) is needed and should be provided by overhead electric light bulbs.

Locusts and crickets thrive in something like a disused aquarium tank. Provide a tightly-fitting lid and adequate ventilation. This is particularly important, for excessive condensation inside the container will cause the insects to deteriorate very rapidly. Bran, sand or newspaper can be used in the base of the tank. Grass and bran provide the main diet items. Water is best given in a container which has its neck plugged with cotton-wool.

Crickets usually breed fairly readily if they are provided with roughly screwed-up pieces of newspaper. Locusts, on the other hand, need shallow containers of fine, and preferably sterilised, moist sand in which to lay their eggs. The egg pods usually hatch in about 10–14 days. They are best removed, for hatching and eventual rearing of the young hoppers, to separate small containers.

Wax moth larvae are rapidly gaining ground as extremely useful additions to the aviculturist's armoury of live food. Apart from the fact that the larvae become large and fat they are soft-skinned and therefore easily dealt with by even the smaller insectivorous birds. Although they can be cultured fairly readily in a variety of home-made mediums, ready-to-use breeding concentrates are now marketed for these and other live foods (*Drosophila*, house flies, crickets are among those in the Nekton range). They certainly simplify the process, and produce good results.

Gentles, or maggots, in their various forms, are still widely used in aviculture and many people have fed them to their birds over several years without mishap. Nevertheless it *is* a risky business and I know of collections of exotic species that were all but wiped out by botulism almost certainly introduced through this type of live food.

If you *must* use these items it is vital they spend up to 5 days in clean bran before they are offered to birds. Better still, let the maggots pupate and feed what is effectively inanimate live food to your stock. Initially, at least, many species of birds show no interest in the pupae; before long you may find they become almost addicted to them and indeed prefer them to the wriggling larvae. Three types are available, mainly from angling shops: maggots, which are the large larvae of the blowfly; 'pinkies' which are the larvae of the smaller 'greenbottle', and 'feeders' which are the very tiny larvae of the housefly.

The garden and countryside can provide a host of valuable additional items of insect food. Beating a tree or section of a hedge (spread a sheet of some kind on the ground under the area to be assaulted and then give the branches above a few good whacks with something like a stout walking stick) can produce a wonderfully varied selection of tit-bits for your stock. Remember, though, the almost constant threat in areas of cultivation ranging from farms and smallholdings to orchards and allotments, of contaminated items resulting from the use of toxic sprays.

HOUSING

Birds are surprisingly adaptable creatures in matters of housing. Whether their quarters are a simple box-cage or a compartment in a lavishly planted tropical house, as long as other requirements are properly taken care of they usually thrive in their own particular patch of territory. That is not to suggest, of course, that foreign finches in a wire budgerigar cage, or a trio of pheasants in a bantam house, is acceptable. It most certainly is not. Nevertheless, lack of space on the part of a would-be aviculturist need not be an insurmountable barrier in the way of owning a few birds. Many bird-keeping enthusiasts are adept at making the best possible use of whatever materials and space they have available to create a comfortable environment for their stock. Sadly, though, there are still far too many examples of birds struggling to exist in conditions clearly alien to their needs. More often than not the root cause of such problems is housing rather than diet or other factors. Often they arise because of lack of knowledge on the part of embryo aviculturists, and are compounded by inadequate advice if birds are purchased from a non-specialist retailer. If a pair of foreign finches is bought from a pet shop there is every chance that the proprietor, if asked, will recommend an ornate, all-wire cage as a suitable home for them.

Now, such contraptions do not seem to have an adverse effect on the health of canaries, budgerigars and other domesticated types of cage birds but are definitely not suitable for foreign species. Designed more with interior decor in mind than the requirements of their eventual tenants, there is little doubt that some of the more expensive of these ornamental cages will not look out of place in the most expensively-furnished lounge or living room ... but preferably as a home for house plants rather than birds. Box-cages should always be used if an aviary is not available.

Cages

Cages provide one of the most basic, but at the same time important, types of housing for a huge array of birds. They also allow bird-keepers with no more than a spare room or a garden shed at their disposal to keep and breed birds ranging from finches to small parrakeets. To be of real value, however, they must be much more functional than the kind sold for pet birds. Box-pattern cages (or 'box-cages') are best. These unlovely, utilitarian homes for birds provide space, protection from cold and draughts, easy maintenance and the sense of seclusion and security an open cage can never offer.

Box-cages are exactly what that short and entirely accurate description implies—boxes with wire fronts (see Fig. 1) Of course they have other necessary refinements: drawer-trays to facilitate cleaning, doors (usually placed in the wire front), perches, food and water vessels and, in some cases, built-in (or -on) nest-boxes. They are one of the most unglamorous pieces of avicultural furniture imaginable, and also one of the most useful.

I would go so far as to say that it is impossible to maintain *any* collection of birds (other than such things as waterfowl, the bigger gallinaceous species, Ratites and so on) without the ubiquitous box-cage being pressed into service on numerous occasions. Box-cages are invaluable for isolating new stock, acclimatising birds just out of quarantine, segregating sick or injured specimens, providing emergency accommodation for individuals being persecuted by companions or mates; in other words helping cope with the sort of every-

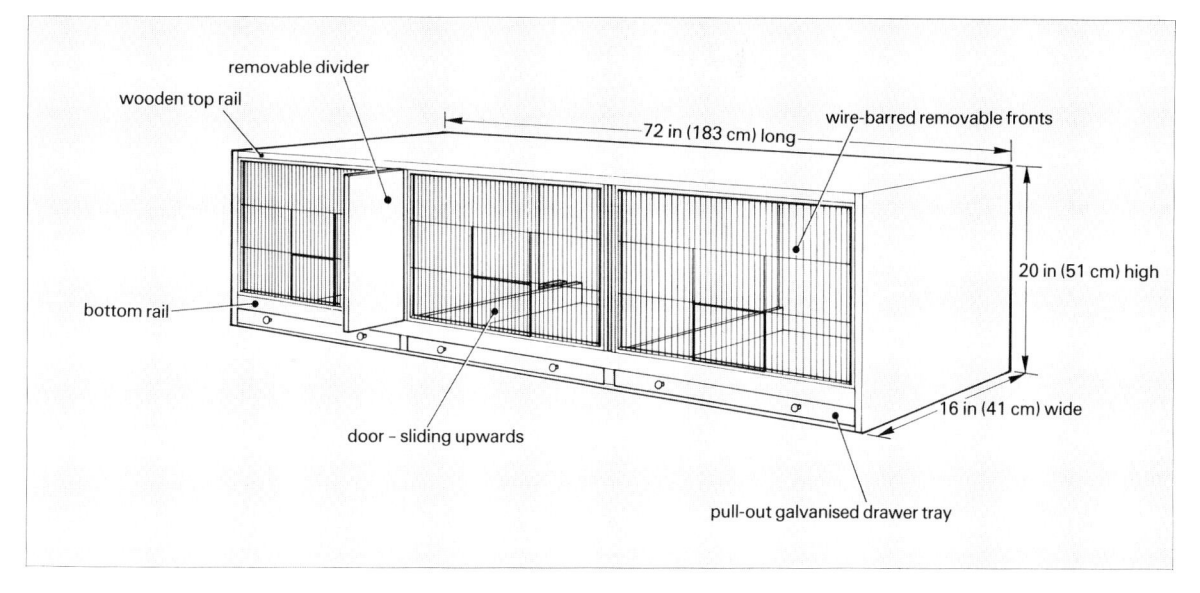

FIG. 1 Treble cages similar to this permit considerable flexibility. They can be utilised as small units for breeding pairs of finches. With both dividers removed they provide excellent accommodation for bigger and more active species up to the size of Shamas, barbets, pairs of Pekin Robins, mesias and sibias.

day problems experienced aviculturists are all too familiar with. For bird-keepers who simply do not have any outdoor space for aviaries, box-cages (provided they are of suitable dimensions) offer scope not only for long-term housing but also for breeding projects with many species.

Do not be persuaded that birds will not live in perfect health and reproduce successfully in the relatively close confines of a cage. Almost all of the perching birds, as well as many parrotlike and other species, easily achieve the first objective. Many, an increasing number of species in fact, breed regularly and freely in cages.

In the USA, for example, cages (and all-wire cages at that!) provide accommodation for breeding pairs of many parrotlike birds such as conures. The birds' output speaks volumes for the efficiency of the system, even though it eliminates any aesthetic value and takes aviculture into the realms of production-line efficiency.

In many minds space equates with freedom, or something approaching freedom. For the human mind that is a perfectly reasonable analogy. Birds, however, do not have our capacity to think and reason. Their way of life does not—cannot—base itself on anything other than the sheer practicalities of feeding, evading enemies, resting, and breeding. If these functions can be carried out within a small territory (which means a lower expenditure of energy) so much the better.

Birds housed permanently in unsuitable cramped quarters will not, as those critics of aviculture with an anthropomorphic turn of mind would have us believe, simply retire into a quiet corner, pine for their lost freedom, and die. What they *will* do, though, is fall away in condition and develop behavioural aberrations that will inhibit or prevent 'normal' activity.

So what is a 'spacious' cage? Opinions obviously vary but my personal opinion is that many of the minimum dimensions that have been handed down through generations of bird-keepers are often inadequate. For example, a cage measuring 36 in (91 cm) long is usually recommended as being suitable for a cock Shama or pair of Pekin Robins and it is a fact that many species of Australian grassfinches breed regularly in cages no more than 24 inches (60 cm) in length. Both of these examples serve mainly to illustrate that there is no all-embracing yardstick and that opinion among aviculturists as to what constitutes an acceptable amount of living-space for birds ranging from Shamas to Gouldian Finches will always provoke argument.

Part I: General Care

My most recent Shama lived for many years and was a picture of health and vitality, as well as an able songster and mimic. His home was a purpose-built treble box-cage measuring 72 × 20 × 16 in (183 × 51 × 41 cm). Why a treble? Because it provides excellent shut-off facilities for cleaning, painting, repairs etc. A three-compartment unit of this kind allows considerable flexibility and can be quickly pressed into service to provide temporary quarters for all sorts of birds ranging from small seedeaters to starlings and even some of the smaller members of the crow family.

It is possible that the Shama in question would have been just as much at home in a cage only half as long as the one he occupied. I believe, however, that these intelligent birds merit as much space as one can give them. This one looked 'right' and was always taking an inquiring interest in what was going on around him. The fact that he was not cramped for space meant that he never showed any of the awful stereotyped mannerisms that arise so often when active birds have only sufficient space to move between two or three perches and their feeding dishes. I much prefer to see Gouldians and other grassfinches housed in cages half the length but of corresponding height and depth to that suggested for a Shama.

If you are reasonably proficient at handywork there is no reason why cage construction should be beyond you. One big advantage to making your own units is that you can tailor them to fit precisely into available space. Cage fronts are readily available from dealers and pet shops: punched-bar fronts are more expensive than the welded type, but the latter are perfectly satisfactory provided they are of good quality.

Hardboard on a light timber frame is a satisfactory combination to produce built-in compartments (see Fig. 2); although ¼ in (63 mm) plywood will make a better job and is easier to work with.

Whether building or buying cages it is best not to sacrifice either quality or generous dimensions for the sake of saving money. Many of the cages sold by pet shops and bird dealers lack both depth and height to make them of value for a wide range of species.

An even bigger problem with many of these mass-produced units is the wooden drawer-tray fitted to most of them. Healthy birds may take a bath two or three times a day (a few, it is true, seem always to be uninterested in water, but most are avid bathers). Wooden trays very quickly become affected by all the splashing. They swell and are hard to slide out of the cage; in extreme cases they may even show mould growth. Galvanised metal trays, although not cheap, will prove a worthwhile investment in the long term.

If you are not planning to make your own cages

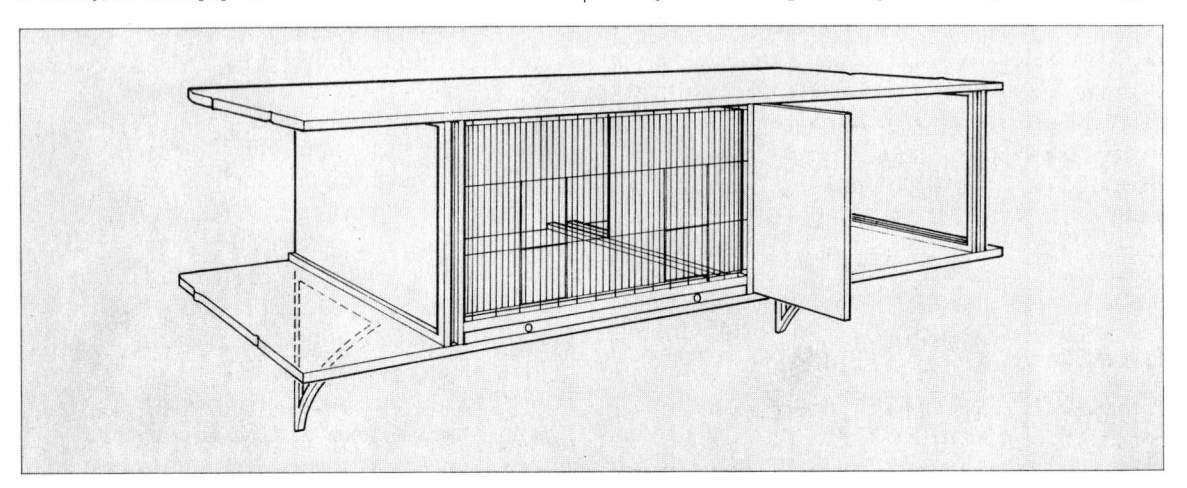

FIG. 2 Compartments are easily constructed in either a spare room or a suitable shed or other portable building. Hardboard or plywood supported on brackets provides the main structure. Heavier materials can also be used, including laminate-faced chip-board. For various reasons, however, thinner, lighter units, properly braced and supported where necessary, are more popular.

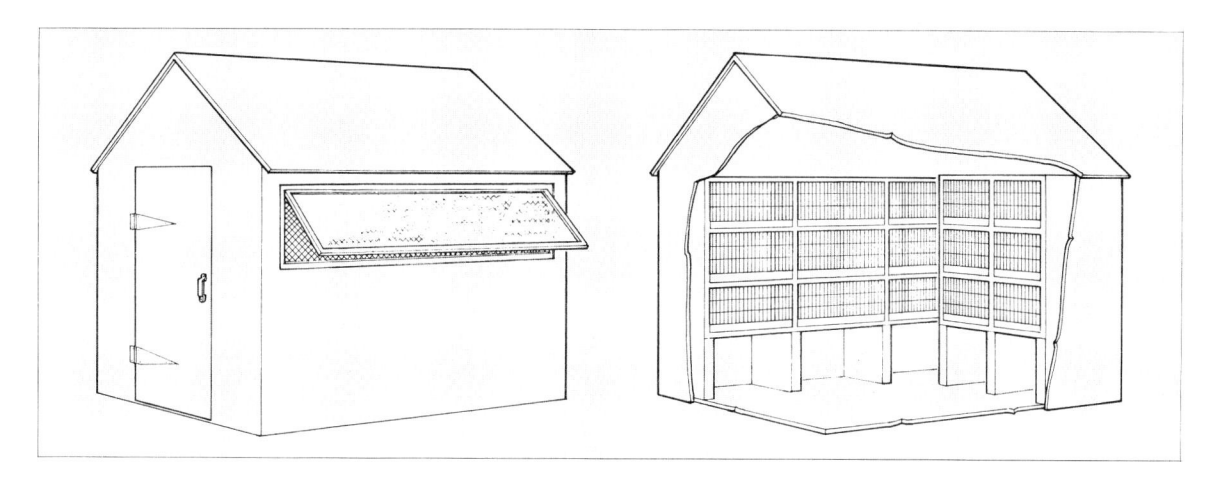

it certainly will pay dividends to consult a specialist manufacturer about your equipment. Do not be persuaded that double breeding cages 36 in (91 cm) long but only 8 or 9 in (20–23 cm) deep, or cages with flimsy hardboard drawer-trays, are what you want, however. It is far better to aim for something measuring around 20 in (51 cm) high and 16 in (41 cm) deep. Depending on the type of birds that will occupy the cages they can usually be acquired in various lengths ranging from 24 in (61 cm), 30 in (76 cm), 36 in (91 cm) all the way up to 72 in (183 cm).

Do pay particular attention to doors. There are two basic types: the large swing-open doors that allow a budgerigar-type nest box to be put inside the cage; and a smaller sliding type. The latter are usually standard on cages to house foreign finches and canaries.

The bigger doors can lead to endless trouble if used for quick-moving little foreign finches, for example. Remember, too, that spacing between 'bars' varies somewhat. Foreign fronts have much closer spacing than either canary or budgerigar fronts. Canary fronts also have bob-holes through which the birds can feed from external dishes; excellent for these domesticated songsters but a recipe for disaster for many exotic species.

Bird Rooms

Although there are many firms specialising in the construction of custom-built bird rooms, for many aviculturists the same term covers a multitude of sites ranging from a shed, a converted site-hut and a portable dwelling to a spare bedroom in the house. I have even seen a disused railway carriage beautifully converted to provide five-star accommodation for a stud of canaries. Whatever type of building is used, two considerations must prevail: that it admits enough light, and that it will not be subject to massive fluctuations of temperature during summer and winter.

Typical garden sheds often lack sufficient window space to be used for birds; although they can be modified, of course. Much better are some of the chalet or summerhouse types of garden building on the market. Both usually have ample window space (see Fig. 3).

Very small portable buildings always present problems with leap-frogging temperatures, although modern insulating materials do much to help overcome such difficulties. The interior should be lined with hardboard and the cavity between outer and inner walls or roof packed with suitable insulation filling.

It is hard to see any kind of garden building measuring less than about 8 × 6 ft (2.4 × 1.8 m) providing a suitable home for a collection of

FIG. 3   An 8 × 6 ft (2.4 × 1.8 m) shed is the minimum size recommended for use as a bird-room. Temperature fluctuations, especially increases influenced by strong sunlight, sometimes create insurmountable problems in very small buildings of this kind. Ensure there is adequate window space and ample ventilation. Lining and insulating a shed will do much to stabilise temperature levels, except in really small portable buildings.

birds, unless its purpose is to provide a series of simple shelter areas connecting with attached outdoor flights.

Buildings built of brick or precast concrete (such as garages) are a different proposition and can be converted to make first-class bird accommodation. Once again attention should be paid to window space: sometimes it is possible to fit rooflights, thus leaving more wall space available for cages or indoor flights.

A spare room in the house can easily be turned into an excellent little bird room and it usually offers one distinct advantage over most outdoor structures in providing a stable temperature. Usually only one window is incorporated and if need be electric lighting may supplement available natural light. Here it is worth investigating some of the excellent fluorescent tubes marketed for tropical fish and reptile enthusiasts, but offering many benefits in the bird room.

No matter what sort of room or building provides a home for your collection, it is strongly recommended that good quality floorcovering is fitted and that interiors are painted in a suitable (light) shade.

If heating is necessary that is very much a decision for individual owners and what they can afford. In my opinion thermostatically-controlled electric heating is best; either tubular or fan heaters can be employed. Oil-filled radiators, many of them with built-in thermostats, provide another convenient and neat means of maintaining modest temperature levels in small and medium-sized rooms. Other possibilities include gas heating as well as appliances that run on paraffin or solid fuel. Various installations are on the market for heating greenhouses and conservatories. Most need little or no adaptation for use in birdrooms.

Remember to give as much attention to ventilation as you do to a means of maintaining a correct temperature for your birds.

Aviaries

A natural-looking aviary in which growing plants complement colourful birds is a favourite way of housing many species. There is a great deal to be said for such an arrangement which works well for many species.

In Europe, for non-hardy species, the climate tends to reduce the period during which this kind of accommodation is occupied; unless it includes suitable bad-weather and roosting quarters. Although plants provide an attractive backdrop against which to display birds, some species (notably parrotlike but including a number of others) will speedily destroy anything from small trees and shrubs down to herbaceous subjects.

Very small aviaries (say 6 × 3 × 6 ft; 1.8 × 0.9 × 1.8 m) provide useful accommodation for single pairs of birds which are better segregated for breeding (see Fig. 4). If they are for spring/summer occupation only there is no need to add a formal shelter, although that does *not* mean no protection at all. Nevertheless, it is easier and less costly to board and felt part of the end and roof than to add a very tiny shelter-section which, in most instances, the birds do not use.

It should also be borne in mind that aviaries with limited floor areas simply will not support plants (even grass) for any length of time. That is not to say that various climbers should not be employed to cover part of the exterior; but putting something like a *Cupressus* or *Lonicera nitida* inside as a potential nesting/roosting site is a waste of time and money.

Cover of some kind, however, is an essential prerequisite to nesting for most passerine birds, and many others too. It is not difficult to provide it by means of bunches of spruce and other evergreens which can be wired into place in corners and the sheltered part of the flight. Remember to carry out this sort of work well before potential breeding pairs are introduced. Spruce provides excellent 'thickets' in which nest-boxes or baskets can be hidden for many small species. Naturally it will turn yellow as it dries out; but it is surprising how long it lasts before finally disintegrating.

Constructing ranges of compartment aviaries (see Fig. 5) saves on space and material costs. It is a system that can work well as long as pairs of the same or closely-related species of birds known to be especially combative are not housed side by side.

Housing

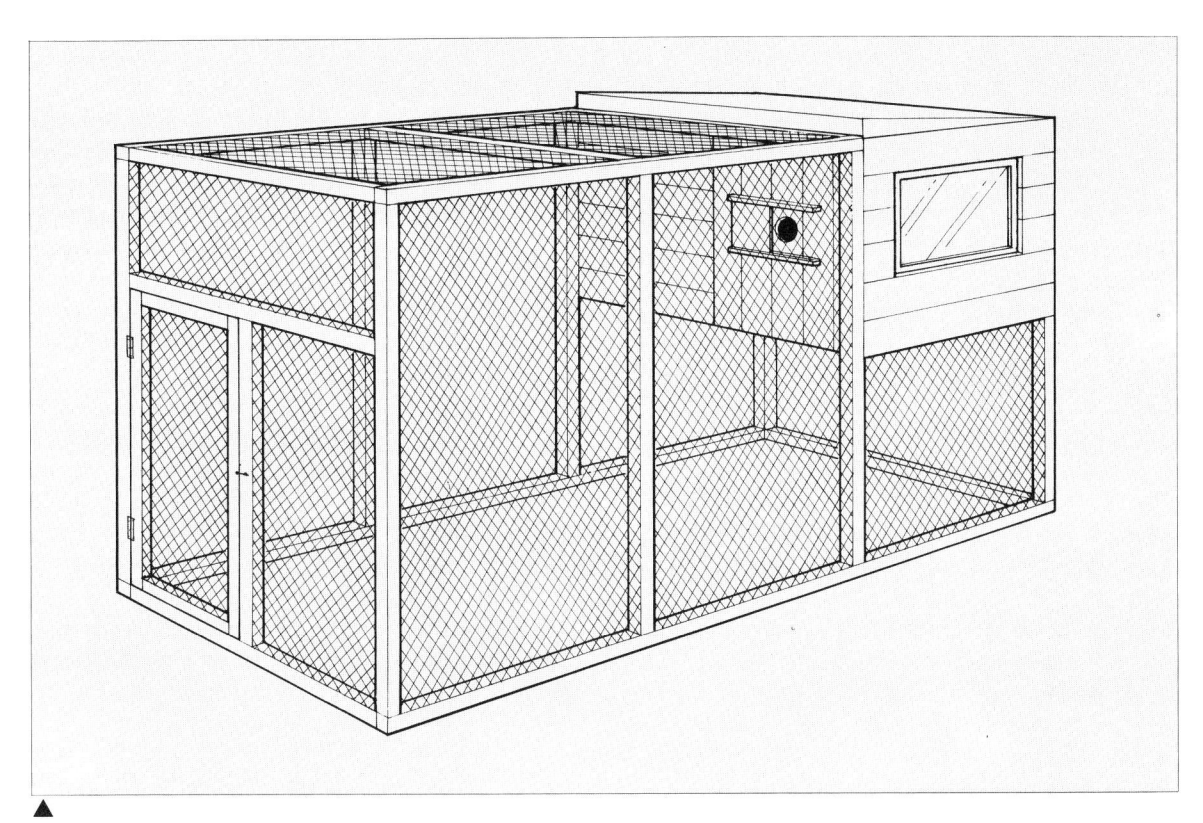

FIG. 4 Single, free-standing aviaries of this kind are built in a variety of shapes and sizes, although the design of this one is among the most popular. A half-shelter of the kind shown represents a considerable saving of expensive timber.

FIG. 5 Compartment aviaries are useful for birds that can be housed alongside, or in close proximity to, each other without fear of combative inclinations causing problems. They are often used to house small parrotlike species including lovebirds, parrotlets and some of the grass parrakeets. For lovebirds it may be necessary to double wire divisions between the flights to prevent fighting. A similar design but with longer flights can be utilised for many other parrotlike species including lorikeets and most *Polytelis* species.

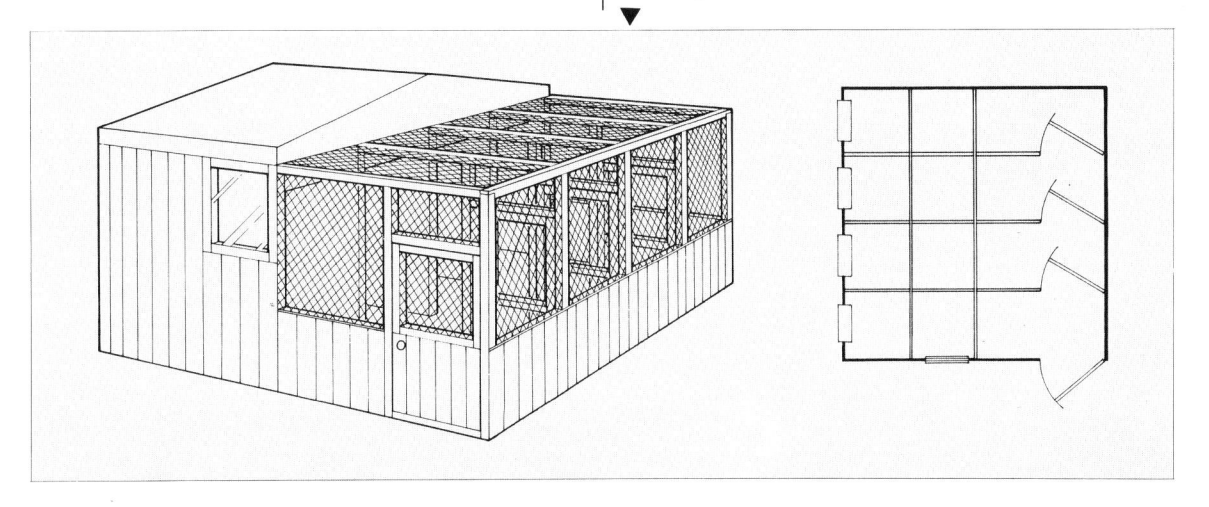

Part I: General Care

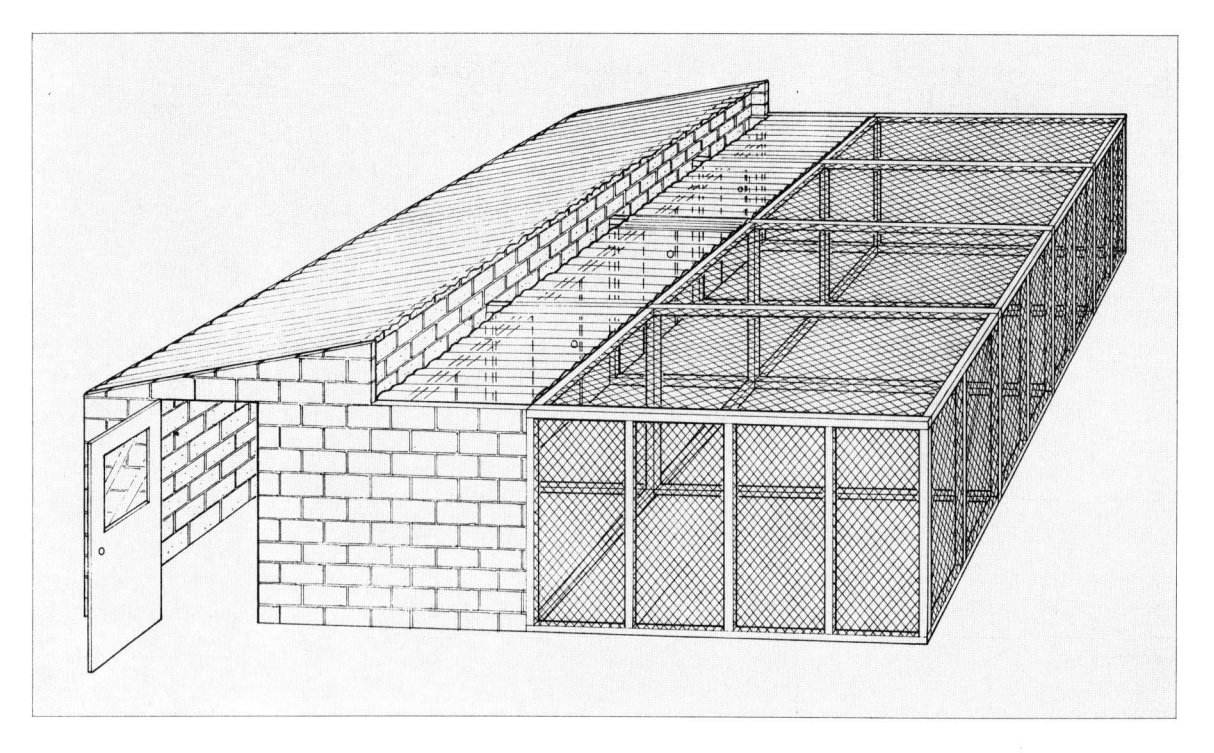

FIG. 6 A range of aviaries to house large parrots (and some of the smaller species known for their ability to destroy wood and lightweight mesh) must be constructed of heavy-duty materials including brick, breeze block or pre-cast concrete with heavy gauge welded mesh.

Quite often a range of flights or flights and shelters is used to house parrotlike species, particularly lovebirds and grass parrakeets (see Fig. 6). With some of these birds (especially the former) in mind it will be wise to ensure a double 'skin' of wire-netting on divisions; some individuals can injure each other quite seriously if this sort of precaution is not taken when aviaries are being constructed.

In view of the difficulty of keeping not only growing plants but also the ground itself in anything approaching good condition over a season or two, it is worth considering erecting all small single unit or compartment breeding aviaries on a cement base.

For many parrotlike species the threat posed by a combination of natural floors and *Ascaridia* worms makes such a foundation essential. It is desirable for many other species kept over long periods in small outdoor aviaries. If the birds concerned are not destructive to plant life much can be done to brighten things up by introducing shrubs and even herbaceous plants in portable containers.

Large aviaries, provided they are not overstocked, are a very different proposition. In a largish unit, say something not less than 300 sq ft (28 m²), the bird-keeper with a flair for landscaping can create a most attractive addition to the garden.

The trouble with aviaries, unfortunately, is that a combination of gaunt timber or metal uprights and cross-members supporting great rafts of wire-mesh can hardly be regarded as things of beauty; but then, neither is a pergola until the roses start to twine over it, or a wigwam frame of canes until it is smothered in sweet peas. Sensibly landscaped and planted aviaries (see Fig. 7) *can* be both attractive and functional. Do remember to get your priorities right, however, and keep in mind that the feathered occupants of such a structure must have priority over some desirable, but potentially dangerous, aspects of your creation. There are, for example, some kinds of plants

FIG. 7 Large outdoor aviaries permit considerable scope for artistic landscaping and subtle planting schemes. A small pool is a possibility, although it should be shallow and even then it may be a hazard for birds prone to night fright. Remember that a really spacious outdoor flight, assuming it houses a number of birds of mixed species, needs to be linked to a good-sized shelter.

which could prove fatal to birds in the close confines of even a very large aviary. It would not be wise to include yew or laburnum in aviaries, and even less so in enclosures or paddocks to house pheasants, cranes and the like.

Think carefully, too, about the possible inclusion of water. There is not the slightest doubt that a shallow pool, even a modest waterfall powered by an electric pump, adds a very desirable extra dimension to a planted aviary, just as it does to a garden. It can also represent a serious hazard to birds whose normal way of life is terrestrial or arboreal. It is not so much that they are likely, in broad daylight, to come to a watery end among the iris and lilies; that is unlikely, although it can happen if birds become over-ambitious taking a bath. Night-frights, when birds take fright *and* flight after dark, are the real problem. Even in water only a couple of inches deep a surprising number of species can come to grief. Landing in an alien element, and almost certainly completely disorientated in their panic, they simply become more and more waterlogged through struggling, and quickly drown.

So it is a question mainly of knowing the habits and temperament of your stock before deciding to add any kind of pond. Remember that in making a judgement you simply cannot predict the likelihood or otherwise of these nocturnal panics. They are brought on by a variety of causes, some real, some imagined. Well-planted aviaries which provide a variety of secluded roosting sites may help reduce the problem. It is worth noting that some species are more susceptible than others to night fright; quail and other small gallinaceous species, and tinamou, are among the bad offenders.

There are many attractive and valuable plants available from nurseries and garden centres. Time spent consulting a local plant specialist, who knows what will and will not do well in local conditions and soil, is usually worthwhile.

Remember to leave adequate flying space in the aviary for the birds. Good natural cover for them is one thing, but if they cannot indulge in plenty of aerial manoeuvring half the value of their environment will be lost.

Upright-growing *Cupressus* provide useful year-round cover in aviaries. Remember that with one or two exceptions they are fairly slow-growing, and the sort of effect you have in mind may take some years to achieve. One of the quickest-growing of them in *C. leylandii*. Even though it can be cut back annually, however, it will probably prove too much of a handful for most roofed bird enclosures. It can though provide a useful windbreak hedge in a variety of situations. There are plenty of other good-looking

Cupressus available including several varieties with silver, gold and blue-green foliage. Some of the similar, but slower-growing, *Juniperus* are also worth considering.

Prostrate-growing *Cupressus* and junipers are good ground-cover subjects. *Cotoneaster horizontalis* is also useful in this direction. These and similar subjects provide the sort of ground-hugging cover appreciated by small waterfowl and others.

Additional valuable and reasonably-priced shrubs and bushes for aviaries include *Lonicera nitida*, *Berberis*, box, *Cotoneasters* (upright-growing types) and *Rosa rugosa* or *rubiginosa*. Several, including the two last-named, can be used to create an invaluable hedge inside a large flight.

There are many tough and attractive perennial plants that can be used to enhance the appearance and usefulness of aviaries. Among them are various bamboos and ornamental grasses. Unless you are really conversant with the characteristics of these two useful groups of plants, do take advice from a specialist grower, for many examples are strong-growing and can prove invasive.

Raspberry canes provide an unusual subject for growing in an outdoor flight. They produce good summer cover, seem to attract insects, and have the additional advantage of a useful berry crop. Of course if you want to cultivate these, and other soft fruits, along conventional lines you should not plant them in bird accommodation where regular disturbance would undoubtedly have an adverse effect on the inmates.

Aviary exteriors should not be forgotten. No matter how attractively landscaped the interiors are the overall effect will be lost unless something is done to disguise the stark outlines of the structure itself.

Clematis is a popular choice for clothing uprights and wire-netting. Several species and varieties are available from plant centres, many of them flowering at different times so that by making a careful choice of two or three different types it is possible to have colour over a long period of the summer.

Russian Vine, *Polygonum baldschuanicum* to give it its full, tongue-twisting botanical name, is one of the quickest-growing climbers of all and will clothe much of a small aviary within little more than a season. Like raspberry canes it seems to be of great interest to small insects; this is another advantage in a flight containing breeding pairs which feed their youngsters either wholly or entirely on small live food. Nevertheless, as is so often the case with what appears at first sight to be a vice-free subject, *P. baldschuanicum* provides one major disadvantage in that it is massively rampant in good growing conditions.

Although the plant's ability to clothe wood and wire with attractive, leafy vines in a matter of only a few months can be a big advantage for aviculturists needing quick, natural cover, the big snag is its invasive nature which has to be seen to be believed. Constant trimming back during the growing season works well (although it is probably only producing stronger growth in the final analysis), but this sort of activity often disturbs breeding birds. Another problem linked with the plant's strong-growing tendencies is the girth achieved by stems only a few years old, which are certainly enough to split any wire-netting they happen to be growing through. It is such a valuable cover-producing plant, however, that it is well worth growing. Constant vigilance is needed to keep it in check, and to ensure that it does not liberate some of your most prized feathered specimens by forcing wire-netting links apart when you are not looking!

There is no reason, of course, why an aviary should not be used for year-round occupation by a wide variety of species provided that suitable shelter accommodation is attached. Many birds are hardy enough to be housed in such facilities without heat. There are many enthusiasts who claim to be able to over-winter small exotic species in similar fashion. That may well be so; but I am totally unconvinced that it is other than harmful (if not immediately, certainly in the medium to long-term) to a number of small seed-eaters and softbills.

There are few warm-climate species I would personally be willing to subject to winter temperatures below 7°C (45°F) and so far as most small birds are concerned I much prefer them to be in indoor flights where artificial light (to lengthen the short winter days) as well as moderate warmth is more easily provided.

A winter shelter should be as spacious as possible and, if possible, lined and insulated to prevent substantial fluctuations of temperature, and heat-loss, for any means of providing warmth nowadays means using increasingly expensive energy. It should also have ample window space, for few birds are keen to enter a dimly-lit covered area. If cats or other predators are a problem it will be best to have non-clear glass or one of the opaque substitutes that are increasingly popular.

Aviary panels are a fairly recent introduction to the equipment market. Most are to the standard size 6 × 3 ft (1.8 × 0.9 m); some are plain and others have a door fitted. They are almost always of standard welded mesh on a light (1½ × 1 in; 4 × 2.5 cm) wooden frame. Certainly they provide a convenient and easy way to assemble a flight, and the only limiting factor appears to be that the finished structure has gone together in 3 ft (0.9 m) or 6 ft (1.8 m) segments.

A small garden shed can be attached to provide a shelter and you have the nearest thing to an instant aviary. It is as simple a means of putting an aviary together as has been devised at present, although by no means the cheapest. That is undoubtedly the self-build system and there are many excellent books on the market that will guide the do-it-yourself enthusiast along the right lines.

Wire-netting, indispensable to generations of aviary-builders, is now being replaced by welded-mesh; although there are many, myself included, who prefer the greater flexibility and 'give' of the former. Welded-mesh, however, is undoubtedly a valuable addition to the aviculturist's armoury of construction materials. For apart from coming in a variety of mesh-sizes and widths, it is also available in a range of gauges, the strongest of which will enclose large parrotlike species which have shown an ability to destroy many other wire-mesh materials over the years.

On the subject of large parrots, and some not so large, it is quite pointless providing an aviary built of wood and anything other than heavy-weight mesh for many of these birds: they will simply dismantle it, sometimes in a matter of hours if they really give their minds to the task. Angle-iron is probably the best material for the framework of a parrot flight whereas shelters must be constructed of pre-cast concrete, bricks or similar materials. Any exposed woodwork will quickly be reduced to shavings unless protected.

A final point on the subject of aviaries. Do remember to paint all mesh (whether it be wire-netting, welded-mesh or chain-link) with a good quality, non-lead, matt-black paint. Apart from providing an extra measure of protection for the mesh, it is surprising how much easier it is to see the birds once the job is done. An old paint roller, or paint-pad, will allow what might otherwise be a tedious and time-consuming task to be completed in a matter of a few hours.

Tropical House or Conservatory

Tropical houses are now being used by a growing number of aviculturists to house birds that thrive in the sort of warm and humid conditions that can only be produced under glass. Hummingbirds, many tanagers, flowerpeckers, broadbills and the like are just a few of the kinds that are particularly suited to this kind of accommodation.

Of course the cost of erecting a suitable building from scratch is likely to be extremely high. A more realistic proposition is to adapt an existing lean-to greenhouse or conservatory (see Fig. 8). Occasionally one is able to purchase, at reasonable cost, a second-hand commercial greenhouse of modest dimensions. Such a building can, with careful but not too extensive modification, make a superb home for tropical birds.

On the question of size it is perhaps worth pointing out that although some of the biggest commercial glasshouses are well beyond the scope of amateur aviculturists very small conservatories or greenhouses are also of no value for conversion to tropical houses. The reason is that they are subject to massive temperature fluctuations, even over a period of a few hours during summer or winter. Although a back-garden crop of tomatoes might be able to cope with this a collection of birds most certainly cannot.

There is virtually nothing that can be done about it in a house too small to allow the sort of

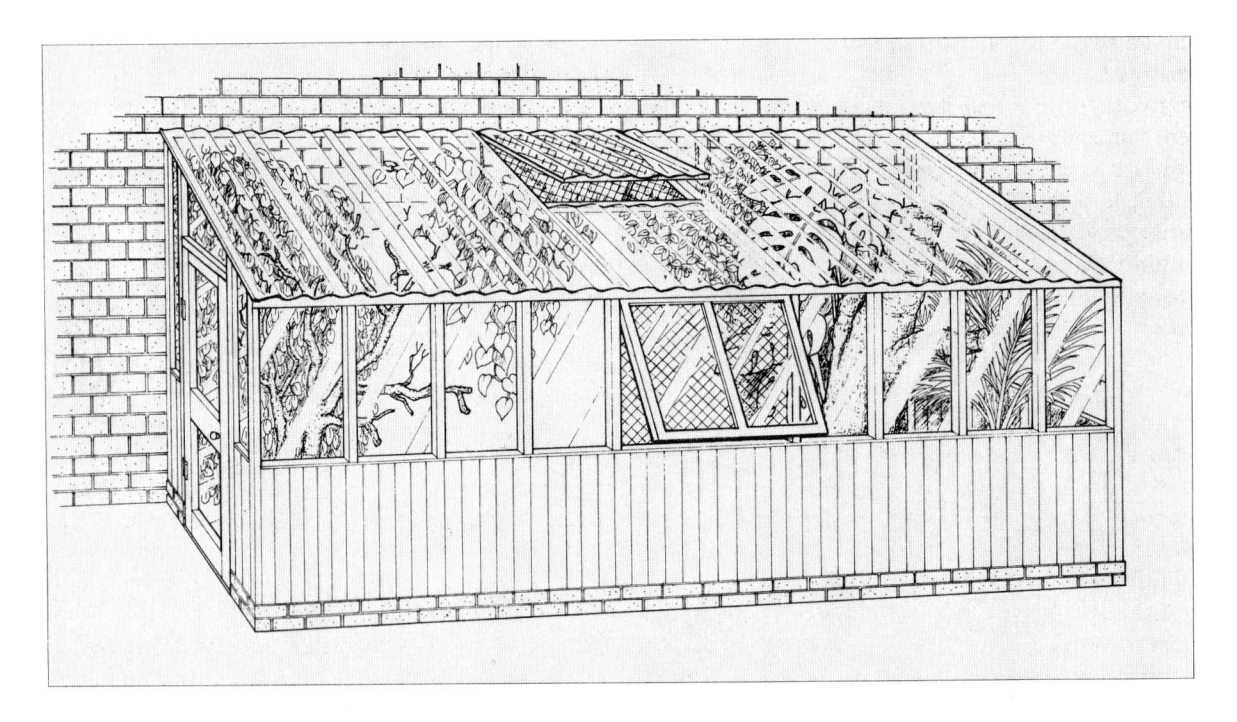

FIG. 8 Many conservatories and sun-lounges can be easily converted to provide valuable accommodation for many exotic birds, especially those at home in the world's jungles and rain forests. Like other portable buildings, however, they must be spacious to be of real value for birds: temperatures fluctuate enormously in any small glazed or part-glazed unit. Remember, also, that glass is an especially fragile material and a single broken pane can lead to the loss of an entire collection of birds. It is well worth considering replacing glass with one of the modern 'substitute' materials, some of which are almost unbreakable. Twin-wall polycarbonate sheets combine strength with good insulating properties and are strongly recommended for cladding a conservatory to be used for birds.

ventilation or throughput of air that would help stabilise things. I would not, personally, believe it possible to make sensible use of a greenhouse or conservatory measuring much less than 20 × 8 ft (6 × 2.4 m) and of corresponding height.

If possible glass should be dispensed with in a building of this kind destined to house birds. The reasons are fairly obvious. It is an extremely fragile material which shatters easily. Although a broken pane means no more than a chill for the gardener's specimens it may mean a mass escape for those belonging to a bird-keeper. Lining the interior with small mesh wire-netting is one solution to the problem. Another and better way is to replace all glass with one of the excellent substitute materials now on the market. As well as being lighter and stronger than glass, many have improved insulating qualities.

Ventilation is extremely important in a building of this kind. As well as being efficient it must also be escape-proof. It is surprising how small an aperture will provide a means of escape for something like a determined Zosterop.

Heating can be provided most easily by one of the greenhouse systems marketed for gardeners. The best of them are efficient and energy-saving with reliable thermostatic controls. The main fuel sources for medium or larger installations are electricity, gas or solid fuel. Smaller heaters that operate on paraffin or calor gas are also available. Electric fan-heaters can prove fairly expensive to run in a large building; but they perform the important task of circulating air as well as providing warmth and are worth considering even as a supplement to the main system.

When the conservatory or tropical house is erected on a concrete base it is important to provide some form of efficient drainage if the soil or other medium in which plants are grown is to remain in good condition. A build-up of noxious parasites is all too easily achieved, albeit unwittingly, in these warm, humid conditions, and diligent management must be practised if such problems are to be avoided.

If a large garden aviary allows scope for the gardener-aviculturist to use his or her imagination to create an attractive addition to the garden, a tropical house will allow that same imagination and flair to run riot. For, depending on the amount of space available (height as well as floor area) and the likely level of temperature to be maintained, a wide range of tropical and subtropical flowering and foliage plants can be grown. Some may be best confined to containers plunged up to their rims in one of the beds; planted out and with a free root-system many will rapidly outgrow their welcome.

Although a tropical house can prove one of the most rewarding and interesting means of housing a collection of exotic birds, it is also fairly demanding, for it is vital to maintain very high standards of cleanliness. Even small left-over particles of food can rapidly promote growths of mould and the condition of all items at food stations should be checked two or three times a day.

Pens and Paddocks

It is hard to attach a firm definition, in avicultural terms, to either a pen or a paddock. Broadly speaking, however, most people would regard 'pens' as the province of pheasants and similar birds whereas a 'paddock' is more likely to be occupied by species like rheas or cranes. Either way, what we are concerned with is an enclosed area of ground with a few necessary refinements such as shelters, as well as trees and shrubs to provide shade and shelter from the elements.

Most pheasants, whether they originate from areas in the snowfields of Tibet and western China or the damp forests of Malaysia and southeast Asia, adapt well to alien climates. Nevertheless, this does not mean that prolonged exposure to bleak, wet conditions is acceptable. *Dry* cold is one thing; bone-chilling damp is something different. If captive pheasants are to remain healthy it is important that they are housed on well-drained land that does not become waterlogged after a shower or two. There are few more unhappy sights than a bedraggled cock pheasant squelching his way round the muddy perimeter of a pen; but that is the lot of many of these birds whose owners fail to differentiate between definitions of hardiness.

Although such birds as Golden, Lady Amherst's, Silver and Swinhoe's Pheasants appear to thrive in fairly small enclosures with a floor area of about 250 sq ft (23 sq. m), the more space that can be allocated to them the better they will be. If a number of pairs or trios are kept, a house opening onto a range of adjoining pens is a popular way of accommodating these birds. Most need little other than an open-fronted shelter (something measuring about 6 × 6 × 7 ft, 1.8 × 1.8 × 2.1 m, suits most of them) opening onto their pen (see Fig. 9).

The latter should be turfed but it is desirable to put down a walkway of pea-gravel around the inner perimeter. Something about 12 in (30 cm) wide and with a 6 in (15 cm) depth will prevent the birds, even on good land, churning up the ground as they patrol their territory.

A single perch, fairly high up and towards the back of the shelter, provides a roosting place. The pen area should be furnished with a few suitable shrubs to give shade during periods of strong sunshine, but more importantly somewhere for the females to take refuge if the cock's attentions become too pressing during spring.

It is sound practice to bury the wire-netting up to 18 in (45 cm) down into the ground around the area of the pens to help prevent mice gaining entrance. This precaution also stops some of the digging species (monals, for example) gaining their freedom.

Although some people confine their birds behind 2-in (5-cm) mesh wire-netting, a smaller mesh is strongly recommended, not only as a rodent deterrent but also to prevent sparrows and starlings gaining entrance to the pens. Apart from the inroads these birds make into the pheasants' food they are also capable of carrying disease.

Paddocks come in a variety of shapes and sizes. One of the best I ever saw (used to house a pair of rheas, among other things) was an old orchard

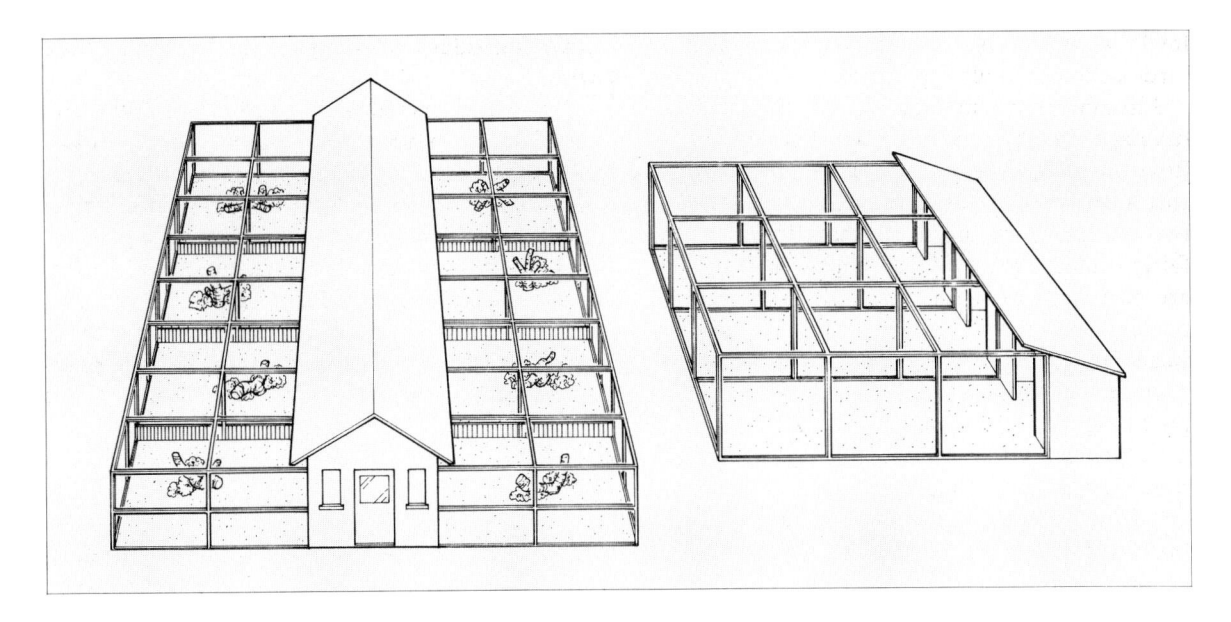

FIG. 9 Here are examples of two types of shelter-pen layouts to house pheasants or other gallinaceous species. Note the solid divisions (which should be 2–3 ft, 60–90 cm, high) between pens to minimise aggressive behaviour between cock birds (left). The layout shown on the right is more modest, but equally effective, combining open-fronted shelters with roomy pens.

completely enclosed behind a six-foot sandstone wall. Few of us are lucky enough to have such an excellent facility available and most paddocks consist of a small fenced area of grassland, or are perhaps an uncultivated part of the garden.

Rheas and large wildfowl (such as geese, cranes and occasionally peafowl, although the latter tend to use their 'official' quarters as no more than a base from which to wander freely if they get the chance) are the usual occupants. The three species of eared pheasants also do well in more spacious quarters of this kind.

A 6-ft (1.8-m) high fox-proof fence should enclose the entire area. Incidentally, if a walled garden, orchard or similar existing feature is being adapted to provide a bird paddock, remember the need for similar anti-fox security measures. These resourceful animals will lose no opportunity of making a meal of surprisingly large birds. Some sort of shelter is essential, of course, and it should be sufficiently large to allow birds to be housed in it during periods of severe weather.

Most of the ground area will probably be down to grass but some trees (for shade) and, unless walled or otherwise well sheltered, a hedge to act as a wind-break are desirable.

Enclosures for Waterfowl

At its most basic an area to house a collection of waterfowl need be little more than a modified paddock, modified, that is, to provide an area of water. On the other hand you may be lucky enough to have a natural pond available. Again a fox-proof fence to surround water and a suitable portion of land is essential.

Many waterfowl (especially some of the well-known duck species such as Carolina and Mandarin ducks, various teal and pintail) will live quite successfully in less ambitious accommodation. Naturally enough, water is an essential part of a duck's environment, but some of the smaller kinds will be perfectly happy with a tiny pond only a few square feet in area. Small pre-formed garden pools (the best are fibre-glass) are sold by water garden specialists and many can be simply adapted to provide a tiny duck pond. They come in a great range of shapes and sizes and can be installed in a garden in little more than an afternoon. Ensure that one to be used for small waterfowl has a suitable ledged shallow area so the birds can leave it at will.

Although they are not fitted with any sort of drainage facility it is well worth undertaking a little bit of extra work to provide one. It is likely that a tiny expanse of water, even if used by only two or three very small ducks, will become foul fairly quickly. Baling out or use of a small pump are two ways of emptying these small moulded pools. It is a fairly simple matter, however, to fit a waste-pipe and plug. The most difficult part is probably cutting or drilling an aperture in the pool base to take the waste-pipe. With care it can be achieved without mishap. Then, in digging out the area to take the pool itself, go 2 or 3 ft (60–90 cm) further down, put in a good depth of broken brick and hard-core to provide a soak-away and ensure that the waste-pipe is sited correctly.

Bigger pools used to be constructed of concrete, a laborious and time-consuming process as well as being one that did not always produce good results if the builders were unskilled. Nowadays even ponds with a surface area covering several hundred square feet can be created fairly quickly using plastic pool liners. Small areas can be dug out with a spade. More ambitious schemes may require the use of a mechanical digger: they come in various sizes, and can be hired (with operator) on a short-term basis. An excavation may be of any size or shape desired. The liners are completely flexible and will follow the contours and shape of the excavated area as the pool is filled.

One of the best materials to use is butyl rubber, which is tougher and much longer-lasting than more inexpensive PVC liners. The latter are fine for most medium-sized layouts but not suited to really big waterways.

Few waterfowl will willingly enter any kind of man-made shelter; so there is little point in going to the expense of buying a shed for them to sleep in. Nevertheless, although they are extremely tough and apparently impervious to the most severe weather, they *must* have plenty of natural cover in their enclosure. *Cupressus*, including prostrate-growing types, are excellent in a waterfowl enclosure. A compact hedge is very important if the pen is in an exposed situation and the birds will use it as a wind, rain and snow-break, changing sides according to the direction of the wind.

Various ornamental grasses, bamboos and reeds should find a place in the pen. If space allows, the large and spectacular *Gunnera* is worth growing. A selection of strong-growing marginal plants can be obtained from specialist growers and many waterside plants can be included. Remember that an area of good grass is as important for many species of waterfowl as a stretch of water. Such birds as geese, Wigeon and others are avid grazers and will not thrive without the opportunity to crop grass regularly.

Equipment and Utensils

As well as cages, aviaries, tropical houses, pools, pens and paddocks the aviculturist will assuredly amass a vast amount of ancillary equipment over a period of time, most of it probably recognisable, despite much modification, as items that inexplicably disappeared from the kitchen and other parts of the household some time previously.

There is, of course, a small industry producing all sorts of avicultural equipment ranging from split rings to ready-made aviaries. Most of the merchandise on offer is well made and useful. Whether much of it is essential is another matter. For example it is possible to spend a fairly considerable sum (especially if a large number of birds is involved) on such mundane items as food and water dishes. They come in a variety of shapes, sizes and materials. Most are functional and fulfil, admirably, the purpose for which they were designed and manufactured. So do plant-pot saucers, however, which are available in a good range of sizes, and at a fraction of the cost, from any garden centre or nursery. I use nothing else and find them excellent for species ranging from finches to flycatchers. A small waxbill can enjoy a thorough soaking in the smallest of them; and at the other end of the scale I have watched a Blue Magpie use the biggest size to get everything wet except the end of its tail.

This modern world bestows all kinds of *free* equipment on aviculturists and other livestock hobbyists. For instance, where would we be without the marvellously useful rectangular polythene tubs in which ice-cream is sold by freezer specialists? I use them to culture a variety of live foods, as storage containers for insectivorous food and individual seeds, for mixing, sifting, stirring and many more bird-related operations.

Knives, forks, spoons and mixing bowls can be borrowed as needed from the kitchen. If you keep the sort of birds that require its use regularly, it may be worth considering the purchase of a blender to be used exclusively for bird food. They really are extremely useful, particularly if you are concerned with such birds as lories and lorikeets. Another piece of kitchen equipment I find invaluable is a small hand-operated shredder, normally used for herbs. Mine performs valuable service reducing prawns and shrimps before they go into the insectivorous mixtures.

If a large number of seedeating species are kept, a winnowing machine may prove a good investment. The only disadvantage I have found with one of these appliances is that, quite often, the discarded seed (which the winnower reclaims) will again be rejected by the same birds who refused it first time round!

Although some theorists question the wisdom or value of hospital cages, I regard them as indispensible items of equipment. I have two, of different sizes and of a well-known make. Over the years I do not have the slightest doubt that they have helped prevent losses of stock. A metal cage, which can be easily disinfected, is preferable to one made of wood. Modern hospital cages with accurate thermostats are easy to control and the cost of operating is negligible (see Fig. 10).

Another manufactured item, which is virtually indispensable if extra artificial light is needed during winter, is a dimmer. There are a number on the market and all seem to be efficient and reliable. Remember that the majority will not operate fluorescent-tube lighting.

For waterfowl and pheasant enthusiasts a few more specialised pieces of avicultural hardware come into the picture. One is almost certainly an incubator, and the choice of models is wide, indeed: gas, paraffin and electric models, still- or forced-air, with or without automatic turning. There are a number of decisions to be made before the correct choice of machine is finally decided. If electricity is available most people opt for that type of incubator. As to whether still- or forced-air is the better system, this is a subject of considerable debate and each method has its devotees. If the number of eggs for hatching is likely to be substantial, that factor alone makes acquisition of a forced-air model desirable. For the hobbyist with only a few eggs to hatch each year the simpler, still-air incubator will almost certainly prove adequate.

I would not, personally, consider purchasing an incubator that lacked an automatic turning facility. In a busy day it is easy to forget the all-important need to turn a batch of eggs at a given time. An automatic machine does *not* forget and can amply repay the small initial extra outlay over a non-automatic model within a single season.

Coops and runs are an integral part of waterfowl and game bird rearing programmes. The handyman will find the construction of a traditional slat-fronted, weatherproof coop, and a suitable run to attach to it, fairly straightforward (see Fig. 11). Alternatively these and other items can be purchased from one of the game bird farms that advertise regularly in the sporting press.

If more detailed information about the construction of aviaries, bird rooms, pheasant pens, pad-

FIG. 10 A hospital cage is an invaluable addition to an aviculturist's basic equipment. Warmth is an essential element towards retarding the progress of many minor bird illnesses and a cage of this kind is a recommended acquisition.

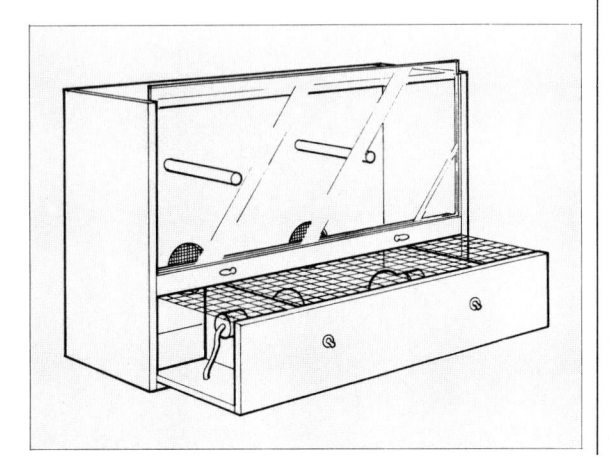

Housing

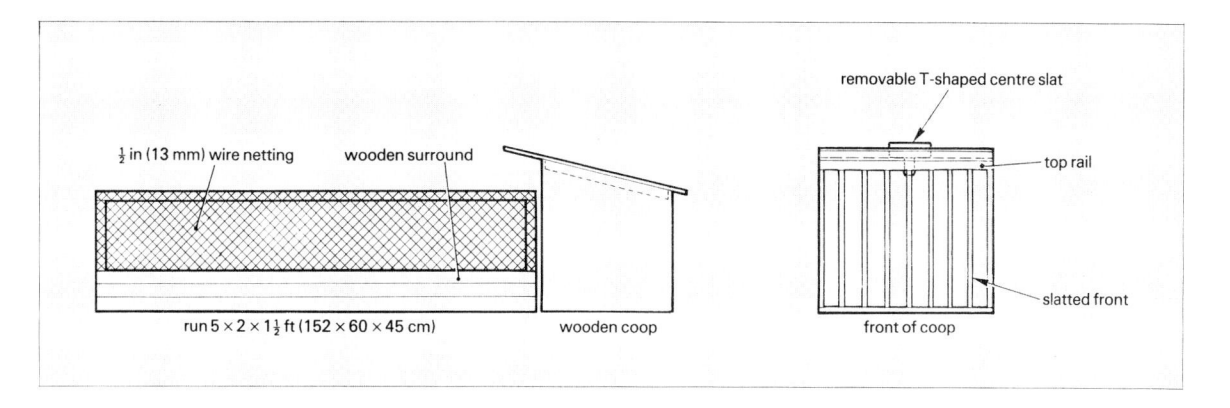

FIG. 11 Coops and runs of this kind can be purchased from game farms specialising in rearing equipment for pheasants, partridge, Mallard, etc. They are also easily constructed by anyone with any pretensions towards being a handyman (or woman). Tongued and grooved timber, weather-boarding or marine ply can be used. Although it is usual to set broodies on a hollowed-out turf, remember to include a solid base in the coop to deter vermin.

docks and all the other traditional and more innovative kinds of bird accommodation is needed there are good specialist books available which will take the uninitiated through a maze of construction details (see Bibliography).

Do remember, though, the importance of housing your birds in the best possible way; whether it is finches and small softbills in a garden shed or aviary, or larger birds occupying a leafy paddock. Space and expense are not necessarily the main criteria in such matters. The person with flair and 'feel' for what he or she is doing is capable of creating a good avicultural environment in a back yard; whereas an individual with acres of ground and a deep pocket may fail through lack of those two important attributes.

[47]

BREEDING

If there was ever a time when collections of exotic birds could be kept solely for the pleasure of their company, or perhaps for ornamental value, it certainly is not now. With many species becoming increasingly rare in aviculture, an inactive breeding life is extremely undesirable. The dictionary defines aviculture quite simply as 'the rearing of birds; bird fancying'. The need to take the first part of that brief definition at face value has never been more necessary than at present.

For some unknown reason there is a tendency to regard people whose bird-keeping interests are connected with canaries, budgerigars, mules, hybrids and the like as 'fanciers'; whereas those who opt for more unusual foreign birds, waterfowl, pheasants etc are more likely to be given the title 'aviculturist'. Whether a 'fancier' or an 'aviculturist', if foreign birds are the main interest it is vital that all concerned, no matter what the size and scope of their collection, pay heed to present trends, and undertake to do everything possible to persuade the birds they own to breed. The range available nowadays has contracted alarmingly. A glance at dealer price lists of some of the foreign species being offered for sale only a few years ago quickly confirms that fact.

By now there can be few whose interest, particularly in exotics, is more than superficial who are not aware of the dread possibility that within a few years their hobby might be starved of all replacement stock unless it can be proved to have been captive-bred. So it is safe to assume that the more dedicated people, 'fanciers' and 'aviculturists', are making a supreme effort to produce home-bred birds of all kinds. Indeed there is increasing evidence of this welcome trend. Aviary-bred birds are appearing in considerable numbers, and certainly in greater variety, on the showbench than was the case a few years ago.

Although it is still only a trickle, the rate at which surplus captive-bred stock is advertised in the bird-keeping press is beginning to accelerate.

Those concerned are to be applauded for their efforts. After half a lifetime 'flirting' with exotic species I have more hard-luck stories in my armoury than any angler. Despite much-improved breeding results with many species, success with other desirable exotics continues to be elusive. Years ago, however, the same remarks applied to many species whose breeding, even in cages, we accept as something of no great consequence nowadays. There was plenty of heartbreak and frustration surrounding early attempts to establish the likes of Gouldian Finches and other popular and presently widely-kept grassfinches.

A surprisingly wide range of exotic species will breed in the relatively close confines of a cage or indoor compartment if they are provided with a quiet, comfortable environment. Indeed, cage breeding can prove a more efficient method than if the birds concerned are housed, with others either of their own kind or other species, in a spacious garden aviary. For many people aviary breeding somehow represents a more natural, and therefore presumably palatable, approach to the task; but precious fledglings face all sorts of hazards in such a situation and some of them are far too important to the future in aviculture of their species to be exposed to undue risks.

Cut-throats, African and Indian Silverbills, Magpie Mannikins, Red Avadavats and Golden-breasted Waxbills are among a number of species which have demonstrated their willingness to raise families in indoor compartments in my own collection over the years. There is little doubt that many others are potentially successful cage or compartment breeders.

What is important at the moment is that people

make a concerted effort while many popular seedeaters are still available in large numbers through exports. It may not always be so; and if enough of us had decided a few years ago that such things as Green Cardinals, Olive and Cuban Finches, to mention only three species, were worthy of such attention they might not have all but disappeared from most collections by now (although I readily acknowledge that Cuban Finches seem to be making something of a comeback thanks to careful breeding programmes carried out by a number of enthusiasts).

Specialisation

Some degree of specialisation by individuals is desirable if we are to achieve the best possible results with the wild stocks still available. There are, unfortunately, far too many 'postage stamp' collections still in being; the avicultural equivalent of Victorian zoos which gathered together diverse and unwieldy accumulations of animals with little regard to their breeding potential.

Of course, for many people the very thought of specialising is too constricting to receive serious consideration. So far as sustained and serious breeding programmes are concerned, however (and these are the only kind that will save aviculture from possibly terminal affliction), some degree of specialisation is essential. There is no need, of course, to restrict bird-keeping to just one or two species in order to specialise. It almost certainly is sensible, however, especially where space and resources are limited, to concentrate on, say, a favourite group or family.

Planning

Remember, also, the importance of a *planned* breeding programme. Of course when one is dealing with unpredictable foreign birds, all the meticulous planning in the world does not mean they will give their full support and co-operation; far from it, quite often. One sort of planning that is important, no matter how the birds behave, is record-keeping: maintaining notes of matings (to avoid the possibility of in-breeding), incubation periods, food choices when rearing etc.

If at all possible establish and maintain contact with other enthusiasts involved with the species you are attempting to breed. Apart from exchanging and comparing notes such liaison allows interchange of stock as and when necessary.

Breeding Accommodation

Small outdoor aviaries are useful for many species, but to generate consistently good results they need to be in a quiet situation and of a design that provides plenty of seclusion for the occupants. People who have achieved good results with species such as owls and birds of prey often house their birds in aviaries which have solid back, sides and most of the roof; only part of one end is wire-netting covered in some cases. I am not suggesting that this is necessary when setting out to breed most passerine types; but a modification of the idea may well produce better results with known, nervous species.

Outdoor aviaries can be either free-standing or built as small ranges. The latter are labour and material saving and can be used for many species, seedeaters, small parrotlike and softbills, so long as they are not likely to antagonise neighbours through the dividing wire.

Some species are not at home unless efforts are made to produce something resembling a 'natural' environment, which means cover of some kind. Many softbilled species seem unwilling to settle to breeding unless these needs are attended to. On the other hand it is worth remembering that (and this is just one example) one of the real connoisseur's softbills, the Yellow-winged Sugarbird, has been successfully bred in a small, all-wire cage.

Pairing

Whereas a number of popular and presently widely-kept species are gregarious in habit, others need strict isolation in pairs when breeding. Several waxbill species can be kept on something resembling a colony system. One of the advantages of this in a species where there is no sexual dimorphism is that pairs will self-select from a larger group and the level of compatibility is usually better than in 'arranged marriages' where two birds are housed together with more optimism than conviction on the part of their owner.

The whole question of compatibility between potential (or hoped-for) breeding pairs is more important than many people realise. Some insectivorous species are notoriously difficult to introduce; although this usually has much to do with a particular species' natural aggression and the fact that one of the pair is ahead of the other in terms of readiness to breed.

Pair-bonding is an essential prelude to successful mating, laying and rearing; although it varies in its intensity, complexity and duration. Much of it is to do with displays (including singing, of course) and, if they are happy together, the two birds preening each other. Some birds simply do not get on at all, however. If both are believed to be in breeding condition but their behaviour suggests partial or total incompatibility, the only thing to do is part them and try to provide each with a new mate. It is surprising how many times the new union leads to the rapid establishment of bonding, followed by successful breeding.

Even when a group of tiny and normally good colonists such as some of the waxbill species are housed together, problems may arise with odd, unmated birds within the group. Their behaviour can sometimes prove sufficiently disruptive to reduce the breeding potential of companion pairs that may be either incubating or rearing broods. If nesting is already under way the level of disturbance caused by such activity can create even bigger problems.

Difficulties of this nature can be minimised by getting a group of birds together well before the breeding season when, hopefully, they will have sorted themselves out in sufficient time for any unpaired examples to be removed.

Mixed Community Breeding

Compartment aviaries or indoor flights allow a good measure of management of the inmates. Larger aviaries, or tropical houses, where a mixed community of species is housed, obviously permit much less control of individual pairs.

If, for any reason, the intention is to stock a single unit of accommodation with a varied, but compatible, group of birds, it is important that a careful choice of species is made if successful breedings are to be achieved. With plenty of space and cover at their disposal it is often surprising how the most unlikely combinations of species succeed not only in living alongside each other but even in successfully raising families. It is not necessarily a method to be recommended, but it can produce positive results if known troublemakers are excluded from the community.

Some species with a penchant for disruptive behaviour should be rapidly eliminated from the list of likely occupants: such birds as medium to large-sized barbets and thrushes, several chat-thrushes, toucans, toucanets, aracaris, magpies, jays and other members of the crow family are among those groups that one would not associate with small passerine species. Remember, too, that many species that live in harmony in the right accommodation can prove troublesome when companions lay eggs.

One attractive and widely-kept softbill that can create just as many problems among nesting companions as more notorious egg-thieves is the Pekin Robin. These charming songsters are frequently housed with companions of all kinds ranging from finches and other small to medium-sized softbills to small doves and quail. The robins (or, more accurately, *Leiothrix*) can move like quicksilver and are adept at removing eggs from an unguarded nest in the twinkling of an eye.

Artificial Incubation and Hand-rearing

Many parrotlike birds are now raised very successfully by means of artificial incubation of eggs and hand-rearing of chicks. The same techniques are employed by zoos with an increasingly wide range of species including cranes, ibis and various waders. It is an admirable practice resulting in the production of many young birds which, in the normal course of events would be lost.

One aspect that *does* worry me, though, is the scope of such activities involving many parrotlike species in private collections. It is, I suggest, one thing to manage artificially the reproductive cycle of rare and endangered species when double-clutching clearly increases survival chances; and hand-rearing may prove safer than leaving natural parents to tackle the job. It is something quite different when the same techniques are applied to less vulnerable species. In some establishments the main objective seems to be to produce tame, hand-reared youngsters for the pet market where they are in great demand at high prices.

Do not imagine that such wholesale tinkering will not lead to problems. Moving from high-value parrotlike species to the other end of the avicultural spectrum, anyone who has had experience with Chinese Painted Quail 20 years ago, and now, will know precisely what such manipulation can lead to. There was a time when these delightful miniature game birds were reasonably free breeders, and once having embarked on the process of producing eggs they could usually be relied upon to incubate them safely and rear the resulting brood of bee-sized chicks without mishap. The family life of the Chinese Painted Quail is a delight to observe; but how many bird-keepers have had the pleasure of doing so over recent years? Only a small percentage, set against the huge number of people who keep the species.

Commercial interests have turned these, and other quail, into the nearest thing to an avicultural battery hen you will ever see. Fortunately there are still some people, notably members of the Quail Group of the World Pheasant Association, who are striving to maintain stocks which have not been tampered with.

There is reason for concern, too, about many Australian grassfinches which are being produced

FIG. 12a This type of nest-box will be used by many species of duck, especially those that in the wild lay eggs in tree-holes. Some birds prefer a box of this kind to be at ground-level and screened by tall grass or some kind of rushy cover.

FIG. 12b Nest-boxes of this design provide important seclusion for many small Passerine species: others may prefer boxes with a half-open front.

FIG. 12c An old wire hanging-basket filled with suitable nesting material will often be taken over by nesting softbills. Small finches may also tunnel into a well-filled basket to create nest-chambers. Even live, planted baskets will be used by many birds, when watering will probably have to be discontinued!

FIG. 12d Doves and pigeons are notorious for their inability to build secure nest platforms for their eggs. It is sensible to provide help with easily-made wire-based trays similar to this one. Most pairs take to them quickly, especially if a few blades of grass (for the smaller species) or twigs (for larger birds) are placed in the base.

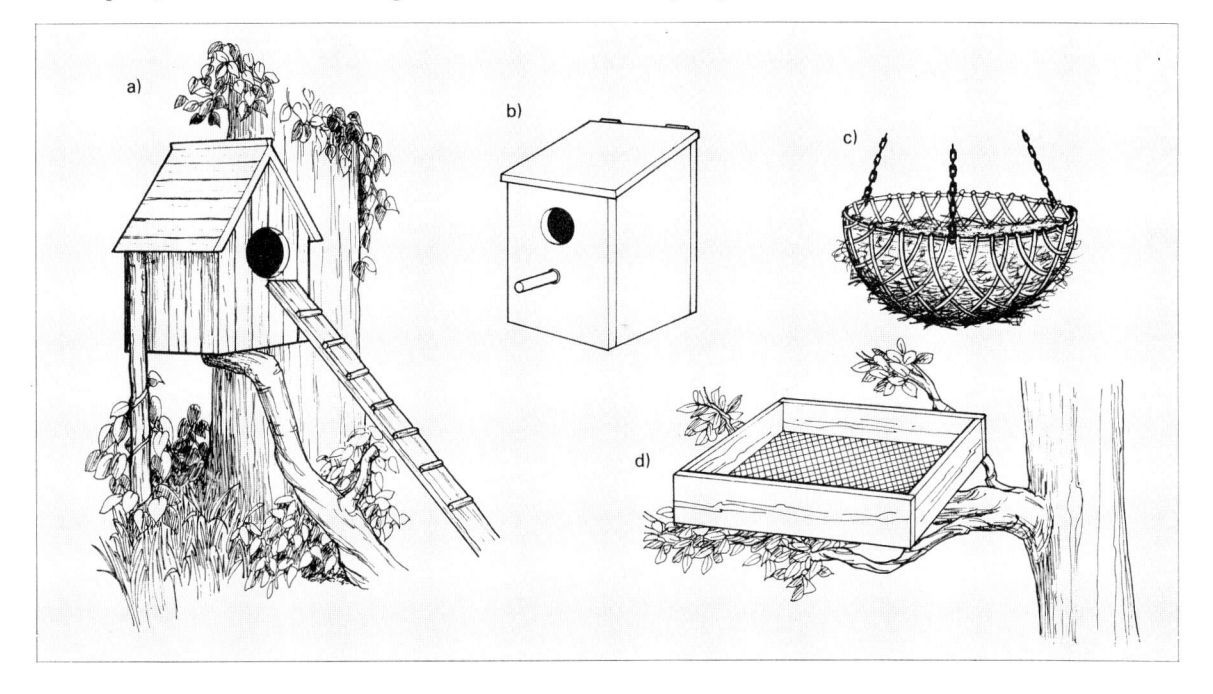

in increased numbers by the simple expedient of removing eggs from rightful parents and having Bengalese undertake hatching and rearing. Although grassfinches are sometimes temperamental parents, on the whole they are no worse than other seedeaters, and a good deal more reliable than many. With luck, however, extra eggs are produced by birds which, like their counterparts in the wild, replace a clutch of eggs lost through accident.

The aviculturists of yesteryear would have marvelled at many of the techniques now perfected by their twentieth-century descendants. Whether they would have approved of *some* of the motivation behind their use is debatable. It would be tragic beyond measure if these new skills were to be debased and abused in pursuit of nothing more than financial gain.

Nesting Materials

Some birds (rheas and many game birds, for example) lay their eggs in the most rudimentary nest imaginable; a simple scrape in the ground. Other species go in for more sophisticated nurseries and are capable of building remarkably intricate structures. In aviculture most of the nest-building birds are provided with a suitable box, basket or other receptacle in which they can lay eggs and bring up a family. Some, such as parrots, have no use for any of the nest-construction materials used by passerine species and deposit eggs on a layer of soft wood chips or shavings.

Many weavers, on the other hand, are indefatigable builders and need masses of suitable plant stems, grasses and other plant materials to weave their remarkable nests.

Nest boxes range from tiny 5-in (12.5-cm) square cubes for small finches to tall 'grandfather clock' parrakeet boxes (see Fig. 12). Ducks of some species utilise boxes or hollow logs. Many are easily home-made; or they can be obtained from specialist manufacturers and bird retailers. Also available are wicker baskets in a variety of shapes and sizes which are favoured by many seedeating species.

Rearing Diets

Remember, always, the importance of providing a suitable rearing diet when eggs are due to hatch. Many young birds are lost simply because their parents cannot obtain essential items for them. These remarks apply especially to various finches which, for several months of the year, consume mainly small seeds and grain, but switch to a near total insect menu during the breeding season.

Various commercial rearing diets are available in crumb form for such birds as waterfowl, pheasants and similar species. It is unwise to modify these complete diets by, for example, adding a percentage of pellet food of some other kind that is not recommended as a rearing diet.

The protein content must be at the correct level, otherwise feather or other abnormalities may arise as the youngsters grow.

DISEASE AND MEDICINE

by A.G. Greenwood MA VetMB MIBiol MRCVS

At first sight, the description of all the disease problems of all the species of birds kept for aviculture would be a hopeless task. Fortunately, however, the various groups of birds have many similar features, with the result that most problems are common to most groups. Thus, for example, egg-binding may be seen in any female bird, from hummingbird to swan, and the causes and treatment are essentially the same.

Active prevention of disease in birds is much more effective than treatment, if only because illness progresses rapidly and its signs are often hard to interpret. Attention given to the advice in this book on selection of birds, housing, feeding and hygiene will be repaid many times in the benefits of healthy stock. However, the serious aviculturist will inevitably encounter disease in his birds at one time or another and it is important that he equips himself to deal with it.

Your Veterinarian

Aviculturists should always try to establish a close working relationship with their local veterinary practice. Many practitioners, but by no means all, take an active interest in avian medicine, particularly the more recent graduates who are much better trained in this field. Those who keep waterfowl or gallinaceous birds will find that specialist poultry practices have much to offer, including excellent diagnostic facilities, but the 'flock' approach to medicine embodied in this specialty does not lend itself so readily to dealing with the problems of the individual bird. Diagnosis and treatment of diseases in birds are undergoing rapid development and modern techniques are making success much more frequent. It is important that the aviculturist is aware of this, as otherwise he may tend to hold back from seeking professional help until it is too late. Not only will this lose him the bird, but it will also give him the impression that treatment is futile, whereas it might well have succeeded if he had acted soon enough.

The establishment of a working relationship with an interested practice is two-sided. The aviculturist should explain quite clearly the features of his collection and what he is trying to achieve with it. He should be prepared to provide simple, clean facilities on his premises for the examination and handling of birds and should have a small hospital room for those needing continuous treatment. It is vital that good records are kept and presented with the sick bird. He should arrange that diagnostic samples and post-mortem material are investigated either by the veterinarian himself or through him by a competent laboratory, so that the vet is aware of all the findings and can act on them quickly.

The value of post-mortem examination of dead birds cannot be over-emphasised. If possible all birds which die, even in certain circumstances chicks and unhatched eggs, should be examined. It is always foolish to assume that you know the cause of death, even where there are obvious signs of injury; disease may have predisposed the bird to being attacked by others, for example.

Many aviculturists wait until the second bird dies and then decide it is time for action. Not only is this one bird too late but, as the spread of epidemics is often explosive, many more may now be in the incubation stages of the disease and beyond help. Equally, the aviculturist should not fool himself that he can 'have a look' and diagnose the cause of death. Pathology is a learned discipline requiring years of training and experience and successful avian pathology is particularly dependent on complex laboratory techniques. In most mortality surveys, no precise diagnosis is achieved in 15 to 20 per cent of bird autopsies and, regrettable though this is, the aviculturist should be prepared for it and not become disheartened.

First Aid

The aim of first aid is to cope with and stabilise emergencies to gain time until skilled help is available. Lives can be saved by the intelligent application of some basic procedures.

The first essentials for any sick or injured bird are warmth, quiet and, for a short period until treatment is available, darkness. Obviously, sick birds need to feed and drink and they will not do so in darkness, with the exception of birds of prey. Nevertheless, a cardboard box kept in a warm room or near a heater makes ideal temporary accommodation for a bird which is very weak, chilled or in shock because of an injury. The box will also do nicely to transport the bird to the vet, but in this case there should be some crumpled newspaper in the bottom to prevent the bird being thrown about during the journey. Sick birds benefit particularly from heat, and a hospital cage which can be held at 80°F (27°C) is ideal for small birds. The heat should not be provided by an ordinary light bulb, as constant light is undesirable. Larger birds should be kept in an ordinary cage in a warm environment.

The major features of an emergency which give rise to shock are severe bleeding, dehydration and pain. Severe bleeding can generally be controlled by simple pressure to the bleeding point. Tourniquets should be avoided as some expertise is needed in their management. The pressure can be applied by hand with a clean gauze or linen pad at first, followed up with a tight bandage. Such bleeding should always receive professional attention as soon as possible. Minor bleeding, as from claws, beaks or feathers, can be controlled with strong ferric chloride solution or potassium permanganate crystals applied to the wound. Dehydration is likely to occur if the bird has been unable to feed or drink for some time or if there is excess fluid loss from diarrhoea. Supplementation with glucose saline (one teaspoon of salt and a tablespoon of glucose to a litre of water) by stomach tube at 10 ml per kilo body weight four times a day will keep up the bird's fluid requirements. Relief of pain in birds and animals is difficult for the amateur as most available products are unsuitable, but fortunately birds seem to have a very high pain threshold and are very stoical. Injured limbs should be gently handled and, if possible, immobilised by taping to the body until they can be attended to. This will minimise both pain and further soft tissue damage from splintered bones. Heavily contaminated wounds can be gently cleansed with mild antiseptic before being presented for surgery, especially if there is to be any delay which may allow infection to set in. Feathers around a wound should be clipped away, not pulled out.

Nursing and Administration of Medicines

Treating an illness or injury with specific drugs or surgery is only the beginning. Nursing care is needed to keep the bird feeding, drinking and comfortable, without which the most expert therapy will be wasted. Few veterinary clinics in Britain are equipped to nurse birds and it is preferable if this time-consuming process can be carried out by the aviculturist himself. A warm cage in a birdroom is needed, and the bird must be offered easily digested food and fluids supplemented with glucose and vitamins as prescribed. Often a course of medical treatment will be needed and, as birds are more effectively treated with most drugs administered by injection rather than by mouth, the aviculturist should ask his veterinarian to teach him simple injection techniques. The use of injections in birds causes much anxiety in many owners and not a few vets but, provided proper fine gauge equipment and the correct drugs are used, such fears are unwarranted. The injections may need to be given several times a day. Drugs which can be given by mouth are best given through a stomach tube made of thin plastic or rubber. With most birds this technique is very simple and easily learned, and it is very valuable for the accurate administration of preventive medicine such as anthelmintics (worming drugs).

If a drug has to be given in feed or drinking water, and it is known to be adequately absorbed by mouth, then the only problem is to know that the correct amount is being taken. When drugs are to be given in water, then sufficient should be dissolved in the amount of water which the bird will drink in a day. This information should be already known, as part of the basic records kept for each bird or group of birds, and is obtained by simply recording the actual volume taken from the water supply each day (with all pools empty) and working out the average. This will obviously depend on the weather, time of year and other factors. Plastic or porcelain pots should be used for administering drugs and treated water should be the only drinking source available.

Mixing drugs with food is more difficult. Many sick birds will refuse food, so this is really only suitable for procedures in healthy birds such as deworming or long term antibiotic therapy for psittacosis. Drugs such as fenbendazole (Panacur) can be mixed with seed or grain to give specific concentrations, but most drugs adhere poorly to food and need to be dissolved in oil first. For birds which dehusk their seed, any drug on the outside will be lost and much research is going into the incorporation of medicines into palatable feed pellets for such birds.

Quarantine

There are statutory quarantine requirements for birds being imported into many countries. In Britain, for example, these requirements are designed primarily to exclude virulent Newcastle Disease (fowl pest). Some birds, such as parrots, present a much greater risk than others from this disease and are accordingly dealt with more severely. Importations of large numbers of birds all now go into a Ministry-approved quarantine station for 35 days. Small numbers of birds (up to ten) which are privately imported can be quarantined under less restrictive conditions, depending on the species and their origin. As such regulations vary throughout the world, and change all the time, the importer should ascertain the import requirements for his birds well in advance.

A valuable side-effect of quarantine is that other severe infectious diseases may appear during this period, often as a result of the stress of transport. Nevertheless, birds entering a collection after official quarantine or, indeed, from any other source should undergo another three-week quarantine period before being mixed with resident birds. They can be acclimatised to new diets in this period, dewormed and sprayed for ectoparasites as well as watched carefully for any signs of disease. Dropping samples can be examined for parasites and *Salmonella*, and psittacines screened for chlamydiosis (psittacosis). Any bird which dies during this period should be examined, and full records kept in case there is any dispute with the supplier.

Infectious Diseases

Birds are subject to a great many infectious diseases and many of them are common between groups of birds. Thus all birds are susceptible to avian tuberculosis and most families have their own strain of avian pox or herpes virus hepatitis. While many parasites are not transmissible between families, each family will have an equivalent parasite type, whose prevention and treatment will be similar. The possibility for transmission of infections between birds depends upon the specific features of their captive management—waterfowl are an obvious example—but the general principles for each type of infection apply to all birds. Infectious diseases are caused by viruses, bacteria, fungi and parasites.

Virus Diseases

Virus diseases have several characteristic features which are important for an understanding of their control. They are mainly very contagious—that is to say they spread very easily from bird to bird, often without the need for direct contact. They are difficult if not impossible to treat at present but they generally confer a good immunity after infection. Recovered birds can often still infect others. The viruses themselves are fragile and cannot survive long outside the body without protection. Consequently the principles of control are isolation of affected birds, immunisation where possible and disinfection of the environment. As treatment has no part to play, prevention is the only defence, ideally by stopping these diseases from entering the collection by quarantine.

Newcastle Disease, or fowl pest, which is caused by a paramyxovirus, illustrates these principles very well. Birds exported from the Far East and South America frequently carry the disease. Consequently, in countries such as Britain a five-week quarantine period has been introduced to prevent its introduction and infected birds are destroyed. The infection can be spread between birds or aviaries by direct contact, by people, by wild birds, in food and even on the wind. Immunisation with vaccines is possible, is widely practised in poultry and has been applied to other captive birds. The virus is killed by heat, desiccation and many disinfectants, but is protected by freezing. There is no effective treatment. The aviculturist faced by this disease must therefore protect his stock either by isolation or immunisation and be careful not to carry the disease in on food, new birds or his clothes. Just such precautions are now having to be taken by pigeon-fanciers in the face of a closely-related paramyxovirus infection of racing pigeons.

Unfortunately, very few virus diseases of aviculturists' birds have corresponding vaccines. Most of the poultry strains of virus, for which vaccines are commercially available, do not affect birds other than galliformes. Some vaccines are available for pigeons and waterfowl infections. Although Newcastle Disease can probably infect any species, the other common viruses, the pox and herpes viruses, have species or family-specific strains.

Avian pox, which causes crusty dry lesions on the bare skin of the eyelids, feet and cere and occasionally spreads into the mouth and throat, is common in canaries, pigeons and wild birds. The viruses are spread by very close contact, or by mosquitoes and other biting insects. In general the different strains of virus do not spread well between birds although there is some overlap. No vaccines are available for these diseases but the infection is self-limiting and usually, if the bird has not been affected internally, the lesions will disappear after a few months and the bird become immune.

Herpes viruses, which cause severe damage to the liver and other organs, are a much more alarming problem. They are frequently the cause of fatal and epidemic infections, and yet can be carried by unaffected birds and spread to others. These viruses are very strain-specific, and infections of cranes, owls, falcons, pigeons, pheasants, waterfowl and parrots have been recognised. Probably the best known of these is Pacheco's Parrot Disease, which is a massive problem in the United States. It has only recently been seen in Britain, as has the falcon virus, but the pigeon, waterfowl and pheasants' strains are well established. Being specific, these viruses can obviously cause devastating effects in specialised collections and must be kept out at all costs by adopting strict quarantine principles. No vaccines are available.

Some of the tricky yet well-known problems of aviculture such as French moult, various 'going-light' syndromes in young finches and canaries and cockatoo feather loss may well be due to virus infections, which explains why they appear to be so intractable. Application of the principles of quarantine may, therefore, help to keep them out of unaffected collections. Active research into many of these problems is going on.

Bacterial Diseases

Disease caused by bacteria have different features from those caused by viruses which lead to different methods of prevention and control. In general they are much less contagious and are only spread by close contact with other birds, animals, people or the environment; they can often be treated with antibiotics; they establish poor immunity in the bird after infection, so that reinfection is always possible and useful vaccines are not readily available; they often originate from within the bird itself when its own bacteria take over because of some other stress which reduces its immunity. From these characteristics we can deduce that the approach to prevention and control involves the elimination of carrier birds, disinfection of the environment, isolation of sick birds, good general hygiene and treatment with antibiotics.

The bacterial diseases are many and cannot all be considered here. However, as they are common to most birds and individual bacteria are much less specific than viruses in that they can cause pneumonia or enteritis or abscesses, for example, discussion of a few of the most important infections which trouble aviculturists will suffice. As many bacterial infections arise as the result of some other stress factor, good husbandry will prevent most of them.

It must be emphasised at the start that although most bird viruses, fungi or parasites are not a risk to normal, healthy people, many of the bacteria are and consequently hygiene in handling sick birds and carcasses is important. The most important of these zoonoses (diseases transmissible from animals to man) are chlamydiosis (psittacosis), pseudotuberculosis, avian tuberculosis and salmonellosis. All of these can cause severely debilitating disease in fit people, and can be fatal to children and the old. Chlamydiosis (psittacosis or ornithosis) is the best known bird disease but the least understood by the general public. Caused by an organism which is very close to viruses but is included here with bacteria because it can be treated by antibiotics, it can affect any type of bird, not only parrots. Indeed it is common in budgerigar aviaries and pigeon lofts and in domestic waterfowl, although parrots seem to be responsible for the majority of human infections which are acquired in aviculture. The disease was once the subject of quarantine regulations in Britain, and still is in some European countries, but it cannot be effectively excluded as birds can carry the organism for long periods without becoming ill. Those countries which use quarantine now combine it with a 45-day period of antibiotic treatment in an attempt to eliminate the infection, but many birds seem to remain infected despite these efforts. The disease can now quite easily be detected by blood and faecal examination, so the possibility does exist to eliminate it from collections with various antibiotic regimes. The problem is, how far do you go? Undoubtedly, many collections would show positive birds if carefully screened, without ever having had a clinical case of the disease. Similarly many wild birds such as doves and gulls are probably carriers and can easily reinfect an aviary as soon as it has been cleared.

Although birds generally catch the disease from other birds and become ill, leading to outbreaks in crowded dealers' premises, it is much more common for a bird carrying the disease to become ill after some stress such as chilling or transport. Another problem is that the manifestations of the disease are varied, from enteritis to nasal discharge to sudden death, so even when the aviculturist and his veterinarian are constantly aware of the possibility the disease may easily be missed. Because of the risk to man, clinical cases should only be treated if they are particularly important or valuable birds, and, of course, they must be isolated from other birds and handled with gloves and mask or respirator. Fortunately, as with most bacterial infections but unlike viruses, infectivity is at its highest when the bird actually becomes ill, so that prompt action can limit spread to other birds. Good ventilation will also limit spread from the dust of dried droppings within a bird room.

Pseudotuberculosis is a major disease of outdoor collections. It is not related to tuberculosis but gets its name from the similar lesions found at autopsy. Like avian tuberculosis, which is common in waterfowl and galliformes but only occurs sporadically in other birds, it is highly persistent in collections and causes seasonal losses every year. Both diseases probably enter collections through contamination by wild birds and then persist in both carrier birds and the environment. Pseudotuberculosis can and does affect most species of birds, being particularly troublesome in touracos and toucans. It is uncommon in waterfowl. Like avian tuberculosis, there is no effective treatment, because the disease is usually too far advanced when diagnosed. There are no commercially available vaccines against either disease, but vaccines specially made against pseudotuberculosis have seemed to be effective in some collections in reducing the incidence of the disease.

A whole host of bacteria cause enteritis in birds. Some, such as *Salmonella*, are acquired from outside sources, occasionally man but more often wild birds or contaminated food, but many are normally present in the intestines of healthy birds. Some stress factor, such as bad food, moulting or severe weather upsets the normal balance in the bird and diarrhoea ensues. Once the bacteria is identified in the laboratory and the correct antibiotic prescribed, most birds can be successfully treated provided attention is given to nursing and fluid replacement as outlined previously.

Respiratory infections may be caused by a similar wide range of bacteria, but are more difficult to treat because of the peculiar nature of the bird's respiratory system. Instead of breathing air in and out of an expandable lung like ours, the bird inhales air *through* its lungs into a system of air sacs, which are like transparent cellophane bags, and then exhales back again through the lungs a second time. These air sacs, and the complicated sinuses around the bird's nostrils, provide blind cul-de-sacs with poor blood supply where infections can be very hard to reach with antibiotics. Consequently, surgery to remove pus and debris is often needed to clear sinusitis and similar, more drastic surgery, is now being tried to clear air sac infections. Improved delivery of drugs, by inhalation therapy with nebulisers, also gives hope for improved success. The bacterial infections in general are those which most benefit from *rapid* initiation of proper treatment.

Fungal Diseases
Most fungal infections in mammals are superficial, involving the skin and hair, whereas internal systemic infections are rare. In birds, however, the reverse is true and fungal infections are often life-threatening to individual birds, although they rarely spread under avicultural conditions. The two major fungus infections of birds, aspergillosis and candidiasis (thrush) are well known to aviculturists. Aspergillosis is typically a wasting respiratory infection affecting parrots, raptors, hole-nesting waterfowl, pheasants and most other birds from time to time. It is generally recognised as fatal and untreatable although new advances in diagnosis and treatment hold out real promise if the disease is caught in the early stages. The fungus grows well on damp, rotting organic material and spreads by spores in the air, so hygiene and good ventilation are important.

Candidiasis (infection with *Candida* yeasts) generally affects the surface of the mouth and crop and typically occurs in young birds being artificially reared. The cause seems to be a combination of vitamin A deficiency reducing the integrity of the surface tissue, a high carbohydrate diet favouring the yeast and failure of the normal bacterial inhabitants of the bird to become established, which suppress the growth of yeasts. Treatment with antifungal drugs is generally effective provided the problem is quickly identified when feeding becomes difficult.

Parasitic Diseases
All wild birds carry parasites of one kind or another. In general, they live in balance with one another and it is not in the parasite's own interests to overwhelm its host. When parasites do cause problems in wild populations, this is usually as a result of overcrowding, in nesting colonies for example. In captivity this balance can too easily be tipped in favour of the parasite so that a bird may have reduced resistance to infestation, or may more easily come into contact with a heavy parasite load, through faecal contamination of its environment. Thus, if parasites can be removed safely from captive birds by treatment or management, they should be.

Ectoparasites live on the outside of the bird, infesting its skin or feathers and are usually mites or lice. Most of these can be killed easily with insecticidal powders or sprays and birds should be examined routinely and treated if necessary, particularly on first introduction into the collection. Most of the proprietary and veterinary products used for dogs and cats are highly toxic to birds and only those recommended for birds should be used. For those parasites which are able to spend most of their life off the host (such as the red mites *Dermantssus* of cage birds) the environment should also be thoroughly treated. The 'scaly-leg' mites (*Knemidocoptes*) are more difficult to attack as they burrow into the skin around the beak, eyelids and feet. They can be treated either by dissolving an insecticide in an oil base and painting it onto the affected areas, or by injecting the bird with a systemic insecticide (ivermectin). Ectoparasites are easily transmitted between birds, but usually only become numerous if the bird is debilitated for some other reason. The quill and feather follicle mites which live within the feather structure are much more difficult to remove, but their involvement in feather loss syndromes is much rarer than generally believed.

Some mites have adapted to living inside their host, the most important being *Sternostoma* which inhabit the respiratory tract and cause breathing problems and sneezing. Canaries, lovebirds and Gouldian finches are traditionally affected, although the infection occurs in a wide spectrum of exotic softbills. Treating the birds in an enclosed space with dichlorvos strips is generally very effective at reducing the number of mites in a collection.

Endoparasites, the roundworms, tapeworms and flukes, can cause serious problems in aviculture, particularly among waterfowl, galliformes and parrots.

The roundworms (nematodes) are most important, being found in the gut, trachea and air sacs. The gapeworm species can affect a wide number of species but *Syngamus* is particularly important in pheasants, crows, starlings, mynahs, birds of prey and even birds such as snow buntings. The parasite is carried by wild corvids and starlings and can live as infective larvae in earthworms. Consequently any aviary established near a rookery or starling roost is doomed to constant infection if the inhabitants have contact with the soil. Earthworms dug up in a garden frequented by wild birds present a similar hazard if used for live food. *Cyathostoma*, the gapeworm of waterfowl, is transmissible to waders, storks and cranes and is, like other waterfowl parasites, spread by wild ducks. The treatment of gapeworm infections has been much improved by the use of newer anthelminthics such as fenbendazole and nitroxynil.

Of the many worms which inhabit the stomach and intestines of birds, those with direct life cycles are the most troublesome. These worms, such as *Ascaridia*, the scourge of parrakeets, lay eggs which pass out with the bird's droppings and can then develop on the ground to produce infective larvae, without the need to pass through an intermediate host, such as an insect. These infective larvae, which are microscopic, can then be picked up by a bird foraging on the ground. In the case of *Ascaridia* itself, the larvae are infective whilst still in the egg, where they have more protection from the elements and can even stick to wire. A moist, warm environment favours the development of these parasites.

Most of the intestinal parasites of birds can be diagnosed by faecal examinations and the birds then treated by anthelminthic drugs. These should be administered individually if possible, although some can be diluted in water or mixed with food. The environment will have to be treated to remove eggs and larvae and this is best done by scrubbing with a strong solution of washing soda (sodium carbonate) or by using a blow-lamp or flamethrower on non-flammable materials. The best preventive measures are to separate the birds from the ground by suspended aviaries, to change the substrate frequently, or to reduce the number of birds in the aviary or paddock and regularly treat the stock to prevent a build-up of infection in the ground. Where the worm has an intermediate host which is known, such as *Daphnia*, the host of the gizzard worm (*Acuaria*) of waterfowl, measures can be taken to reduce the numbers of this host or minimise contact with wild birds which may introduce infection into the collection.

In general tapeworms (cestodes) and flukes (trematodes) are little trouble to aviculturists and they can now be readily treated with the veterinary products niclosamide or praziquantel.

Some of the best known parasitic diseases of birds are caused by protozoa—the simple-celled organisms. These include coccidiosis, histomoniasis (blackhead), trichomoniasis (frounce or canker) and malaria. The importance which these infections adopt in each group of birds varies considerably. Thus coccidiosis and histomoniasis are chiefly diseases of gallinaceous species, trichomoniasis of pigeons and raptors, and malaria of penguins and canaries. A number of other organisms, among them *Giardia*, are emerging as intestinal parasites of budgerigars, parrots and passerine birds causing chronic diarrhoea.

Apart from the blood parasites, such as malaria, which are spread by mosquitoes and other insects, most protozoan diseases are spread by contamination of food and water by other birds, or wild species. As the parasites are generally very delicate and survive poorly outside the bird, good hygiene and careful siting of feed and water pots are the main preventive measures. Some parasites are passed from parents to young, so screening of breeding stock is advisable if a problem arises.

Except under crowded conditions, or where extensive husbandry is practised, the aviculturist should not be unduly troubled by parasites. He should only be aware of measures needed to minimise infection and that there are now a whole range of safe and effective drugs for their treatment.

Non-Infectious Diseases

The main non-infectious diseases encountered in aviculture are due to nutritional disorders or involve the skin and its appendages (claws, beak and feathers). The two are often inter-related. Most birds have the same requirements for basic nutrients although some appear more sensitive to certain deficiencies in their diet. In general, if the dietary advice provided in this book is followed and well stored and carefully prepared foodstuffs are used, few problems should occur. However, some birds which are offered a broad diet can be very selective in their feeding and may thus induce deficiencies themselves. Vitamins, minerals and trace-elements are most important in this context, together with some amino-acids such as lysine. If these are not already incorporated into a prepared diet, as for waterfowl or pheasants, then they should be added as a supplement in drinking water so that they cannot be avoided by the birds. There are many proprietary preparations available.

Vitamins A, D_3, E and K are most commonly lacking in birds, particularly parrots which may select themselves a very poor diet of sunflower seed only. Vitamin A deficiency results in white plaques on the tongue and palate and eventually in the typical tongue abscesses seen in caged parrots. Green food is the only adequate material source of this vitamin. Vitamin D_3 is available directly to birds which are in sunlight, but indoor birds will need a supplement and young birds in the nest can often suffer from a combined vitamin D_3 and calcium deficiency causing leg deformities if adequate supplements are not provided for the parents.

Iodine, which is a mineral required in only trace quantities, is deficient in many seed diets, and iodine blocks or fortified seed are available. Deficiency will upset the bird's thyroid gland and cause feather growth problems.

Some of the skin and feather problems of birds are easily remedied, such as overgrown beaks, whereas others like feather-plucking often seem insoluble. Every aviculturist should have a sharp pair of nail clippers (not scissors) and assorted small files for trimming beaks and claws. To avoid bleeding, it is best to proceed cautiously taking a few small 'bites' rather than one severe cut. Haemorrhage can be stopped with ferric chloride solution. Badly misshapen beaks need veterinary attention under anaesthetic and often powerful tools in the case of big birds like macaws. Some of the newer hoof and dental repair acrylics have a place in the filling and shaping of damaged beaks.

Feather loss in birds is often very difficult to deal with. Each case seems to have a different cause and treatments which are successful on one occasion are useless on another. In general, aviary and flock birds suffer these problems much less than caged specimens. It is important to observe the nature of the feather loss—are they being pulled out by the bird or by another, or are they just falling out and, if so, at what stage of growth? Has the damage stopped but feather regrowth failed? Is this a deviation of the bird's normal moult?

A full examination of the bird's environment, social behaviour, and diet is required. Sometimes a particular feature will be clearly involved and can be quickly remedied, but often the bird will need to be separated, studied and treated individually. As a last resort, if self-mutilation is a problem, an Elizabethan collar can be applied.

Genetic disorders are not uncommon in birds, especially those which are intensively bred. Conditions that appear to have a hereditary disposition include cataract in Yorkshire canaries, feather cysts, bill deformities and abnormal feather colouring. It is up to specialist societies to prevent the perpetuation of these problems although there are difficulties if the disease arises (such as cataract) well into the breeding life of the bird.

Diseases of Breeding

Disorders of the reproductive system are of considerable importance, and are able to affect all groups of birds. Infertility plagues the aviculturist, and should be investigated. Surgical sexing of monomorphic species by laparoscopy is gradually eliminating wrong pairing of birds, perhaps the greatest cause of 'infertility' in parrots. If a more serious problem is suspected and eggs laid which fail to hatch, then advantage should be taken of the vast amount of research which has been directed at infertility in poultry. A number of laboratories examine eggs as a routine, and eggs from persistently infertile birds or young which fail to hatch or to pass the early stages of rearing should be submitted for investigation.

A serious group of problems which affect hen birds are connected with abnormal egg-production, including egg-binding and egg peritonitis. Egg-binding is a complex syndrome, associated with oversize eggs, muscle weakness in the oviduct and low blood calcium and glucose at laying time. It may be related to lack of exercise. The condition should be suspected when any laying bird becomes depressed and inactive and needs immediate attention. Keeping the bird warm and lubricating the cloaca is often sufficient but if the egg is not expelled within 24 hours professional help should be sought and surgery may be required. Egg peritonitis results from abnormal passage of the soft egg, so that it is expelled into the abdomen and the subsequent infection can be rapidly fatal. Obviously a close watch on laying birds is essential.

Sex changes occur in some species, particularly pheasants, and the change is usually from female to male. As the gonad becomes regressed in this condition it is unlikely for the bird to breed in its 'new' state and the condition cannot be prevented.

Conclusion

It has been the aim of this chapter to introduce the aviculturist to the general nature of the disease problems which may occur in his birds, whatever they may be. It has been assumed that he has sufficient knowledge to recognise sickness in his birds and that his main interest will be in prevention. No attempt has been made to cover medical and surgical treatment specifically, as this is the province of the veterinarian and, in this field, a little knowledge is of no real value. Rather it has been emphasised that the establishment of a good relationship with a veterinary practice, perhaps by a group of aviculturists or a specialist society, will pay dividends in long-lived and healthy stock.

THE LAW AND THE ETHICS

Many of the birds included between the covers of this book are familiar and widely-kept, but others are now much less so. The likes of quetzals and birds of paradise now—rightly—have no place in commercial aviculture. 'Rightly' because few, surely, would argue that endangered species of any kind, whether representative of the world's diminishing fauna or its flora, should be subjects for financial gain.

Nevertheless, that is not to say that the time may not arise when loss of habitat and other pressures reduce wild populations of some species to little more than remnants. In such unhappy circumstances judgements will have to be made about establishing captive-breeding programmes to prevent the extinction of the species concerned.

A far-fetched theory? Well, hardly. There are plenty of examples of birds—some of which have even been reintroduced into wild haunts from which they had previously vanished—whose futures have been secured by the painstaking efforts of professional and amateur aviculturists.

The best known of these 'recovered' species is the Hawaiian Goose, now not only abundant in waterfowl collections but also reintroduced into wild haunts in the Pacific, thanks—in great measure—to the efforts of the Wildfowl Trust in Britain. In the Channel Islands the Jersey Wildlife Preservation Trust is achieving similar success with the endangered Mauritius Pink Pigeon. Now there is a growing captive-bred reservoir of these birds in several collections and examples have been returned to their island home in the Indian Ocean. There are many other species which are now, sadly, declining in the wild but well established in aviculture. Among them are Australian grassfinches and parrakeets, lovebirds, ornamental pheasants and waterfowl.

If the beautiful Rothschild's Starling is to be saved for future generations to admire it is likely that the most crucial conservation work will take place in aviaries rather than in the bird's natural environment on the island of Bali.

Quetzals, birds of paradise, and others that have the dubious privilege of heading many a list of species most in need of protection, may yet need to fight their last vital battle for survival in a man-made environment; and that, I believe, is reason enough to set down details of feeding and management techniques for species unavailable to aviculturists at present but which may need all the help they can get in the future.

I regard myself as fortunate in having served part of my avicultural 'apprenticeship' in an era when a few of the great collectors were still operating—Wilfred Frost and Cecil Webb among them. I was on the staff of the now extinct Belle Vue Zoo in Manchester in those days. From time to time one or other of them would appear, as likely as not from some remote corner of the (then) Empire, to offer for sale various avicultural treasures brought back from a recent collecting trip. Whilst Frost's wares tended to be from the Far East and Pacific regions, Webb was probably more at home in the forests of West Africa; although both men travelled more extensively in their lifelong quest for unusual species.

Birds of paradise, bower birds, sunbirds, kingfishers, drongos and all manner of other rare and delicate birds were tended meticulously during long sea voyages to Britain, and most arrived not only in excellent health but also feather-perfect. For the Webbs and Frosts, and others like them, were unique in the varied skills they brought together. The fact that they were able to earn a living through collecting and selling exotic birds suggests that they were gifted entrepreneurs. There was undoubtedly a touch of the explorer in their make-up; they were expert trappers, capable of outwitting the most alert and sharp-eyed denizens of the forest, and, not least, they were highly-skilled aviculturists. They had to be, for they dealt in quality as opposed to quantity and the loss of a rare and valuable specimen could represent a substantial financial disaster for them.

I treasure my relationships with such people for they helped instil in me an awareness of the need to work constantly to improve and update my knowledge of the species I was caring for. Whatever else *they* cared about, the collectors of yesteryear didn't believe in waste, least of all of their precious charges.

Several years ago, on a visit to London Zoo, I spent an afternoon in the company of Cecil Webb and the late John Yealland (another legendary figure in avicultural circles). Much of the conversation revolved around various West African sunbirds, which, over the years, Webb had successfully brought back to England. We moved on to talk about the almost mythical *Picathartes*—the Bare-headed Rock Fowl—that he had also collected for the zoo.

Then we were discussing a bird which, at the time, I confess I had never even heard of. Cecil Webb had first come across the Oriole Babbler on a pre-World War II expedition to West Africa's Gold Coast (now Ghana). He described the bird vividly, and went on to remark that it was one of the most stubbornly insectivorous species he had ever dealt with. For several weeks his first specimens could be persuaded to take nothing other than small live insects.

That almost forgotten conversation came into its own nearly 25 years later when I was able to acquire a pair of ... Oriole Babblers. Although they eventually learned to accept a more inanimate diet, to begin with at least they were just as addicted to live food as Webb had suggested.

Nowadays, with the world's wild places disappearing at an alarming rate and with many wild creatures labelled 'endangered,' 'threatened' or 'vulnerable' in various parts of the world it would be impossible for collectors to ply their trade, and they have vanished from our planet rather more rapidly even than some of the sorely threatened creatures they used to pursue. In this case, however, it is not the now departed pursuers who can be held responsible for the more recent demise of their quarry. Responsibility for that sad situation cannot be laid at any one door.

One sinister threat to endangered species is that posed by illegal trade; smuggling, the black market, call it what you will, it threatens conservation programmes overseas and damages the reputation of honest and law-abiding aviculturists when the more acquisitive and unscrupulous are prepared to pay high prices for birds they know *cannot* have have been imported legally.

Although some people regard certain aspects of the legislation governing imports as irksome, whether they are right or wrong in harbouring such views is neither here nor there. What is vital is that both national and international law must be obeyed. Any kind of 'black market' is to be deplored for such activity leads inevitably to horrendous losses as live birds are packed in often appalling conditions in order to evade detection by customs and other law enforcement officials.

The Convention on International Trade in Endangered Species of Wild Fauna and Flora (CITES) came into force in 1975 when ten countries ratified it. At the time of writing 91 states are party to the Convention with several further countries poised to join. The species covered by the Convention are grouped into three categories.

Appendix I. Taxa threatened with extinction which are or may be affected by trade.

Appendix II. Comprises taxa which although not necessarily now threatened with extinction may become so unless trade is strictly regulated to avoid utilisation incompatible with their survival. Also included in this appendix are species which are listed because of their visual similarity to other Appendix I and II species.

Appendix III. Contains species which any party identifies as being subject to regulation within its own jurisdiction for the purposes of preventing or restricting exploitation, and as needing the co-operation of other parties in the control of trade.

As well as live specimens the Convention also

applies to eggs, feathers and any other parts and derivatives which can be identified as belonging to a listed species. In some instances species are only listed on CITES with respect to certain sub-species or geographically separate populations. The species texts in this book are annotated with the relevant CITES appendix where the species involved is covered by the Convention. Imports or exports of Appendix I specimens are only authorized in exceptional circumstances. These exceptional circumstances under which licences *may* be granted fall into four main categories.

   a) Personal or household effects.
   b) Captive-bred specimens.
   c) Non-commercial loans, donations or exchanges between registered scientists or scientific institutions.
   d) Travelling zoos or circuses.

All of these are themselves carefully defined by the text of the Convention or the various resolutions passed by conferences of the parties. Imports of Appendix II and III specimens are generally permitted provided that the appropriate export documentation is presented. These are the basic principles governing imports and exports of CITES species, however many states take further domestic measures regarding the capture, collection, import, export, sale and possession of bird species including many which are not covered by CITES at all. The appropriate authorities in the countries involved should be contacted for advice regarding restrictions within their jurisdiction. A list of principal authorities responsible for issuing import and export licences in the main states importing and exporting birds is presented in the Appendix.

The party states meet at a conference every two years to decide on increased or decreased protection for certain species and to pass resolutions clarifying particular points of the text of the Convention or introducing new procedures to assist in the conservation of species covered by CITES. It follows therefore that all the information given here and the annotations in the text are liable to alteration from time to time.

An increasingly wide range of exotic species is being bred under controlled conditions and wherever possible stock should be obtained from a reputable breeder. This will not, however, be possible where many of the rarer and less free-breeding kinds are concerned. Fortunately commercial importers are also mainly honest and trustworthy people who want no part of the activities of crooked fringe elements. Regular advertisers in bird-keeping journals are a good source of stock.

It is time to start asking questions when previously unheard-of individuals offer for sale high-priced rarities which rarely breed in control and whose importation might be regarded as unlikely to obtain official approval.

PART II
THE BIRDS

RHEIFORMES

Rheidae (Rheas)

There are two species of these small ostrich-like birds from South America. Popular and widely-kept zoo and bird garden exhibits, they are easy to feed and house in paddock-type accommodation. Properly-managed adults are usually prolific egg-layers; the chicks are sometimes difficult to hatch and rear successfully in an incubator and brooder. Wild rheas live in small groups; there are occasionally larger parties in a few areas where they are still plentiful.

In aviculture they need a generous area of grass with a minimum 48-in (122-cm) high perimeter fence. Rheas need an open-fronted shelter for use in severe weather; trees and shrubs also provide shade and shelter.

PLATE 1 *Rhea americana*, Common Rhea

Rhea americana [PLATE 1] (**III**, ssp. *albescens* **II**)
Common Rhea, Greater Rhea

---

60 in (152 cm). The plumage is mainly grey with black on the top of the head, the chin and neck. The sexes are similar but the males are usually taller.

Range Lowland areas of northern and eastern Brazil, parts of Uruguay, Bolivia, Paraguay, central Argentina.

Status They are becoming scarce in many parts of the range.

Avicultural rating Not difficult, but they need plenty of space. Having been bred successfully in many public collections for many years, they are now more readily available to private aviculturists.

Feeding Diet T, plus wholemeal bread, green and root vegetables. Suitable-sized grit is essential.

Breeding This is not difficult, but it is best if small groups (one cock to two or more hens) are housed together. The females lay up to 10 eggs. The incubation period is 35–40 days, and is undertaken by the cock. The chicks may not feed for three or four days after hatching—then suitable pellet foods will be taken (see p. 25) plus chopped green food; small grit is essential. Natural hatching and rearing is best.

General management They are easily managed. One cock to two or more hens is usually best, but larger groups of mixed sexes may be compatible. They are hardy and adaptable, and can be housed with other paddock-type stock of appropriate size (deer, wallabies etc).

Pterocnemia pennata (I)
Darwin's Rhea, Lesser Rhea

---

40 in (101 cm). Brownish-coloured plumage with some white spotting. The sexes are similar.

Range Western areas of South America from south-east Peru southwards to Chile and southern Argentina.

Status Endangered in the wild and now absent from many former areas of range.

Avicultural rating Not difficult to manage; but rarely available. Although now extremely rare in the wild, Darwin's Rhea is being bred in a number of zoological collections, notably in the USA.

Feeding/Breeding/General management Similar to Common Rhea.

TINAMIFORMES

Tinamidae (Tinamou)

There are more than 40 species of tinamou, but only a few have been bred successfully in aviaries. Superficially, they are like partridge-type game birds, but are not related. Primitive, seemingly lacking in intelligence, they are poor flyers. Tinamou are ground-dwelling, but some species roost in trees. They have a wide distribution, from Mexico and Central America, to most of South America. Tinamou eggs are among the most beautiful in the bird world; according to the species, they may be blue, green, chocolate, deep red, buff, yellow, grey, and all have a porcelain-like glaze. They are hunted for food and sport in many parts of their range. Females are usually dominant in courtship; the cocks may mate with several hens, or the reverse situation may apply. Once established, tinamou have proved generally easy to manage. They are hardy but need winter shelter. Provide plenty of natural ground cover in the aviary. They do best in spacious quarters. When frightened, they take off vertically in a panic-flight like quail or partridge; and may damage their heads on rigid tops. Wing-clipping or pinioning reduces this problem.

Nothoprocta perdicaria
Chilean Tinamou

10–12 in (25–30 cm). The plumage is mainly a uniform, streaked brown; the neck is long. The females are usually bigger than the males.

Range Chile, western Argentina.

Status There are much reduced populations in some areas, due to hunting.

Avicultural rating Fairly easy, but they are nervous, take fright easily and may upset their companions.

Feeding Basic diet B, with the addition of coarse-grade insectivorous mixture, chopped hard-boiled egg, fruit, berries (in season), green food and some live food. Provide grit.

Breeding Established birds usually lay well during the breeding season; but laying is often sporadic and the clutch size is variable. The eggs are a deep chocolate colour with a characteristic glaze. The incubation period is 20–22 days. They are often uncertain sitters; so a broody bantam or incubator can be used successfully. Artificially-hatched chicks are often difficult to rear and may refuse food; housing with similar age domestic or pheasant chicks may stimulate interest in food, but not always.

General management They can be kept in groups in large, well-planted aviaries; but watch for disputes between individuals, especially in the breeding season. Shelter is needed in severe weather, but heat is not usually required for captive-bred, acclimatised stock. Frost can affect their feet.

CICONIIFORMES

Ardeidae (Herons, Bitterns)

There are more than 60 species in this family, with worldwide distribution in both temperate and tropical climates. All are wading birds by habit and mainly predatory on smaller denizens of waterways, marshes etc. Most are essentially zoo or bird garden subjects; a few may be regarded as fringe avicultural species if suitable accommodation is available. Many are exceptionally hardy after acclimatisation; the nature of their diet requires meticulous attention to cleanliness (particularly in warm weather) if the aviary or enclosure is not to give off offensive odours very quickly.

Bubulcus ibis (III)
Cattle Egret, Buff-backed Heron

20 in (51 cm). The area of bare skin on the face is greenish-yellow (pink to red in the breeding season); the plumage is mainly pure white but feathers of the crown, breast and back are elongated and warm buff in the breeding season. The bill is yellow (pink to red in the breeding season); the legs and feet are dull yellow (bright yellow to pink-red in the breeding season). The sexes are alike.

Range Almost worldwide; southern Europe, much of Africa eastwards to Iran, India to southern Japan, Philippines, Moluccas, eastern USA, Central America, northern areas of South America.

Status A very successful species; abundant in many parts of the present range and extending. Its spread is usually connected with the advance of cattle ranching.

Feeding They usually do well on a diet of freshly-chopped raw meat, fish (whiting or mackerel etc) and whatever items of live food can be procured for them. If only a small number are kept, occasional small dead mice, locusts etc can be offered. The balance of the diet should be mainly fish in the approximate proportions 3–1.

Breeding Surgical sexing techniques now make pairing a more positive proposition. The Cattle Egret is a colony breeder and nests in trees as well as among reedy ground-cover; it may use open boxes with mesh rather than solid bases fixed securely above ground-level. Clutch 4–6; incubation 22–24 days. The staple diet is usually sufficient for rearing, providing that it includes a good proportion of live food; the young are fed by regurgitation.

General management They need a lofty, spacious aviary, or they can be kept in a grass, marsh and pool area if pinioned. They are good occupants of a temperate bird house; but tropical house conditions are not ideal. They need some warmth in winter (45°F; 7°C) and do best in frost-free accommodation.

Threskiornithidae (Ibis, Spoonbills)

The Threskiornithidae contains more than 30 species, which are mainly warm-climate birds from Africa, South-east Asia, Australia and the Pacific region, Central and South America. Ibis have long, down-curving bills; those of spoonbills are long, straight and spatula-ended. They are mainly birds of freshwater regions; they are gregarious and live and breed in colonies. Few can be regarded as good subjects for aviculturists, but only because of the space needed to house them. Generally, they are easily managed, hardy and often free-breeding exhibits in many zoos and bird gardens.

Eudocimus ruber [PLATE 2]
Scarlet Ibis

22 in (56 cm). The plumage is entirely scarlet except for the glossy-black tips to the primary feathers. The bill is red-brown with a black tip to all-black; the area of bare skin on the face, the legs and feet are red. The sexes are alike.

Range Coastal areas of northern and north-eastern South America, Trinidad.

Status Protection in Trinidad has helped the species recover; there are good populations now in this area.

Avicultural rating Occasionally available and fairly easy to manage, they are occasional breeders.

Feeding Freshly chopped raw meat and fresh fish in the proportions 1:2 suits these birds. They also enjoy prawns, shrimps and other crustaceans as well as locusts etc. The red colouring fades unless a colouring agent is included in the diet; several suitable preparations are on the market. Offer all food in a shallow pan of water.

Breeding If space permits they are best kept in small groups. They build a loose platform nest of twigs or sticks; so provide suitable bases in the aviary 6 ft (2 m) from the ground. Clutch 2; incubation 16–17 days.

General management Spacious accommodation is needed with some lofty perches and an area of shallow water for wading. Usually fairly hardy when acclimatised, they can be housed with large softbills.

Phoenicopteridae (Flamingos)

Although there is still some debate about classification, most authorities now accept five species of flamingo. In the New World there are the Caribbean (*Phoenicopterus ruber*), Chilean (*P. chilensis*), Andean (*Phoenicoparrus andinus*) and James' (*P. jamesi*). The Old World species is the Lesser (*Phoeniconaias minor*), with the Greater (*Phoenicopterus ruber roseus*) regarded as a subspecies of the Caribbean. They are mainly white or pink; one species is distinctly red. Instantly recognisable, they have long legs and necks, and an oddly-shaped down-curved bill. They are very gregarious; breeding colonies of several thousand birds occur in some parts of their range. Flamingos have specialised feeding habits; the bill and the tongue combine to work with a suction-pump action. The comb-like filters and tongue 'pump' water into and out of the bill; the filters strain off algae, molluscs etc, which are then swallowed. This action is virtually continuous while feeding—pumping then swallowing. They are good avicultural subjects which need plenty of space; a paddock with a pool is best. A shed is essential for winter housing. The birds become hardy, but are prone to frostbite; they can break their legs in ice traps. They are easy to feed: modern, ready-mixed diets provide good nutrition and help to retain their colour. Flamingos are long-lived (more than 40 years) when properly housed and fed.

PLATE 2 *Eudocimus ruber*, Scarlet Ibis

Phoenicopterus chilensis [PLATE 3] (II)
Chilean Flamingo

48 in (122 cm). Mainly pink and white, they have grey legs with distinctive pink joints and feet. The sexes are alike.

Range Estuaries and lakes of temperate South America.

Status Good populations in some parts of the range but generally reduced.

Avicultural rating Not difficult, but they need carefully planned accommodation. They are easy to feed with modern diets. These are spectacular and beautiful occupants of suitable pool and paddock accommodation, and are long-lived and fairly hardy.

Feeding Formerly difficult and time-consuming when soaked bread, boiled rice, wheat, shrimps and colouring agents were used. Now specially-formulated Flamingo Food (see p. 27) in pellet form is widely used with great success. The birds may also obtain crustaceans and other live food in a carefully-managed, not over-populated, stretch of water.

Breeding This is infrequent, although results are improving in some collections as the imbalance of male/female stock is corrected through surgical sexing. The birds need deep water in the pool in order to mate properly (depth equal to leg length). Reduce the depth when eggs or chicks are present, to minimise the risk of chill or drowning. Wild birds build cone-shaped nests of mud with a shallow depression at the top for a single egg. They are usually sited in shallow water or on islands. Various materials can be used to help create artificial nests; if accepted the birds usually finish these off. A muddy area in the enclosure is essential. They breed at about five years of age. Incubation is carried out by both parents for 29–32 days. The chicks enter the water a few days after hatching; they are fed by their parents for several weeks on regurgitated food. They are

PLATE 3 *Phoenicopterus chilensis*, Chilean Flamingo

difficult to hand-rear, but it is possible. If providing artificial nest cones, allow plenty of space between them to prevent squabbles among sitting birds.

General management Despite their specialised behaviour and requirements, flamingos usually thrive if fed and housed correctly. They need protection from frost and ice in winter. Some species require only an unheated shed (in mild districts); otherwise slight heat may be necessary according to the species and quality of climate. Do not allow birds outside in freezing conditions. Pools must have gradually sloping sides for ease of entry and exit. Pinioned birds are essential. Feather-clipping is not recommended; it needs at least annual attention which necessitates catching, causing a risk of stress, leg fractures etc. Pinioning should be carried out by a veterinary surgeon. In winter, the shelter can be carpeted with a good depth of peat, wheat straw etc; rubber mats (under the bedding) help prevent slipping. Clean out the bedding regularly. When outside, the birds will spend most of the time in the pool, resting on one leg when not feeding. Enclose them behind a high fence if possible; even pinioned birds can clear low barriers with the aid of strong winds.

Phoenicopterus ruber roseus (**II**)
Greater Flamingo

---

60 in (152 cm). The plumage is mainly pale pink in colour with black flight feathers. The legs and feet are pink. The sexes are alike.

Range Southern Europe, parts of Africa and Asia.

Status Abundant in some parts of its range but declining in a number of areas.

Avicultural rating See Chilean Flamingo. Widely kept in public and private collections, although the Chilean is now more often seen.

Feeding/Breeding/General management See Chilean Flamingo.

Phoeniconaias minor (**II**)
Lesser Flamingo

---

34 in (86 cm). It is easily distinguished from the other flamingo species by its smaller size. The plumage is pink, deeper in colour on the wings and back. The legs and feet are pink. The sexes are alike.

Range Eastern and southern Africa, Madagascar, north-west India.

Status Still the most abundant flamingo species.

Avicultural rating See Chilean Flamingo. This is a surface-feeding species (Chilean and Greater are both bottom-feeders). It is now less often seen in collections.

Feeding/Breeding/General management Similar to Chilean Flamingo.

ANSERIFORMES

Anatidae (Ducks, Geese and Swans)

The Anatidae contains nearly 150 species with worldwide distribution (except polar regions). Many species are widely kept and popular avicultural subjects; most are hardy, easy to feed and manage. Generally good breeders, most captive waterfowl are now bred in collections. They are gregarious; in the wild many species gather in large flocks prior to migration. Most need a staple grain, pellet or vegetable matter diet; a few are more specialised feeders. Pinioned birds are easy to accommodate, but must be protected from foxes and other predators. An area of water (natural or man-made) is essential for the establishment of a waterfowl collection. Most are good community birds; a few are best housed alone and others need segregation during the breeding season. Very few waterfowl use sheds or artificial accommodation, even during the most severe weather; therefore good natural protection is necessary. Trees, shrubs etc provide shelter, shade and windbreaks.

Dendrocygna viduata [PLATE 4] **(III)**
White-faced Whistling Duck, White-faced Tree Duck

---

19 in (48 cm). This species has a conspicuous white face and front of the head; the back of the head and neck are black. The upper surfaces are brown; the neck, breast and back are chestnut. The flanks are barred black and white. The birds have an upright carriage. The sexes are similar but the female's white face may show a suffusion of brown.

Range This is an Old and New World species, which is widespread in tropical South America, Africa south of the Sahara and Madagascar.

Status Abundant in many parts of the range.

Avicultural rating Easy to manage, this species needs indoor accommodation during frost or

PLATE 4 *Dendrocygna viduata*, White-faced Whistling Duck

severe weather. Like other members of the genus, it is extremely popular in mixed collections of waterfowl. It has an attractive appearance and a distinctive whistling call.

Feeding Use Diet N, plus some animal and vegetable matter.

Breeding A fairly reliable breeder, this species can usually be trusted to hatch and rear its own broods; otherwise use a broody bantam or incubator. Usually 6–10 eggs; incubation 28 days, mainly by the female. Provide a box or a hollow log nest screened by vegetation. Ducklings need care to begin with, especially if artificially hatched.

General management Although not especially shy, they prefer an enclosure offering some seclusion; also an ample area of water. They are easy to care for but not completely weather-resistant in severe winters; so are best housed in sheds during periods of frost and ice. They are frequently seen perching or occupying a high vantage point.

Dendrocygna autumnalis (III)
Red-billed Whistling Duck, Red-billed Tree Duck, Black-bellied Whistling Duck, Black-bellied Tree Duck

---

20 in (51 cm). The plumage is mainly black and brown; the sides of the head and neck are ash-grey. There is a white ring round the eye. They have an upright carriage. The sexes are alike.

Range Central and South America; also southern areas of USA.

Status Abundant in many areas of the range.

Avicultural rating See White-faced Whistling Duck. This is a very popular and widely kept species, despite its fairly muted plumage colours and patterns. It is generally a good mixer.

Feeding/Breeding/General management See White-faced Whistling Duck.

Dendrocygna bicolor (III)
Fulvous Whistling Duck, Fulvous Tree Duck

---

20 in (51 cm). Mainly various shades of brown with darker barrings on mantle and wings. The distinctive flank feathers are buff-white. They have an upright carriage. The sexes are alike.

Range Southern USA, Central and South America, India, parts of East Africa.

Status Still fairly widespread.

Avicultural rating See White-faced Whistling Duck. This is the most widely-kept member of the family.

Feeding/Breeding/General management See White-faced Whistling Duck. Incubation period 25 days.

Cygnus atratus
Black Swan

---

40 in (102 cm). The plumage is entirely black except for the conspicuous white flight feathers and red bill. The sexes are similar, but the female is usually smaller and slighter in build.

Range Australia and Tasmania; introduced into New Zealand in the mid-nineteenth century.

Status Populations have been reduced in a number of areas, but there are still good numbers in some places.

Avicultural rating This is the most frequently kept member of the swan family; they are easy to manage. Generally easy to feed and house, they need a good stretch of water and, like other swans, are best housed alone. They can be kept with other birds in a large enclosure. Young birds can be kept together to establish a colony; they will nest successfully in colonies, but it is difficult to introduce adults to each other for group exhibition.

Feeding Diet N, with the addition of extra green

food including aquatic plants when available.

Breeding A fairly free breeder, the Black Swan makes a bulky nest of rushes, bits of reed, twigs etc; usually with a down lining. Up to 6 eggs, occasionally more; incubation, by both parents, 34–36 days. Breeding pairs are best segregated away from other species; they are often aggressive at this time and capable of killing small or medium-sized companions. They are monogamous i.e. they pair for life.

General management Easy. Although at best in a large paddock with good water, the Black Swan is the best of the genus if space is limited.

Cygnus melanocoryphus [PLATE 5] **(II)**
Black-necked Swan

39 in (99 cm). The plumage is pure white except for the contrasting black head and neck; a red knob at the base of the upper mandible is present in both sexes. The females are usually slightly smaller; otherwise the sexes are alike.

Range Southern half of South America; Falkland Islands.

Status Declining in many parts of the range.

Avicultural rating Easily managed but not good mixers; this is one of the most handsome and elegant swans.

Feeding Diet N, with the addition of ample green food; aquatic plants are especially favoured.

Breeding Most pairs are extremely aggressive when breeding. Established adults are usually fairly prolific; clutch 4–7 eggs. Incubation by female only; 37 days. A rudimentary nest is built, usually a pile of twigs, rushes and bits of weed.

General management Not difficult, but keep in mind their aggressive disposition when deciding on accommodation or companions. They are quite capable of drowning even medium-sized

PLATE 5 *Cygnus melanocoryphus*, Black-necked Swan

waterfowl if the opportunity arises. They may need protection in severe weather if frostbitten feet are to be avoided.

Cygnus buccinator
Trumpeter Swan

---

64 in (163 cm). The plumage is all white; the lores and bill are black. The sexes are alike but the female is usually smaller and of slighter build.

Range Now restricted mainly to wildlife refuges in northern North America.

Status Declining in the wild.

Avicultural rating This species does well under controlled conditions; but its aggressive nature and large size make it a handful. A single pair of these magnificent white swans should be the sole occupants of a large pen.

Feeding Diet N, with additional green food; grass clippings can provide useful bulk, but take care that they have not been contaminated with toxic sprays.

Breeding Adult pairs breed fairly readily; 4–7 eggs. Incubation, by the female, 35–37 days.

General management Hardy and easily managed; but they need a really large area to be at their best. They can be savage (both with companions and humans) especially when breeding. They have a strident call (hence the name) which might annoy neighbours in close proximity.

Cygnus bewickii
Bewick's Swan

---

48 in (122 cm). The plumage is entirely white; the black bill has an extremely variable yellow area at base. The sexes are similar.

Range It breeds in Russia and Siberia; and is a winter migrant to Europe, Japan and China.

Status There are still substantial populations but pressures are mounting in some parts of the range due to habitat loss.

Avicultural rating Attractive and easily managed, it is smaller than other white swans, but still needs ample space to do well.

Feeding/Breeding/General management Similar to Trumpeter Swan, but it breeds only rarely in confinement. Usually 4–5 eggs; incubation, by the female, lasts approximately 32 days.

Coscoroba coscoroba [PLATE 6] **(II)**
Coscoroba Swan

---

35 in (89 cm). Although superficially similar, this is not one of the *Cygnus* swans. It is small and all white with a coral-pink bill. The sexes are alike.

Range Southern South America.

Status Variable; probably declining.

Avicultural rating Not difficult in the correct conditions.

Feeding Diet N, plus supplies of green food, including duckweed and other aquatic plants; some animal matter.

Breeding An uncertain breeder, it often lays well, but is a nervous sitter. Hatch and rear artificially

PLATE 6 *Coscoroba coscoroba*, Coscoroba Swan

if there are doubts about the ability of the breeding pair. Usually 4–8 eggs; incubation up to 35 days, by the female.

General management It still has a reputation for being delicate in some collections. It does best on running water; some protection is needed in severe weather, so provide plenty of natural shelter and seclusion in the enclosure.

Anser fabalis brachyrhynchus
Pink-footed Goose

24–30 in (61–76 cm). The head and neck are a dark earth-brown. The upper surfaces are grey-brown, the underparts lighter. The bill is black with a red band, the legs and feet pink. The sexes are alike. It is usually regarded as a subspecies of the Bean Goose (*A. fabalis*).

Range It breeds in Greenland, Iceland, Spitzbergen; and winters in western Europe.

Status Still abundant.

Avicultural rating Popular and widely kept. These birds are gregarious but need ample space when breeding if fighting is to be prevented.

Feeding Diet N, with additional green food if grazing is poor or they are housed in a small enclosure; grass clippings are useful.

Breeding These geese are fairly good breeders which appreciate some natural cover. Usually 4–6 eggs; incubation, by the female, 27–28 days.

General management Hardy and easy to manage; generally good mixers, but they can be truculent when breeding.

Anser erythropus
Lesser White-fronted Goose

21–25 in (53–64 cm). The upper surfaces are dark brown, the underparts greyish. There is a light edging to the feathers of the upper parts and some black barring on lower breast and belly. The bill is pink with a conspicuous area of white extending from the forehead to the crown. The legs and feet are orange. The sexes are alike.

Range Northern Siberia and Scandinavia; it winters southwards into Europe and parts of Asia.

Status Numbers have been reduced in recent years.

Avicultural rating See Pink-footed Goose. This is a smaller version of the White-fronted Goose (*A. albifrons*) and needs similar treatment. It has long been a popular subject with wildfowl enthusiasts.

Feeding Diet N, with ample grazing, or extra green food.

Breeding They are fairly free breeders when established. Clutch 3–8; incubation, by the female, 26–27 days. Some natural cover in the pen is desirable.

General management Their small size makes them suitable for a limited space, but watch companions; they can be spiteful in mixed company. Individual birds often become tame and confiding.

Anser caerulescens
Lesser Snow Goose

---

27 in (69 cm). The adult plumage is pure white with contrasting black primaries. The bill is orange-pink, the legs and feet pink. The sexes are alike. There is also a blue colour phase in which the head and neck are usually white, but the rest of the plumage shows varying amounts of blue-grey.

Range Islands off USA and Siberia; it winters in California, Mexico etc.

Status There is a substantial but declining population.

Avicultural rating Popular, hardy and usually good mixers.

Feeding Diet N, with ample grazing; they may need extra grain during the winter, but do not allow them to become over-fat.

Breeding They are usually good breeders. Clutch 3–8; incubation, by the female, 22–25 days. Provide some ground cover in the pen.

General management They are easily managed and usually mix well with other waterfowl. Full-winged flocks have been kept at liberty and showed no inclination to wander (the provisions of the Wildlife and Countryside Act now make release programmes illegal in Britain).

Anser caerulescens atlanticus
Greater Snow Goose

28–30 in (71–76 cm). Similar to the Lesser Snow Goose, but in this subspecies there is no blue phase. The sexes are alike.

Range Various Baffin Bay islands; it migrates to the east coast of the USA.

Status It may be increasing its numbers.

Avicultural rating See Lesser Snow Goose.

Feeding/Breeding/General management See Lesser Snow Goose.

Anser rossi
Ross's Goose

23–25 in (58–64 cm). The plumage is all white with black primaries. The bill is pink with a grey-blue patch at the base of the upper mandible; the legs and feet are pink. The sexes are similar.

Range It breeds in the Perry River area of Canada (a fact which was unknown until 1938); and migrates to the Gulf of California.

Status Probably fairly stable at present.

Avicultural rating Popular and easy to manage. Small and attractive Snow Geese, they are normally confident and good mixers.

Feeding/Breeding/General management See Lesser Snow Goose, but note that the incubation period can be as short as 21–22 days.

Anser canagicus [PLATE 7]
Emperor Goose

---

30 in (76 cm). The head and back of the neck are a conspicuous white; the chin and throat are black. The rest of the plumage is grey, with black and white lacing giving a barred effect. The bill is pink, the legs and feet orange. The sexes are similar.

Range North-east coast of Siberia and north-west coast of Alaska; it winters in Aleutian Islands and Kamchatka.

Status The population has been reduced due to hunting, but they are still fairly numerous.

PLATE 7 *Anser canagicus*, Emperor Goose

Avicultural rating Perennial favourites, and deservedly so, they are easily managed and hardy.

Feeding Diet N, with extra green food if only poor or little grazing is available.

Breeding They are free breeders and the goslings are easy to rear. Clutch 4–6 eggs; incubation, by the female, 24–27 days. Provide ground cover in the pen.

General management Straightforward. A good beginner's bird; it needs little other than ample grazing and a modest area of water to maintain condition.

Anser indicus
Bar-headed Goose

28 in (71 cm). The plumage is mainly grey in colour with darker and lighter lacings on the feathers; the head and sides of the neck are white. There are two dark brown bars across the top of the head and neck; the back of the neck is black. The bill, legs and feet are yellow. The sexes are alike.

Range It breeds in central Asia and winters in northern India, Assam, Burma.

Status Numbers are believed to be fairly stable.

Avicultural rating Widely kept and easily managed. Outside aviculture, this species is best known for being observed flying over Everest at an altitude of about 27,000 feet (8,200 m).

Feeding Diet N, plus ample grazing.

Breeding Fairly ready breeders; they usually nest among natural cover or will use a kennel-type box. Usually 3–6 eggs; incubation, by the female, 27–30 days. The goslings are not difficult to rear.

General management Hardy and weather resistant; they are good mixers, but watch their behaviour when breeding.

Branta sandvicensis (I)
Hawaiian Goose, Nene

24 in (61 cm). The plumage is mainly ash-brown in colour, lighter on the undersurfaces; the feathers have lighter terminal edges. The face, throat, top of the head and neck are black; the cheeks warm buff. There are dark stripes on the flanks. The bill, legs and feet are black. The sexes are alike.

Range Hawaii and island of Maui (re-introduced in 1962).

Status Slowly improving, following near-extinction.

Avicultural rating Not difficult, if good foundation stock is obtained.

Feeding Diet N, with ample grazing, or extra green food.

Breeding Variable; they can be prolific with good stock, but haphazard inbreeding in some collections has led to the production of inferior birds. Usually 3–6 eggs; incubation, by the female, 29–30 days. They like to nest amid good natural ground cover.

General management Not difficult; they are usually hardy and easily managed. They become tame and confiding, although individual ganders can be pugnacious.

Branta leucopsis [PLATE 8]
Barnacle Goose

24–26 in (61–66 cm). The white face and forehead contrast sharply with the glossy-black crown, neck and breast. The upper surfaces are grey, lighter on the mantle, the underparts grey-white, with some light barring on the flanks. The bill, legs and feet are black. The sexes are alike.

Range Greenland, Spitzbergen and islands north

Anseriformes

PLATE 8 *Branta leucopsis*, Barnacle Goose

of Siberia; it winters in western Europe.

Status Numbers are increasing in some parts of the range but hunting pressures in some areas are having an adverse effect.

Avicultural rating A popular, easily managed species. One of the most widely-kept geese in aviculture; the compact build makes the species a popular choice where space is limited.

Feeding Diet N, with ample grazing, or extra green food.

Breeding Usually prolific; they are easy to hatch and rear, artificially or naturally. Usually 4–6 eggs; incubation, by the female only, 24–25 days. They may use a nest box; otherwise they will nest among ground cover.

General management Arguably the 'ideal' goose for beginners. If space permits they are best in small flocks rather than individual pairs. They are aggressive in the breeding season, but are generally good mixers.

Branta bernicla
Brent Goose

22–25 in (56–64 cm). The head, neck and breast are black, the upper surfaces dark grey. The lower underparts vary in colour according to subspecies: the Pacific Brent (*B. b. nigricans*) is the darkest, the Atlantic Brent (*B. b. hrota*) the lightest. There is a partial white collar round the neck. The bill, legs and feet are black. The sexes are alike.

Range There is circumpolar distribution of various subspecies: Arctic North America, Greenland, Spitzbergen, Arctic Siberia; it winters on the Pacific and Atlantic coasts of North America; Western Europe.

Status There are still fairly large populations.

Avicultural rating Despite the high cost of stock, they are popular and widely kept.

Feeding Diet N, with ample grazing, or extra green food.

Breeding Erratic; some subspecies have bred very infrequently in collections. Usually 3–4 eggs; incubation, by the female, 23–24 days. Some animal matter is useful for the goslings.

General management Small and compact, they look well kept as small flocks. They are hardy and not difficult to manage. Despite their small size, they do best in a spacious pen with a good area of grass.

Branta canadensis [PLATE 9] (ssp. *leucopareia* I)
Canada Goose

22–24 in (56–61 cm). The size varies according to the subspecies. They have generally brown or grey-brown upper surfaces and flanks; the head and neck are black, sometimes with a conspicuous white patch extending from the cheeks to the top of the neck. The breast and lower underparts are white. The bill, feet and legs are black. The sexes are alike.

Range It breeds in northern USA and Canada, Aleutian Islands, and winters southwards to Mexico and the Gulf coast. It is now established

[81]

PLATE 9 *Branta canadensis*, Canada Goose

as a resident breeding species in other countries, including the UK.

Status Widespread.

Avicultural rating Popular and widely kept; some subspecies are among the most inexpensive of waterfowl. There are 10 subspecies which vary greatly in size, weight and plumage.

Feeding/Breeding/General management See Barnacle Goose. Some subspecies breed readily, others less so. Incubation 25–28 days; a shorter period for smaller subspecies. The goslings develop quickly.

Branta ruficollis [PLATE 10] (**II**)
Red-breasted Goose

---

22 in (56 cm). The upper surfaces are black; there are white wing bars and tail coverts. The top of the head and rear of the neck are also black; a black band extends from the crown to the chin. There is a white patch between the base of the bill and the eye, and an area of chestnut on the side of the head bordered with white. The upper breast is chestnut, the lower undersurfaces black, with a white band separating the two colours. The bill, legs and feet are black. The sexes are similar.

Range It breeds in northern Siberia; and winters mainly in the Black Sea area, but also in parts of western Europe, although not usually in large numbers.

Status The wild population is declining.

Avicultural rating Popular and easily managed, but expensive.

Feeding Diet K, plus ample grazing, or extra green food.

Breeding This is often erratic. Clutch 3–8; incubation, by the female, 23–25 days.

General management They are hardy and mix well. Extremely decorative, they often become

PLATE 10 *Branta ruficollis*, Red-breasted Goose

tame and confiding. They need good accommodation with grazing and water.

Cereopsis novae-hollandiae
Cape Barren Goose, Cereopsis Goose

40 in (102 cm). These large and handsome birds are mainly French-grey in colour with a white crown; there are black tips to the primaries and secondaries. The bill is black, with a horny yellow sheath extending from the base to near the tip. The legs and feet are orange. The sexes are similar but the female is usually smaller.

Range Islands off the southern coast of Australia.

Status Protected in the wild, numbers are now slowly recovering after a marked decrease due to excessive hunting.

Avicultural rating They are handsome, hardy and easy to manage, but strong and aggressive.

Feeding Diet N, with good grazing, or extra green food.

Breeding Some pairs are prolific; others less so. Clutch 4–10; incubation, by the female, 34–36 days. They usually nest in mid-winter, and often use a kennel-type box; this is best if the eggs are to survive severe weather. Ground cover is appreciated.

General management Their size, strength and temperament make these handsome birds something of a handful. They should be housed alone, and they need secure fencing; they are good diggers and will uproot an unburied wire-mesh perimeter. They rarely enter water, so they can be kept without a pool, although ample drinking water is essential.

Chloephaga rubidiceps
Ruddy-headed Goose

20 in (51 cm). The head and neck are chestnut; the upper surfaces, breast and flanks are finely barred grey and black with a suffusion of red on the underparts. The bill is black. The legs and feet are yellow. The sexes are similar but females are smaller and less brightly coloured.

Range Southern parts of South America, Falkland Islands; some winter movement northwards into Argentina.

Status Fairly numerous.

Avicultural rating Hardy; they need clean, ideally running, water and good grazing. They are less widely kept than a few years ago.

Feeding Diet N, with good grazing, or extra green food.

Breeding Generally difficult; they need the seclusion of a kennel-type box amid good ground cover. Clutch usually 4–6 (sometimes up to 10); incubation, by the female, 30 days. They are usually early spring nesters.

General management Like other Sheldgeese, they can be extremely savage, especially prior to and during the breeding season. They are best housed alone.

Chloephaga poliocephala
Ashy-headed Goose

22 in (56 cm). The head and neck are light grey; the lower neck, breast and nape are chestnut with some fine black barring. The mantle is brown; the back and tail are black. The undersurfaces are white with distinctive black bars on the flanks. The bill is black. The legs are black inside, orange outside. The sexes are similar.

Range Southern Chile and Argentina.

Status Fairly widespread; individual populations vary in size.

Avicultural rating See Ruddy-headed Goose.

Feeding/Breeding/General management See Ruddy-headed Goose. This species usually proves a more willing breeder than the foregoing one.

Tadorna ferruginea
Ruddy Shelduck

24 in (61 cm). The plumage is mainly tawny brown in colour; the head and neck are cream with a narrow black collar. The primaries are black, the coverts white; the speculum is green. The bill, legs and feet are black. The sexes are similar but the females show a paler head colour and lack the black collar.

Range North Africa, areas of eastern Europe to Mongolia and China; the species winters southwards to India, southern China, north-east Africa.

Status The population appears fairly stable.

Avicultural rating They are widely kept and popular, despite their savage disposition.

Feeding Diet N, with some grazing, or extra green food.

Breeding They are usually prolific. They appreciate cover in the enclosure. They nest in boxes buried in the ground with entry by a drainpipe; ensure good drainage. Clutch usually 8–12; incubation, by the female, 28 days.

General management These are straightforward subjects, except that they are not good mixers. Single pairs can be satisfactorily housed in small enclosures with a small pool and good grazing. They are not recommended for mixed groups.

Tadorna tadorna [PLATE 11]
Common Shelduck

24 in (61 cm). The male in breeding plumage has the head and neck a glossy bottle green; the lower neck and upper breast are white. The mantle and lower breast are chestnut. The undersurfaces and wing coverts are white. The bill is red with a knob at the base. The legs and feet are pink. The females are smaller, less brightly coloured and lack the knob at the base of the upper mandible. Both sexes are duller after the summer moult.

Range Europe eastwards to central Asia and China; some southward movement outside the breeding season.

Status Still widespread, despite pressures in some parts of the range.

PLATE 11 *Tadorna tadorna*, Common Shelduck

Avicultural rating Widely kept, easy to manage, but aggressive.

Feeding/Breeding/General management See Ruddy Shelduck. Incubation period usually 28 days.

Callonetta leucophrys [PLATE 12]
Ringed Teal

---

16 in (41 cm). The face and sides of the head are fawn; the crown, nape and tail are black. The mantle is grey-brown. The wings are glossy-green with some chestnut, the speculum greenish-gold. The fawn-pink breast has some black spotting; the flanks and undersurfaces are grey with black vermiculations. The bill is blue-grey; the legs and feet are pink. The female is easily recognised; her dark-brown head and other markings are less clearly defined. There is no eclipse.

Range Central South America.

Status Uncertain, but populations are believed to be stable in many areas.

Avicultural rating Easy to manage and a good mixer, which becomes tame and confiding.

Feeding Diet N, plus some animal matter if possible. Duckweed or chopped green food are also enjoyed.

Breeding Usually a ready breeder, it prefers a raised box with duckboard. Usually 6–8 eggs; incubation 25–26 days. Both parents participate in rearing. The ducklings are not difficult to rear by natural or artificial means.

General management Usually of a quiet disposition, hardy, easy to house, feed and breed, this is a good community bird, but it can be bullied by more aggressive species.

Aix sponsa [PLATE 13]
Carolina Duck, Wood Duck

---

18 in (46 cm). The mantle and upper surfaces are a glossy bottle-green with purplish reflections; the head and backward-sweeping crest are similar. The chin and throat are white; the breast is chestnut with a purple suffusion and white spots, shading to white on the lower breast and abdomen. The flanks are buff-brown. There is white on the sides of the breast, in the stripe over the eye and on the lower edge of the crest. The bill is red with a black tip and a narrow yellow border at the base; the legs and feet are pink. The female is mainly a combination of olive and grey-brown with some iridescent areas; her chin and throat are white. Her bill is grey; her legs and feet are yellowish. During the eclipse, the male resembles the female.

Range United States, southern Canada, Cuba.

Status Numbers have been reduced through excessive hunting; they are now recovering in many areas.

Avicultural rating This duck vies with the Mandarin Duck (*Aix galericulata*) as the most popular and widely kept species.

Feeding Diet N, plus the usual extras.

Breeding Straightforward. They are hole-nesters

PLATE 12 *Callonetta leucophrys*, Ringed Teal

PLATE 13 *Aix sponsa*, Carolina Duck

in the wild, so provide a raised nest box. Clutch 8–12; incubation, by the female, 30–32 days. The ducklings are easily reared, either naturally or artificially; provide chick crumbs and hard-boiled egg.

General management Hardy and very easy to manage, they can be housed in modest-sized enclosures with a small pond. A good mixer, this bird can be promiscuous and hybridise with other species.

Aix galericulata
Mandarin Duck

---

18 in (46 cm). The male in breeding plumage is probably the most exotic-looking of all waterfowl. The upper surfaces are a dark glossy green; the wings are dark brown with upright, chestnut-coloured 'sail' feathers. The upper breast is purple-maroon, with black and white stripes at the rear; the lower breast and underparts are white. The forehead and crown are dark glossy green merging into bronze-chestnut as the feathers extend into a sweeping crest; the sides of the head and crest are white, and the cheek feathers warm buff. The tawny feathers of the front and sides of the neck form a ruff. The flanks are orange-brown. The bill is red, the legs and feet yellow. The female is mainly brown with lighter streaking and a greyish head, with a white ring round the eye. Her bill is grey, her legs and feet yellowish. The male is similar in eclipse.

Range China, Japan, north-east Asia; it is now established as a breeding species in many European countries and is extending its range in Britain.

Status It is believed that there are reasonable numbers in a few parts of the range.

Avicultural rating See Carolina Duck.

Feeding/Breeding/General management See Carolina Duck. Incubation 29–30 days. The ducklings are often nervous and easily stressed when young.

Anas platyrhynchos laysanensis (**I**)
Laysan Teal

22 in (56 cm). The head and neck are sooty-brown. The remainder of the plumage is dark brown; the feathers are margined with buff-brown. The green speculum is bordered with black and white. There is a white ring around the eye. The bill is grey-green; the legs and feet are orange. The female is similar, but the markings are more clearly defined and the speculum is darker.

Range Laysan Island.

Status This was formerly one of the world's most endangered birds, and it is still vulnerable in the wild due to severe climate changes; but from a total world population of fewer than 10 in 1912 good captive populations have now been built up and the subspecies can probably be regarded as having a secure future.

Avicultural rating It is remarkably tame (the partial cause of its downfall in the wild); it lacks bright colours, but has an attractive, friendly personality.

Feeding Diet N, plus the usual extras.

Breeding This is a good breeder. It will nest in a box or among cover. Clutch 5–7; incubation, by the female, 26–28 days. The ducklings are easily reared by the female, or artificially.

General management Its tameness and easy-going manner are the greatest attractions. It is a fairly good mixer, and does not need a large area of water.

Anas acuta [PLATE 14] (**III**)
Northern Pintail, Common Pintail

23 in (58 cm). The whole of the head, chin and throat are chocolate-brown with a greenish gloss. There is a blackish stripe down the nape and the rear of the neck; the rest of the neck and the breast are white. The mantle, back and flanks are vermiculated fawn-grey and white; the speculum is bronze-green bordered with brown and white. The underparts and under tail coverts are white; the tail is grey-black. The bill is lead-grey, the legs and feet grey-green. The female is smaller than the male, and browner with darker streaking and mottling.

PLATE 14 *Anas acuta*, Northern Pintail

Range Europe, Asia and North America; this species winters in North Africa, Egypt, Ethiopia, south-east Asia, Central America, West Indies and Hawaii.

Status An abundant species in most parts of the range.

Avicultural rating Easily managed and good mixers; they are usually quiet but the drakes will mate readily with other species.

Feeding Diet N, with some animal matter; small molluscs and crustaceans are taken in the breeding season.

Breeding They are generally good breeders. The nest is usually made among cover, but they will use a ground-level box. Clutch 6–9; incubation, by the female, 23–24 days. The ducklings are not difficult to rear either naturally or artificially.

General management These handsome and easily maintained ducks have been long established in aviculture, and are known to have been kept by the ancient Egyptians. They fare best on a fairly large stretch of water, as they are less confident in a restricted space.

Chenonetta jubata
Maned Goose, Maned Wood Duck, Australian Wood Duck

---

18 in (45 cm). The head and neck are dark brown with the feathers at the back of the head extending downwards to form a crest or 'mane'. The mantle is grey, and the rest of the upper surfaces mainly black; the wings, sides and flanks are grey, the latter with fine black vermiculations. The breast is grey-fawn with darker lacing. The bill, legs and feet are greenish-black. The female's colours and markings are generally more muted, and there are conspicuous buff-white stripes above and below her eye.

Range Australia and Tasmania.

Status Much reduced populations in many areas where the species is regarded as a crop pest.

Avicultural rating Easily managed, the Maned Goose mixes with other species but can be aggressive to smaller ducks. It is hardy.

Feeding Diet N, with a good area for grazing; additional green food may be offered if the grazing is poor, but they do best on a good plot of grass.

Breeding They are fairly good breeders. Established pairs usually hatch and rear their own broods successfully, with the female incubating and the male guarding the nest site. Clutch usually 7–9; incubation 28 days. A ground-level box is used.

General management They enjoy shallow water with ground cover up to the pool edges. They are often early nesters. Some pairs are spiteful in mixed company, especially if space is limited.

Anas penelope (III)
European Wigeon

---

18 in (45 cm). The head and neck are chestnut with a gold-buff crown and forehead. The upper surfaces are grey with black vermiculations; there are black flight feathers and a green speculum. The breast is light chestnut, the undersurfaces white. The bill is grey, the legs and feet grey-brown. The female lacks the brighter colours and is mainly brown above with paler underparts. The male in eclipse is similar, but distinguished by his white wing coverts.

Range Northern Europe and Asia; this duck winters in north Africa, Middle East, India, Japan.

Status Widespread, and still good populations.

Avicultural rating This species has a long association with man; it is said to have been kept by early Egyptians. It is easily managed if conditions are right.

Feeding Diet N. Grazing is essential; the birds do not do well unless a permanent good area of grass is available to them.

Breeding They are fairly easy to breed. Provide good natural cover in the pen; they may use a ground-level box or nest among the vegetation. Clutch usually 7–10; incubation, by the female, 24–25 days. The ducklings enjoy duckweed and small aquatic insects.

General management These grazers are not suitable for release into cultivated areas, as are some other small ducks; they will damage most low-growing plants. They are hardy and good mixers, and frequently hybridise with other *Anas* species.

Anas americana
American Wigeon, Baldpate

---

18 in (45 cm). They are similar in general appearance to European Wigeon but have different colours and patterning. The forehead and the front of the crown are white; there is an area of glossy green from the eye to the nape. The rest of the head and neck are buff with fine black spots. The upper surfaces and breast are vinous-brown; the flanks are similar with fine black vermiculations. The bill is grey with black at the tip and base; the legs and feet are grey-green. The female is mainly reddish-brown; the head is whitish with spotting on the throat and crown.

Range USA, western Canada; this species winters in Mexico, parts of Central America, Cuba, West Indies.

Status There are good populations within the main distribution areas.

Avicultural rating Fairly widely kept; see European Wigeon.

Feeding/Breeding/General management See European Wigeon.

Anas sibilatrix [PLATE 15]
Chiloe Wigeon

---

19 in (48 cm). They have a conspicuous white face and forehead; the rest of the head and neck are black with a glossy-green area on the sides, and there is a white patch below the eye. Most of the upper surfaces are black; the feathers are edged with white. The breast is white with heavy black barring; the undersurfaces are white with chestnut on the flanks. The bill is bluish-grey; the legs and feet dark grey. The female is similar but her markings are usually less distinct.

Range Southern South America.

Status They are numerous in many parts of the range.

Avicultural rating Popular and widely kept. See European Wigeon.

Feeding/Breeding/General management See European Wigeon. This species is a free breeder and will use either a ground-level or an elevated nest box, or it may nest among ground cover. Incubation 26–28 days. The drake is protective and assists in rearing the young.

PLATE 15 *Anas sibilatrix*, Chiloe Wigeon

Anas formosa [PLATE 16]
Baikal Teal, Formosa Teal, Spectacled Teal

PLATE 16 *Anas formosa*, Baikal Teal

17 in (43 cm). The forehead, crown, nape, chin and throat are black; the face is buff-coloured with a black dividing line extending from the eye to the throat. There is a crescent-shaped metallic-green area behind the eye; and a narrow white border. The upper surfaces are grey with fine black vermiculations. The breast is warm brown with black speckling. The flanks are blue-grey with fine vermiculations; there are white bands below the bend of the wing and tail. The scapulars are elongated, and cinnamon-brown with a black centre stripe. The speculum is metallic green. The bill is grey, the legs and feet yellow-brown. The female's upper surfaces are mainly tawny-rufous with dark striations. There is black on her crown, nape and sides of the face; buff on the cheeks. The male in eclipse is similar.

Range North-east Asia; this species winters in China and Japan.

Status Uncertain, but declining in many areas of the range due to excessive hunting, utilisation of habitat for agriculture etc.

Avicultural rating Hardy and easily managed. They are poor breeders and therefore extremely expensive.

Feeding Diet N, with usual extras.

Breeding They are extremely unreliable breeders; the reason may be the difficulty of bringing male and female into condition at the same time. When eggs are laid most aviculturists prefer to hatch and rear them artificially; but there is no reason why the parents should not rear the young successfully. Clutch 8–10; incubation 25–26 days. Provide plenty of natural cover in the breeding pen and rushy margins to the pool. They will use a ground-level nest box amid cover.

General management Like most Northern Hemisphere waterfowl, they are very hardy. Quiet and secretive in habit, they need good cover, and ensure that they are not bullied by more aggressive companions.

Anas erythrorhyncha
Red-billed Teal, Red-billed Pintail

18 in (45 cm). The crown, nape and rear of the neck are dark brown; the cheeks (from the base of the bill to the side of the neck) are buff-white. The upper surfaces and wings are dark brown with light-coloured edges to the feathers; the breast and underparts are grey-brown with darker markings. The bill is red; the legs and feet are dark grey. The sexes are alike.

Range East and South Africa, Madagascar.

Status Widespread and fairly common.

Avicultural rating Hardy and easy to manage, this is a popular subject which has an attractive appearance, with the bonus of no eclipse.

Feeding Diet N, with the usual extras.

Breeding They are easy to breed, but like a secluded area for nesting; they usually lay eggs among natural ground cover rather than in a box. Clutch 5–10; incubation 25–26 days.

General management This bird co-exists with other species quite well; breeding pairs may do better in small pens of their own, in common with related small species.

Anas crecca [PLATE 17] (**III**)
European Green-winged Teal

PLATE 17 *Anas crecca*, European Green-winged Teal

15 in (38 cm). Much of the plumage is grey with black vermiculations; the head and neck are chestnut with a white-edged area of metallic green from the front of the eye down the sides of the neck. The upper breast is creamy-buff shading to white on the lower underparts; there is some darker speckling on the breast. There is a buff patch on each side of the rump. The bill is dark grey, the legs and feet grey. The female is mainly brown, many of her feathers with lighter margins; the face, chin and throat are paler. The crown and nape are darker. The male in eclipse plumage is similar but has darker coloured upper surfaces.

Range Europe and Asia; this species winters in North Africa, India and China.

Status Common in many parts of the range, but their secretive behaviour sometimes misleads census-takers!

Avicultural rating Hardy and easy to manage, they are represented in many collections, but are not among the most popular species.

Feeding Diet N, with the usual extras.

Breeding They are generally difficult to breed; shy and secretive, they need plenty of rushy cover, and reed-fringed channels in the pen. Success is more likely if pairs have no competition from other species. Clutch usually 8–10; incubation, by the female, 21–22 days. Many aviculturists prefer to entrust the eggs and ducklings to a bantam.

General management Not difficult, but this is a nervous, easily stressed little duck. It is not especially showy and is usually active at night when feeding.

Anas crecca carolinensis (**III**)
American Green-winged Teal

---

15 in (38 cm). They are very similar to European Green-winged Teal, but the vertical white band on the side of the breast distinguishes the drakes. The females are almost identical to the European species, as is the male is eclipse.

Range They breed in Alaska and northern Canada, and winter in southern areas of the USA, Central America, West Indies.

Status Still extremely numerous in many parts of the range, despite their popularity with sportsmen.

Avicultural rating See European Green-winged Teal.

Feeding/Breeding/General management See European Green-winged Teal. This subspecies seems a little less secretive and shy, but is still far from a free breeder. It needs natural cover in the pen and may use a concealed ground-level box.

Anas flavirostris
Chilean Teal, South American Green-winged Teal, Yellow-billed Teal

16 in (41 cm). The plumage is mainly grey, darker on the upper surfaces, paler below; there are lighter margins to the feathers of the upper parts and large black spots on the breast. The bill is yellow with a black tip; the legs and feet are dark grey. The female is smaller and duller.

Of the four subspecies described, the Sharp-winged Teal (*A. f. oxyptera*) is the one most usually seen. There has been some hybridisation between the subspecies in aviculture and pure stock may be hard to find.

Range *A. f. flavirostris*: central Chile, north-west Argentina south to Tierra del Fuego. *A. f. oxyptera*: northern Peru to northern Chile, Argentina. *A. f. andium*: Colombia, Ecuador. *A. f. altipetens*: W. Venezuela, Colombia.

Status Numerous in most parts of the range.

Avicultural rating Undemanding and popular with aviculturists.

Feeding Diet N, with the usual extras.

Breeding These are generally easier to breed than the European and American Green-winged. They like natural cover in the pen and may use either a ground-level or an elevated nest box. Clutch usually 6–8; incubation 25–26 days. The drake assists with rearing.

General management Easily managed birds, they are usually tamer and more confiding than their European and North American relatives. They tend to be more active at night.

Anas castanea
Chestnut-breasted Teal, Chestnut Teal, Australian Brown Teal

22 in (56 cm). The head and neck are a dark glossy-green; the back is dark brown, the feathers with a chestnut edging. The breast and under surfaces are chestnut, boldly spotted and blotched with dark brown. The bill, legs and feet are dark grey. The female is mainly reddish-brown with darker spots and striations. The head and nape are darker, the chin and throat fawn-brown.

Range Tasmania, coastal regions and islands off South Australia.

Status Reasonable numbers, but declining in most parts of the range.

Avicultural rating A popular and easily managed species.

Feeding Diet N, with the usual extras.

Breeding They are fairly ready breeders. They like natural cover and will use either a ground-level or an elevated nest box. Clutch 8–12; incubation, by the female, 28 days. The male is usually attentive during incubation and until the ducklings fledge.

General management Easy to manage and hardy. They mix well with other species.

Anas punctata [PLATE 18]
Hottentot Teal

14 in (36 cm). A delightful small species; only the Cotton Teal (*Nettapus*) are smaller among waterfowl. The crown and nape are dark brown; the throat and cheeks golden-buff with a black 'thumb-print' marking on the sides of the neck. The rest of the plumage is mainly brown; the wings are black with green gloss. The bill is black; the legs and feet are dark grey. The female is similar but usually lacks the deep gloss on the wings, and has paler under surfaces.

Range East and South Africa; Madagascar.

Status They appear to be fairly numerous in many areas.

PLATE 18 *Anas punctata*, Hottentot Teal

Avicultural rating Popular because of their small size; they tend to be shy and secretive in common with other small teal species.

Feeding Diet N, with the usual extras; they particularly enjoy duckweed and floating aquatic vegetation.

Breeding They are now beginning to breed more freely, having had a reputation for uncertainty for many years. They need a quiet, secluded pen, ideally with no other competition for nest sites, food etc. Provide plenty of natural cover; they may use a ground-level or an elevated nest box. Clutch usually 6–8; incubation 23–25 days.

General management These delightful little birds are hardy and easily managed. Confidence builds up with careful management and they then become more 'visible' than related species.

Anas versicolor puna
Puna Teal

18 in (45 cm). The plumage is mainly buff and light brown in colour, spotted and barred with dark brown; the top and sides of the head and the nape are very dark brown, contrasting with the buff-white cheeks and chin. The long bill is blue-grey, the legs and feet dark grey. The sexes are similar, but the female is usually smaller and with less clearly defined markings.

Range Elevated areas in central Peru, northern Chile, Bolivia.

Status Populations are believed to be relatively stable.

Avicultural rating Popular, easy to accommodate and feed, and hardy.

Feeding Diet N, with the usual extras.

Breeding They are usually free breeders. Provide natural cover and a choice of ground-level or elevated nest boxes. Clutch 5–8; incubation 25 days. Both parents participate in rearing the ducklings.

General management Like related subspecies, this teal appreciates seclusion and natural cover in the pen, especially when breeding. It usually has a quiet disposition.

Anas bahamensis
Bahama Pintail, Lesser Bahama Pintail, White-cheeked Pintail, Northern Bahama Pintail, Summer Duck

18 in (45 cm). The head and nape are chocolate brown; the cheeks, chin and throat white. The upper surfaces and wings are chestnut brown, the feathers with lighter edges; the underparts are chestnut brown with heavy black spangling. The long pointed tail is cinnamon-beige. The bill is blue-grey, red at the base; the legs and feet are dark grey. The sexes are similar but the female is usually smaller.

Range Northern areas of South America, Greater Antilles, Bahamas, Cuba.

Status Numbers are being reduced through excessive hunting and predation.

Avicultural rating An attractive and popular species, which is easily managed.

Feeding Diet N, plus extras.

Breeding This species is usually a free breeder. Provide ground cover in the pen; a ground-level nest box is used. Clutch 6–10; incubation 25–26 days. The ducks rear their own young successfully, but resent disturbance.

General management Easy to manage and hardy, these attractive ducks have no eclipse and remain in colour throughout the year. A silver mutation is also available.

Anas querquedula (III)
Garganey, Cricket Teal

---

16 in (41 cm). The forehead, crown and nape are dark brown; there is a white streak from the eye to the nape. The upper surfaces are brown with pale margins to the feathers; the breast is mottled brown. The flanks are grey with fine vermiculations. The elongated scapulars are brown and white. The bill, legs and feet are dark grey. The female is dark brown with lighter and darker streakings; the throat is white. The male in eclipse resembles the female.

Range Europe; eastwards across Asia to Japan; this species winters in Africa, India, Indonesia; some birds reach Australia and New Guinea.

Status The species has become scarce in some parts of the range.

Avicultural rating They have a quiet disposition and are easy to manage, and hardy.

Feeding Diet N, with extras including animal matter (aquatic insects, crustaceans etc).

Breeding This bird is not a free breeder. It needs a quiet, secluded area, ideally with no competition from other species. It usually nests among dense ground cover. Clutch 7–12; incubation, by the female, 22–23 days. The ducklings need care when young.

General management A good mixer, but it may be bullied by other birds. It is hardy, but ensure that there is good ground cover in severe weather; this bird enjoys quiet, reedy accommodation.

Anas clypeata (III)
Shoveler, Common Shoveler, Northern Shoveler, European Shoveler

---

20 in (51 cm). The enormous bill (from which the bird takes its name) is the first point of recognition. The whole of the head and neck are glossy-green; most of the upper surfaces and breast are white. The mantle is brown, the flanks and underparts chestnut. The bill is black, the legs and feet orange. The female's upper surfaces are brown with lighter margins to the feathers; she has heavily spotted buff underparts. The male is similar in eclipse plumage.

Range Europe, Asia, North America; it winters in East Africa, India, China, Japan.

Status Relatively common in many parts of the range.

Avicultural rating Popular with aviculturists; it is easy to manage and has a quaint appearance, but can be aggressive.

Feeding Diet N, plus extras.

Breeding They are reasonably easy to breed. These birds like good ground cover such as tall grasses, rushes etc. Clutch 6–12; incubation, by the female, 24–25 days.

General management They appreciate an area of shallow, muddy water, plus good natural cover in the pen; they are sometimes a problem in mixed company, especially in the breeding season.

Netta rufina
Red-crested Pochard

22 in (56 cm). The head and throat are chestnut; the nape, mantle, breast and under surfaces are black. The upper parts are grey-brown, the flanks and shoulders white; there is a prominent white band across the flight feathers. The bill is red, the legs and feet yellow. The female's head and neck are dull brown; her other colours are also muted. The male in eclipse is similar, but retains the red bill.

Range Mainly eastern Europe, Asia; it winters in North Africa, the Mediterranean, Asia.

Status The population may be decreasing; numbers fluctuate in many areas.

Avicultural rating These handsome birds do well in collections and are usually good mixers. They are hardy, and fairly ready breeders.

Feeding Mainly Diet N, but with some additional vegetable and animal matter.

Breeding They need natural cover in the pen, and may make their own nest or use a surface-level box. Clutch 6–12; incubation, by the female, 24–27 days. They occasionally lay eggs in other occupied nests.

General management Usually good community birds, they appreciate deeper water for diving. They are often kept in small groups rather than individual pairs. Offer food (other than pellets) in the water, so that the birds will dive.

Aythya valisneria
Canvasback

22 in (56 cm). The whole of the head and neck are dark wine-red; the mantle is pale grey with vermiculations. The breast and saddle are black, the rest of the plumage grey. The bill is black, the legs and feet lead-grey. The female is mainly brown with a pale grey mantle. The male in eclipse is similar.

Range Western and central areas of the USA and Canada; it winters in the southern USA and Mexico.

Status There are still good populations.

Avicultural rating Fairly widely kept, Canvasbacks like seclusion, and are hardy.

Feeding Diet N, with additional green food; they are particularly fond of aquatic plants.

Breeding This species is not a free breeder; it needs seclusion and cover such as rushes, tall grass etc. They may build a nest or use a ground-level box. Clutch 6–10; incubation 25–29 days.

General management They like a good stretch of water and a deep area for diving. Although at times uncertain breeders, they have hybridised with several other species.

Aythya fuligula
Tufted Duck

17 in (43 cm). The head, breast and upper surfaces are glossy black with purple and green reflections; there are contrasting white flanks. The long crest extends downwards from the crown. The bill, legs and feet are slate-blue. The female is almost entirely brown and has a shorter crest. The male in eclipse is similar but darker.

Range The Tufted Duck breeds in most of Europe north to the Arctic Circle; eastwards throughout Asia to Japan. It winters in southern Asia, Africa, Europe.

Status Abundant; the population appears to be increasing in some parts of the range.

Avicultural rating One of the most widely kept of diving ducks, it is hardy and easy to manage, provided that a suitable stretch of water is available.

Feeding Diet N, with the usual additions, including some animal matter.

Breeding Easy to breed; it usually builds a nest among ground cover such as reeds, rushes etc close to water. Clutch 6–10; incubation, by the female, 24–26 days. The ducklings are easily reared but need live insects when young.

General management Popular for many years, these ducks are hardy and easy to manage, but need a good stretch of water and an area for diving. They usually mix well with other species.

Somateria mollissima [PLATE 19]
European Eider

---

24 in (60 cm). The crown is black with a central white line to the nape; the upper surfaces, cheeks, throat and breast are white, the latter with a buff suffusion. The nape and patches on the side of the head are pale green; most of the underparts are black. The bill is grey with a greenish wash, the legs and feet greenish-brown with black webs. The females are brown, heavily barred with darker and lighter shades. The male in eclipse is mainly very dark brown with some white areas on the neck and back.

Range This species breeds in Iceland, northern Europe (including the UK and Ireland); it winters in north-west Europe, the Mediterranean.

Status Fairly stable population.

Avicultural rating This species has been in aviculture for many years and was bred as early as the mid-nineteenth century. It is not as widely kept as some species.

Feeding Diet N, with additional animal matter. The latter is important, and the modern higher-protein pellet foods used in commercial fish hatcheries are useful.

Breeding They are fairly ready breeders, but appreciate good ground cover to provide seclusion

PLATE 19 *Somateria mollissima*, European Eider

when nesting; they may use a box or nest in the open. Clutch 4–6; incubation, by the female, 25–27 days.

General management They fare best on a good area of clear water; cool, running water is ideal, but the birds also thrive in other environments including static, shallow water. Suitable grit or stones are important to aid the Eider's digestion; they also help to wear down the bill.

Bucephala albeola
Bufflehead

---

16 in (41 cm). The head, neck, chin, throat and back are glossy black with blue, green and purple reflections; the undersurfaces and two large bands from the back of the eye to the neck are white. The bill is blue-grey, the legs and feet flesh-coloured. The female has a dark brown head and neck with a white patch on the sides of her head; her upper surfaces are darker, and the underparts grey-white. The male in eclipse is similar but has a darker head, and the white areas are larger.

Range Parts of northern and western Canada; the species winters in the USA.

Status Widespread but local.

Avicultural rating Regarded as a 'challenging' species; expensive and not suitable for novices. It is attractive and mixes well.

Feeding Diet K, plus high-protein items such as trout pellets and some animal matter. Food prepared for smaller insectivorous birds may be suitable, but do not feed such items in water.

Breeding This is a rather infrequent breeder, although successes have become more frequent within the last decade. Provide seclusion and a raised nest box. Clutch 6–10; incubation, by the female, 29–30 days. The ducklings may prove delicate; treat them carefully to reduce any possibility of stress.

General management Rarely successful, except when kept on a good stretch of clean, deep water; running water is ideal, plus the opportunity for diving. Do not try to keep this and related species in cramped pens with poor quality water.

Bucephala clangula [PLATE 20]
European Goldeneye, Common Goldeneye

---

18 in (45 cm). The head, mantle and back are glossy black with green and purple reflections; the

PLATE 20 *Bucephala clangula*, European Goldeneye

circle in front of the eye and most of the under surfaces are white. The bill is blue-black, the legs and feet orange. The female is mainly grey with a brown head, white collar and wing patches. In eclipse, the male resembles the female but usually retains a few black head feathers.

Range Northern Europe and Asia; it winters in southern Europe, India, China, Japan.

Status Populations are believed to be fairly stable, although nowhere are they abundant in the range.

Avicultural rating See Bufflehead.

Feeding/Breeding/General management Similar to Bufflehead. A fairly ready breeder, this bird may use a ground-level or an elevated nest box. Clutch 6–12; incubation, by the female, 28–30 days.

Bucephala islandica
Barrow's Goldeneye

19 in (48 cm). Similar in appearance to European Goldeneye, but the white patch at the base of the bill extends upwards; there is a row of white spots on the wings. The bill is black, the legs and feet yellow. The female is identical to the female European Goldeneye. Males in eclipse resemble the drake European Goldeneye in non-breeding plumage.

Range Southern Alaska, Iceland; it winters along western and eastern coasts of the USA, Aleutians.

Status Numbers are believed to be stable, although populations are small in many parts of the range.

Avicultural rating See Bufflehead.

Feeding/Breeding/General management Similar to Bufflehead. Clutch 8–10; incubation, by the female, 30–31 days. Elevated or ground-level nest boxes are used. They need seclusion and cover.

Mergus cucullatus [PLATE 21]
Hooded Merganser

17 in (43 cm). Striking-looking birds with a large erectile crest which is white edged with black; the front of the head, neck, chin, throat, back and wings are black. The under surfaces are mainly white; the flanks are pale chestnut with fine black vermiculations. The bill is black, the legs and feet dark grey. The female is mainly grey-brown; her flanks are mottled, and the under surfaces whitish. The male in eclipse is similar, but with markings that are slightly more distinctive.

Range Eastern and western areas of Canada; it winters in areas of western and southern USA.

Status Numbers are generally declining because of loss of habitat.

Avicultural rating A much coveted species, but their special needs inhibit many prospective owners.

Feeding A floating pellet diet is needed in addition to Diet N, plus animal matter. Useful items range from sand eels to shrimps.

Breeding Established pairs usually breed; they can bully smaller, weaker companions when nesting. Clutch 8–12; incubation, by the female, 31–32 days. The ducklings can be difficult to rear. Artificially-hatched babies are often poor feeders; small live food (*Daphnia*, *Tubifex* worms, freshwater shrimp) helps during the first week or so.

General management This species is well worth keeping if the correct conditions (a spacious, deep, clear stretch of water) are available. They are good mixers but sometimes too exuberant for some companions. Watch their environment, as these and related species often pick up and eat items ranging from nails to marbles, with fatal consequences.

PLATE 21 *Mergus cucullatus*, Hooded Merganser

Oxyura jamaicensis
Ruddy Duck, North American Ruddy Duck

---

16 in (41 cm). The forehead, crown and nape are black; there are large white cheek patches. Most of the upper surfaces are chestnut, the underparts silvery-white. The bill is light blue, the legs and feet blue-black. The female has similar markings but all her colours are more muted. The male in eclipse is similar but has a darker head.

Range Western areas of North America; this species winters in Mexico, eastern and southern United States.

Status Adaptable, and therefore there are still good populations in many parts of the range.

Avicultural rating Becoming more widely kept; they need deep water.

Feeding Diet N, with additional animal matter and some aquatic plants.

Breeding They are fairly ready breeders in a natural setting, with a large, deep area of natural water. Artificially-hatched ducklings present some problems and are not easily reared. Clutch 6–10; incubation, by the female, 23–24 days. Provide rushy cover in the pen and a variety of nest sites; they may nest amid cover, in a ground-level box or among floating vegetation.

General management Not difficult to keep if conditions are right. A natural environment is strongly recommended, and a good depth of clean water is very important. They are hardy birds and usually good mixers if not closely confined. They do well in groups and have an amusing habit of diving *en masse* when danger threatens.

FALCONIFORMES

Falconidae (Falcons, Caracaras)

There are more than 60 species of diurnal birds of prey in the Falconidae, including some of nature's most efficient killers. They are wide-ranging throughout the world (except the Antarctic). Many species are kept (and bred) by raptor enthusiasts. Only one species is described here and is now virtually unobtainable to aviculturists.

Microhierax caerulescens (II)
Red-thighed Falconet, Collared Falconet

6 in (15 cm). The forehead, sides of the head and neck are white; the crown, mantle, back and wings are black. The chin, throat and underparts are rufous-buff, paler on the flanks. There is a prominent black eye-streak; the thighs are rufous. The wings and tail feathers are spotted and edged with white. The bill is lead-grey, the legs and feet black. The female is similar but usually slightly bigger.

Range Himalayas, northern India, Burma, Thailand, Laos, Cambodia.

Status Abundant in many parts of the range.

Avicultural rating It has not been available for some time; it was previously exported in small numbers but fairly frequently. It is a delicate and extremely difficult bird to maintain in good health over a lengthy period.

Feeding One of the most difficult elements in the successful management of these birds is establishing a correct and acceptable diet for them; a mix of conventional live food and dead flesh seems to suit. Locusts, crickets and other large insects are taken; also offer pinky and young (but furred) mice, and day-old chicks of Japanese Quail (*Coturnix japonica*). Various small passerine birds killed through accident are acceptable, but risky if of a wild species because of the possibility of passing on infection or toxins.

Breeding They nest in tree-holes. Clutch 4–5; incubation 26–28 days. There is a strong bond between the breeding pair and they are said to be aggressive in defence of the nest.

General management This species reacts very quickly to a drop in temperature and is then clearly unhappy. It needs comfortable quarters, not a hot-house, but without substantial fluctuations; it is best at around 55°F (13°C) or slightly higher, but keep conditions considerably warmer until the birds are properly acclimatised. A nest box or suitable hollow log is essential for roosting or as a hiding place; although not easily stressed these birds seem to need a refuge of this kind and frequently disappear from view even during daylight hours.

GALLIFORMES

Phasianidae (Pheasants, Grouse, Quail, etc)

This family comprises more than 200 species, including many excellent avicultural subjects. There is a great diversity of size, from the Chinese Painted Quail measuring only 5 in (13 cm) to the Indian Peafowl which, including a magnificent train, is nearly 96 in (244 cm) in length. Pheasants and quail are widely kept; others, such as grouse, less so. They are essentially ground-dwelling birds. Pheasants are confined to the Old World; quail are distributed in both the New and Old Worlds. Most have distinctive plumage, and marked sexual dimorphism is a feature in many species. They are generally hardy, although some small tropical species need protection in winter. Spacious aviary-type accommodation is needed for pheasants; pinioning is not usual, therefore open-topped paddocks are not suitable. Quail breed well in small, portable units but are visually more appealing in natural aviaries. Probably most species of pheasants and quail in aviculture are now artificially hatched and reared worldwide. Many species do well at liberty but foxes and predators can be a problem and in the UK the provisions of the Wildlife and Countryside Act do not permit the release of non-indigenous species.

Callipepla squamata [PLATE 22]
Scaled Quail, Blue Quail

$9\frac{1}{2}$ in (24 cm). The plumage is mainly blue-grey in colour; the feathers with darker edges give a scaled appearance. There is a short, upright crest with some white spotting. The flanks are grey-brown with some lighter streaking; the lower breast and underparts are buff. The females are similar but have dark streaking on the face and throat.

PLATE 22 *Callipepla squamata*, Scaled Quail

Range Mexico and south-western areas of the USA.

Status Fairly common, but the population fluctuates a good deal.

Avicultural rating They are becoming more widely kept, but are not as easily managed as other New World species.

Feeding Diet D, plus recommended additives.

Breeding This bird is often an erratic layer; the usual clutch is 10–14 but they may lay more over an extended period. Artificial incubation is often safest, although some hens prove good sitters and mothers. Incubation lasts 23 days.

General management They need dry accommodation and a frost-free shelter. They are usually good mixers with passerine species; the cock becomes belligerent in the breeding season. Provide high perches for roosting.

Lophortyx californica
California Quail, Valley Quail, Crested Quail

---

10 in (25 cm). These are attractive small pigeon-sized birds, mainly grey in colour. They have a forward-tilting black crest and a distinctive black throat, bordered with white. The abdomen is chestnut; much of the lower plumage has scale-like markings. The female has a short crest and lacks the male's distinctive head patterning.

Range Western areas of the USA and Mexico.

Status Common in many areas and apparently increasing in numbers.

Avicultural rating Hardy and easy to manage, this is a good beginner's bird.

Feeding Diet D, with recommended additives.

Breeding Often a prolific layer; usual clutch 10–15 (some hens lay many more but do not sit). The eggs are often scattered in the aviary. Incubation 22–23 days, by the hen. They are easy to hatch and rear artificially. Very small bantams may be used—or incubators and brooders.

General management They are usually good mixers (but not with other quail species); the cock's exuberance may upset smaller companions in the spring. They will breed in small units, or flourish in larger mixed community aviaries.

Colinus virginianus (ssp. *ridgivayi* I)
Bobwhite, Virginian Colin

9–10½ in (23–27 cm). There is some variation in size according to the subspecies, of which about 20 are described. Bobwhites are mainly rufous in colour with buff and brown spots and lacings; the breast and underparts are buff brown. The eyebrow stripe and throat are white; there is a black eye streak and throat border. The female is generally lighter in colour and lacks the distinctive patterning.

Range Central and eastern USA but some subspecies extend into Mexico and Central America.

Status Generally abundant.

Avicultural rating Easily managed and widely kept; but they do not do well in damp conditions.

Feeding Diet D, plus recommended additives.

Breeding They are free breeders although the hens are often uncertain sitters. Clutch very variable: 6–30. Incubation, by the hen, 23–25 days. When eggs are laid into a specific place, this often indicates an intention to incubate; if they are scattered about the aviary it is less likely. Artificial hatching and rearing is straightforward, but the chicks mutilate each other if overcrowded. Do not mix these with other species of quail chicks.

General management Easy to manage. The Bobwhite does well in small portable units or larger aviaries; it is usually a good mixer with non-quail species. It does not tolerate damp and is best housed indoors during the winter if prolonged wet conditions are likely. A dry, sandy base in outdoor flights is better than damp earth or long grass. Some low, shrubby vegetation is desirable; this may provide the incentive for hens to incubate.

Alectoris chukar
Chukar Partridge

---

14 in (36 cm). The forehead, crown and breast are grey. The upper parts are olive-grey; the flanks are barred with chestnut, black, white and grey. There is a conspicuous white throat patch with a black border. The bill, legs, feet and eye-ring are red. The sexes are alike.

Range Parts of Asia and Asia Minor; introduced into other countries (including the USA and Canada) as a game bird.

Status Fairly common in many parts of the range.

Avicultural rating Hardy and easy to manage; aggressive.

Feeding Diet D, with recommended additives; Chukars are fond of wild berries in season.

Breeding They are often prolific layers, but rarely incubate successfully in aviary conditions. Clutch 8–16 (or more); incubation 24–25 days. Bantam hatching and rearing is usually successful; newly-hatched chicks should have abundant live food at first.

General management They are hardy and easy to manage, but very aggressive. They will attack birds up to the size of a Monal Pheasant. Chukars need dry, well-drained outdoor accommodation; they dislike damp but otherwise are totally weather-resistant. They like to roost above ground.

Alectoris rufa
Red-legged Partridge, French Partridge

13½ in (34 cm). Similar to Chukar Partridge but slightly smaller. The sexes are alike.

Range South-west Europe; introduced into Britain and other countries.

Status Variable, but mainly stable populations.

Avicultural rating Hardy, easily managed and decorative, but aggressive and not a good mixer.

Feeding/Breeding/General management Similar to Chukar Partridge but *slightly* more resistant to damp conditions. Some ground cover in the aviary is desirable.

Francolinus francolinus [PLATE 23] (V)
Black Francolin, Black Partridge

PLATE 23 *Francolinus francolinus*, Black Francolin

12 in (30 cm). The plumage is essentially black with a chestnut collar on the throat. There is a white patch behind the eye, white barring on the lower back and white spots on the upper back, flanks and breast. The female is browner with a white throat. The markings are less distinct.

Range Cyprus; Syria and Iran eastwards to Pakistan, India, Assam.

Status Hunting has reduced numbers in many parts of the range.

Avicultural rating Not widely kept, they are attractive and reasonably hardy.

Feeding Diet D, with recommended additives.

Breeding Some pairs are very willing to lay; others less so. There is a similarly erratic pattern to incubation. Established breeding pairs may raise 1–3 broods per year although 1–2 is usual. Clutch 6–8; incubation, by the female, 22–23 days. Provide ground cover in the aviary.

General management Good aviary birds, but they need dry, well-drained conditions. They are fairly hardy, but do not withstand winter frost and wet. Provide facilities for dust-bathing at all times. Imported stock is usually nervous and easily stressed; aviary birds are better, but need time to settle when in a new environment.

Coturnix japonica
Japanese Quail

6 in (15 cm). The plumage is mainly brown and warm buff with darker and lighter striations. The chin and throat are dull red. The female is similar but usually has less streaking on the flanks and breast. Pure stock is difficult to obtain as hybridisation has taken place to produce birds for the table. Many types and colour variations are available, including white, but these are mainly nondescript and not worth the attention of aviculturists.

Range Parts of south-east Asia, including Burma, northern Thailand and Laos eastwards to Hong Kong; Japan.

Status It appears to be fairly common in many areas, but there may be some confusion with a subspecies of the Common Quail in parts of the range. Numbers have generally been reduced.

Avicultural rating Other than the canary and budgerigar, it is difficult to think of other birds which can be as effectively 'domesticated' within a short period. This species is kept for culinary purposes as well as aviculture. It is easy to manage.

Feeding Diet D, with recommended additives.

Breeding Normal clutch 8–12; incubation 18 days. The females may prove uncertain sitters but artificial incubation and rearing is straightforward. Many modern hybrids lay substantial numbers of eggs, almost like battery-hen production.

General management They are very easy to manage in either small portable units or a larger outdoor aviary. They are hardy but prefer dry to damp conditions. They may be aggressive to smaller birds when breeding. Provide natural cover in aviary for roosting and nesting sites.

Coturnix delegorguei [PLATE 24]
Harlequin Quail, Delegorgue's Quail

PLATE 24 *Coturnix delegorguei*, Harlequin Quail

6 in (15 cm). The upper surfaces are dark brown with prominent buff streaking on the back; the breast and underparts are chestnut but the centre of the breast is black with black spots extending to the throat and flanks. A white stripe over the eye extends down the sides of the neck; the chin is white with a black margin. The bill is dark brown, the legs and feet yellow-brown. The female is a uniform brown with both darker and lighter streaking and feather margins; her throat is grey-white.

Range Savannah and grasslands from Senegal to Ethiopia; South Africa.

Status Variable, but numbers have been reduced in some parts of the range.

Avicultural rating Not widely kept, mainly because of a shortage of stock; they are easily managed.

Feeding Diet A, plus recommended additives including green food and grit.

Breeding Healthy, unrelated stock usually prove fairly free breeders; but they are uncertain sitters. If the hen will not sit, the eggs hatch satisfactorily in an incubator; but persevere with pairs if possible as it is important to retain the normal brooding instincts. Clutch 6–12; incubation 17 days. Like the Chinese Painted Quail, the nest is usually a scrape in the ground but hidden in ground vegetation; so coarse grasses and rushes are ideal sites.

General management See Chinese Painted Quail. This species is easy to manage in indoor or outdoor quarters; it does well in small portable units, or semi-permanent (until winter) accommodation in a garden aviary. Cocks (especially) frequently become very tame and will feed from the hand.

Coturnix coromandelica
Rain Quail, Black-breasted Quail

6 in (15 cm). The upper surfaces are warm brown with conspicuous yellow-buff streaks over the scapulars and back. The breast and underparts are buff with a large area of black in the centre of the breast, and black spots on the flanks. There are two black stripes from the base of the bill to the nape; the eyebrow is white, as are the cheeks and throat, bordered with black. The bill is dark grey, the legs and feet flesh-coloured. The female is mainly brown and lacks all the cock's distinctive patterning.

Range India, Sri Lanka, Burma.

Status Fairly common in many parts of the range.

Avicultural rating Once common in aviculture but now rare; it is not difficult to keep, but is usually more nervous than the Harlequin Quail.

Feeding/Breeding/General management Similar to the Harlequin Quail. They are fairly free layers but uncertain sitters.

Excalfactoria chinensis
Chinese Painted Quail, King Quail, Blue-breasted Quail

5 in (13 cm). The upper surfaces are mainly brown with darker and lighter streakings and a suffusion of blue-grey on the cheeks and back; the underparts and flanks are blue-grey. The lower breast and abdomen are chestnut; the throat and a streak below the eye are black, bordered with white. The bill is black, the legs and feet yellow. The females are a uniform warm brown with darker and lighter streaking.

Range South-east Asia, Australia.

Status Abundant in many parts of the range.

Avicultural rating Popular and easily managed. This is the most widely-kept of all small gallinaceous birds.

Feeding Diet A, plus recommended additives including green food and grit.

Breeding This species is usually a very free layer, but modern 'strains' are less willing to incubate. This situation has almost certainly been brought about by the intensive methods employed by some breeders. Fortunately some aviculturists are maintaining stocks that incubate and rear their own broods; such birds are well worth searching for. Clutch 5–12; incubation 16 days, almost always by the female. The cock assists with the chicks but becomes aggressive towards them at about the time the sexes can be determined visually. Chicks on hatching are very tiny and able to pass through $\frac{1}{2}$ in (13 mm) wire-netting. Provide brick or wood skirting to the aviary.

General management They are easy to house indoors or outdoors, but fare best in a garden aviary during the summer. They are hardy but best kept indoors during the winter. They are good companions for anything that will not molest them. Provide permanent interest at ground level. They can now be obtained in several colour mutations, including pied, silver and fawn.

Perdicula asiatica
Jungle Bush Quail

6 in (15 cm). The upper surfaces are earth-brown with darker markings on the back and feathers showing pale buff shaft-stripes; the forehead, eyebrow, chin and throat are warm chestnut. The breast and underparts are white with well-defined black barring. The bill is brown, the legs and feet orange. The female lacks the barred under surfaces and shaft-stripes but has a chestnut chin and throat like the male.

Range Himalayan foothills, southwards through India to Sri Lanka.

Status Probably still widespread and with good populations. But their jungle habitat is being destroyed in many parts of the range.

Avicultural rating The fact that this species is available at all in the UK is largely due to two men, Bill Thornhill and Gary Robbins of the World Pheasant Associations's Quail Group, who re-established the species from a small imported shipment. From birds bred by them slow progress is being made in building up numbers. This species is recommended, if you can obtain stock.

Feeding Diet A, with recommended additives, including green food and grit.

Breeding They are fairly ready breeders in the right conditions. They appreciate coarse, tussocky grass for cover, and the close proximity of a stout, rotten tree bough. Unlike birds 'tainted' by intensive breeding (Japanese, Chinese Painted Quail and a few others), Jungle Bush Quail usually prove reliable parents. Clutch 5–8 eggs; incubation, by the female, 21 days.

General management They do best in a secluded, natural aviary, but they will breed in smaller, portable units. This species mixes well but the cock is aggressive whilst the female is sitting. These birds have a wonderful, rolling call-note, reminiscent of a Roller Canary, and often become very tame.

Rollulus roulroul
Roulroul, Crested Green Wood-Partridge, Crowned Wood Partridge, Green Partridge

11 in (28 cm). Much of the plumage is green with purple and blue reflections; the wings are chestnut. The head is black, with a white patch on the crown and a long crest of chestnut feathers starting from the back of the head; there is an area of red naked skin around the eye and a red patch at the base of the lower mandible. The bill is black, the legs and feet red. The female lacks the crest and chestnut colouring on the wings.

Range Malaysia, Sumatra, Borneo.

Status Never abundant, but there are reasonable numbers where suitable habitat exists.

Avicultural rating The Roulroul needs specialist accommodation; it is invariably scarce and expensive, and is recommended for experienced aviculturists only.

Feeding Diet A, plus mixed fruit (similar to Diet E), some insectivorous mixture (Diet G), green food and live food. Mealworms should be given in moderation, especially if birds are housed in a restricted area.

Breeding They are not particularly free breeders; although individual pairs occasionally surpass themselves. A willingness to incubate and rear young also varies among individuals and is much influenced by environment and choice of companions. The eggs can be artificially incubated but chicks may be difficult to rear; a small bantam may help but the normal coop and run housing is not suitable if outdoors. Clutch usually 2–4 eggs;

incubation 18 days. The nest is usually well screened in cover; both parents assist in rearing young. Provide fine grade insectile food, small live food, and cracked small seed for chicks.

General management These birds are likely to thrive only in warm, moist tropical house conditions. They can be allowed into a well-planted, sheltered outdoor flight during high summer. They are ground-dwelling birds but need a perch for roosting. Normally good mixers, they are nervous in new accommodation at first but often become tame. Provide an area of dry sand for dust bathing and remember that, like related species, they need grit.

Bambusicola thoracica [PLATE 25]
Chinese Bamboo Partridge

PLATE 25 *Bambusicola thoracica*, Chinese Bamboo Partridge

11 in (28 cm). The upper surfaces are brown (chestnut on the head and neck, olive on the back) with buff and black mottles and striations. The eye-stripe is grey, the under surfaces grey-brown to pale buff. The male has spurs, but the female, although similar in appearance, does not.

Range China, Japan.

Status Uncertain, but probably reasonably widespread.

Avicultural rating Rarely imported, but fairly easy to manage and they would appear to be hardy.

Feeding Diet A, with the addition of green food and some small live food.

Breeding They are fairly free breeders, and the adults usually prove reliable parents. Clutch 4–7; incubation 18–19 days.

General management They are hardy, but need shelter from driving rain, and well-drained ground. The cocks can be belligerent in mixed company. They do best in an outdoor aviary with comfortable winter accommodation. They enjoy dust bathing.

Tragopan satyra [PLATE 26]
Satyr Tragopan

PLATE 26 *Tragopan satyra*, Satyr Tragopan

27 in (69 cm). The face, forehead, hind crown and chin are black; the centre of crown, ear-coverts, nape, throat and underparts are red, but this is less intense on the lower under surfaces. The mantle, back, wings, rump and tail are olive-brown. The breast and underparts, mantle, back, wings and upper tail coverts are liberally sprinkled with black-ringed white spots. The bill is black, the legs and feet flesh-brown. The female

is smaller; her upper surfaces are rufous-brown with darker striations, the chin and throat grey-white, the underparts buff-brown.

Range Central and eastern Himalayas.

Status Uncommon throughout the range.

Avicultural rating Occasionally available; it is hardy but needs spacious accommodation.

Feeding Diet M, but with the main emphasis on pelleted foods; do not include brassicas among green food. Grit is essential.

Breeding This is not among the easiest of pheasants. The hen often lays in a ground-scrape under cover—unlike others of the family which frequently nest in trees. Clutch 4–5; incubation 27–28 days. Artificial incubation and rearing is usual with this species; the chicks need care, and some small live food in the diet is initially beneficial.

General management This species needs plenty of space and should be housed on well-drained land. It is hardy, but provide natural windbreaks (shrub hedges etc) as well as natural shelter from strong summer sunshine. A spacious shelter is desirable to provide refuge from prolonged cold or wet conditions. Tragopans are usually fairly strong flyers.

Lophophorus impeyanus [PLATE 27] (I)
Himalayan Monal Pheasant, Impeyan Pheasant

---

28 in (71 cm). The head, throat and crest are metallic green; the mantle is green suffused with yellow-gold. The sides and back of the neck are copper; the wings are dark blue with purple reflections. The lower back is white, the tail chestnut. The under surfaces are velvety-black; there is an area of blue-green skin around the eye. The bill is dark brown, the legs and feet brownish-green. The females are mainly dark brown, streaked and mottled with buff; they have smaller crests.

PLATE 27 *Lophophorus impeyanus*, Himalayan Monal Pheasant

Range Himalayas.

Status Numbers have been reduced through excessive hunting.

Avicultural rating Popular avicultural subjects; they are hardy, but do not tolerate damp conditions.

Feeding Diet M, and ensure that adequate grit is available at all times.

Breeding They are usually good breeders; the eggs are mostly taken for incubator or bantam hatching, although some people leave the female pheasant to rear. Apart from the safety aspect, the advantage of non-pheasant hatching is that further clutches are usually laid to replace those taken; and as many as 14–16 eggs are possible from one pair. But if the female Monal is a reliable mother it will be sensible to allow her to raise one clutch each year. Normal clutch 4–6; incubation 27–28 days. The chicks are not difficult to rear; small live food is necessary whilst young.

General management They are very hardy, but do not tolerate damp or wet conditions; ideally provide a sandy soil, but it must be well-drained. They are best in large enclosures; the cocks are often aggressive in confined areas and may injure the hens. They usually live in pairs but one cock to three or four hens is possible. They are said to be prone to infection and short-lived; this may be so if they are badly kept, but with correct housing and good ground they can be very successful, living up to 30 years or more. Provide some natural cover in the enclosure and an open-fronted shed for bad-weather roosting. Place the perch well off the ground; they may sleep on a plank or shelf in preference to the perch. They are great diggers; do not underestimate their ability in this direction.

Gallus gallus
Red Junglefowl

26 in (66 cm) including 11–12 in (28–30 cm) tail. The crown and neck hackles are orange-red to golden-yellow; the back is an iridescent chestnut. The under surfaces are black, the rump golden red. Much of the plumage has blue, purple and green reflections. The face, serrated comb and throat wattles are crimson. The bill is dark brown, the legs and feet lead-grey. The male has well-developed spurs. The female is smaller and has a short, rounded tail. She also has a smaller comb and lacks spurs. Mainly various shades of brown with lighter and darker spots and streaks, she lacks the male's iridescent plumage.

Range Kashmir and central India, Burma, Thailand, Malaysia, Sumatra, Java.

Status Populations are much reduced in many areas, mainly through excessive hunting and habitat destruction; the species is maintaining its numbers in only a few parts of its range.

Avicultural rating These are attractive avicultural subjects, which do well at liberty where predators are not a problem (and where the law permits); they are hardy and easy to manage.

Feeding Diet M, but liberty birds find much of their own food; ensure that grit is always available.

Breeding They are easily bred and make good parents. Clutch 5–10; incubation 20–21 days. The cock assists with rearing. The best ratio is one cock to 4–5 hens. The chicks are easy to rear.

General management This species is the ancestor of domestic poultry, with which it will readily hybridise if given a chance. It does best in spacious pens, or at liberty; the birds have powerful feet and will quickly destroy turf by digging, if in a restricted area. Liberty birds tend to be secretive; they like to roost high. If there is plenty of space, they can be housed with other ground-dwelling birds, including pheasants.

Gallus sonneratii [PLATE 28] (II)
Sonnerat's Junglefowl, Grey Junglefowl

PLATE 28 *Gallus sonneratii*, Sonnerat's Junglefowl

30 in (76 cm) including 16 in (41 cm) tail. The long hackles on the neck are black with grey edges, white shafts and spots; there are also wax-like yellow spots created by the fusing of the webs. Most of the body plumage is black with lighter shafting and edges to feathers. There are extensive purple reflections; the tail and coverts are also black with iridescent highlights. The comb is less serrated than that of the Red Junglefowl; the face, throat and wattles are crimson. The bill is dark brown, the legs and feet pink. The female is mainly brown with blackish-brown upper parts. She is smaller in size with a shorter rounded tail, and lacks the male's spurs.

Range Western and southern areas of India.

Status Numbers are much reduced in many parts of the range.

Avicultural rating Less often kept than the commoner Red Junglefowl, it is reasonably hardy but does not appreciate cold, damp conditions.

Feeding/Breeding/General management Similar to Red Junglefowl, but is rarely kept at liberty as it needs more direct management. Use the same ratio of cock to hens; or pairs. The chicks may prove delicate; they need a dry, sheltered environment to begin with. There are usually 8 eggs maximum, but more can be achieved by double clutching. The cock has a distinctive call, not like the crow of the Red Junglefowl or of domestic poultry.

Gallus lafayettei
Lafayette's Junglefowl, Ceylon (Sri Lankan) Junglefowl

---

28 in (71 cm) including 14–15 in (36–38 cm) tail. The hackles are various shades of yellow with darker shafts. The back and rump are orange-red; the under surfaces are mainly the same shade with darker markings. There is some black barring on the flanks. The comb is red, with the central area yellow; the face and wattles are red. The bill is yellow, the legs and feet yellow-brown. The female's plumage is various shades of brown with lighter underparts. She is smaller than the male, and the comb is almost absent; she has a shorter tail.

Range Sri Lanka.

Status It is found throughout Sri Lanka where suitable environment exists, but numbers are much reduced in many areas.

Avicultural rating Infrequently available and kept in only a few private collections.

Feeding/Breeding/General management Similar to the Red Junglefowl but the clutches are smaller: usually a maximum of 4 eggs. They are invariably more shy and wary than Red and Sonnerat's Junglefowl. The male has a distinctive call.

Lophura nycthemera
Silver Pheasant

---

48 in (122 cm) including 24–26 in (61–66 cm) tail. The top of head, including the crest, the chin, throat and rest of the underparts are glossy black. The upper surfaces are white; the feathers show a black lacing pattern, especially pronounced on the wings and tail. There is an area of red bare skin on the face, and the wattles are red. The bill is dark yellow, the legs and feet red. The female is smaller; mainly brown with darker striations and lacking the male's well-developed spurs.

Range Southern China, parts of Burma, Thailand, Laos, Vietnam.

Status Numbers are much reduced, especially in recent war zones; the population is not known in many parts of the range.

Avicultural rating Robust and easy to manage, this is one of the most popular of all pheasants. Old males can be fierce and attack females, other companions, dogs ... and their owner!

Feeding Diet M, and ensure that adequate grit is available at all times.

Breeding They are free breeders. They are usually kept in pairs, but 3–4 hens to one cock is better. The cocks are extremely aggressive in the breeding season; so provide natural cover for the females to escape the male's attentions from time to time. Clutch 6–12; incubation 26–27 days. The hens are good sitters; the cocks are known to hatch and rear if any accident befalls their mate. It is not a difficult task to hatch and rear the chicks in an incubator or under a bantam.

General management They are very hardy, and do best in spacious accommodation; provide high perches 6 ft (1.8 m) from the ground for roosting. They will perch in trees; it is best if an open-fronted shed is available as severe-weather protection, but generally they are resistant to the elements. Watch the behaviour of the cock in spring; they are doubtful companions for other birds most of the time, but quite impossible at this season. As with other pheasants, if they are housed in aviaries adjoining others occupied by related species it is essential to have a close-board or other solid division (to 24 in, 60 cm) between them to prevent fighting through the wire.

Lophura swinhoei (**I**)
Swinhoe's Pheasant

32 in (81 cm). The crest, mantle and central tail feathers are white; the back and rump are black and the feathers have iridescent blue margins. The head, neck and under surfaces are blue-black, the scapulars chestnut. The area of skin around the eye (which can be inflated), comb and wattles are bright red. The bill is horn-coloured, the legs and feet red. The females are mainly brown with darker vermiculations, and considerably smaller than the males.

Range Taiwan.

Status Endangered, although re-introduction programmes have been successfully carried out using captive-bred stock.

Avicultural rating A popular avicultural subject.

Feeding Diet M, and ensure that adequate grit is available at all times.

Breeding They are fairly ready breeders, which are usually kept in pairs. Clutch 8–12; incubation 24–25 days. The chicks are not difficult to rear but must be moved to fresh ground very frequently to prevent the possibility of infection.

General management They are very hardy, but not happy in damp conditions, so well-drained ground is needed. They may use an open-fronted shelter in severe weather; otherwise provide roosting perches in a sheltered part of the pen. Provide natural ground cover in the flight.

Lophura diardi [PLATE 29]
Siamese Fireback Pheasant

PLATE 29 *Lophura diardi*, Siamese Fireback Pheasant

32–34 in (81–86 cm). The head, throat and crest are black. The neck, back and breast are grey with fine black vermiculations; the centre of the back is orange and the lower back is red with iridescent purple-blue edges to the feathers. The tail is black with a suffusion of green; the lower under surfaces are black. The area of naked skin on face and the wattles are red. The bill is black, the legs and feet red. The female is mainly brown and chestnut with some lighter mottling on the back, wings and tail. She lacks the crest and is smaller. She also lacks spurs.

Range Burma, Thailand, Indo-China.

Status Numbers are decreasing in most parts of the range.

Avicultural rating It is becoming more widely kept, although this is not one of the most inexpensive pheasants; it is relatively hardy but does not tolerate damp conditions.

Feeding Diet M, with an adequate supply of grit.

Breeding A reasonably free breeder, this species is monogamous. Clutch 5–8; incubation 24–25 days. The cocks are frequently over-enthusiastic with the hens; so provide plenty of natural ground cover so the female can escape from time to time. The chicks are not difficult to rear but need a quiet, sheltered environment to begin with; small live food is important. Provide shade in the rearing pens.

General management They are hardy, but dislike wet or damp. They are best shut in at night during periods of severe frost, which may affect their toes. Well-drained ground is important; bacterial problems are likely if they are quartered on heavy or moisture-retentive ground.

Crossoptilon crossoptilon [PLATE 30] (I)
White Eared Pheasant

36 in (91 cm). The plumage is almost entirely white, silvery on wings and coverts; the tail feathers are black glossed with green. The top of the head is velvety black. The naked skin on the face is red. The bill is flesh-coloured, the legs and feet red. The female is similar but smaller; most of the females also lack spurs.

Range South-eastern Tibet to north-west Yunnan, central Szechwan.

Status Endangered, but captive-breeding programmes should ensure their survival.

Avicultural rating Still rare in collections; there have been improved breeding results in Europe and USA.

Feeding Diet M, with adequate grit; like other eared pheasants this species enjoys digging for bulbs, roots etc.

PLATE 30 *Crossoptilon crossoptilon*, White Eared Pheasant

Breeding Established adult pairs usually lay. Clutch 6–12; incubation 24–26 days. Artificial incubation and rearing are usual, and it also produces double clutching which is valuable with endangered species. The chicks are not difficult, but watch carefully during the first few days for signs of stress or aggression. Sheltered, dry rearing quarters are important. The young chicks are inclined to toe-pick.

General management Relatively easy to manage; they are usually quiet and less inclined to the vigorous breeding behaviour of many other pheasant species. They need spacious pens; their quiet, almost lethargic, disposition makes them good liberty birds but with the disadvantage that predators find them easy prey. They are hardy but dislike damp; well-drained ground is important. They are great diggers, but with the bill rather than the feet. They are not suitable for restricted quarters as their plumage will be quickly damaged and the birds often pick at each other.

Crossoptilon mantchuricum (I)
Brown Eared Pheasant

---

39 in (99 cm). The crown is black, the neck blackish merging with the brown mantle. The lower back, rump and tail coverts are white. The under surfaces are brown. The feathers of the ear coverts are long, upturned and white; the naked skin on the face is red. The bill is flesh-coloured, the legs and feet red. The female is similar but smaller; she lacks the spurs.

Range North-east China.

Status Probably decreasing to the point of near-extinction.

Avicultural rating Popular and widely-kept for many years.

Feeding/Breeding/General management Similar to White Eared Pheasant. These are willing layers, but fertility is often a problem. Clutch 6–8; incubation 27 days. In-breeding is a major problem with this species; most of the stock in Europe and the USA is descended from one trio exported in the mid-nineteenth century. New blood, which became available in the 1970s, will undoubtedly improve the species' prospects of long-term survival in aviculture.

Crossoptilon auritum
Blue Eared Pheasant

---

38 in (97 cm). Most of the plumage is slate-blue, except for the upturned white ear tufts, chin and throat; the crown is velvety-black. The bare skin on the face is red. The bill is horn-coloured, the legs and feet red. The female is similar in appearance and size, but usually slightly slimmer.

Range Western China.

Status Decreasing.

Avicultural rating A long-established and popular avicultural subject.

Feeding/Breeding/General management Similar to White Eared Pheasant. Clutch 8–12; incubation 27–28 days. The adult cocks occasionally show more aggression than the two other eared pheasant species. Like them, this species is also a poor flyer.

Syrmaticus reevesii
Reeves' Pheasant

---

84 in (213 cm) including a 40–60 in (101–152 cm) tail. The mantle, back and rump are cinnamon-buff and the feathers have black margins; the under surfaces are tawny-buff and the feathers have darker edges giving a laced effect. The crown is white; the forehead, cheeks and upper neck are black. The lower underparts and tail coverts are black. The bill is horn-coloured; the legs and feet are grey-brown. The female is mainly brown and grey; her chin and throat are buff-white and the bird is smaller with a much shorter tail.

Range Northern and central China.

Status Decreasing.

Avicultural rating A well-established and popular avicultural subject.

Feeding Diet M, with adequate grit.

Breeding They are best kept in trios or quartets; a 1 : 1 ratio will invariably lead to considerable persecution for the female. Provide plenty of ground cover in a large enclosure. Clutch 8–12; normally two clutches per year or more if the eggs are removed for artificial hatching; incubation 24–25 days. The females are usually reliable mothers. The chicks are hardy and easy to rear, but quickly show signs of a belligerent disposition among themselves and if they are housed with companions.

General management This species needs *plenty* of space if it is not to suffer damaged plumage, especially the tail which in exceptional cases may reach nearly 72 in (183 cm). There are many examples of Reeves' Pheasants being released into the wild as potential additions to 'sporting' birds; they do not seem to have adapted successfully in the UK but there are still plenty in some other European countries, and there may be 'wild' populations in other parts of the world.

Chrysolophus pictus
Golden Pheasant

42–44 in (107–118 cm) including a 30–31 in (76–79 cm) tail. The forehead, crown and crest, back and rump are a bright golden yellow; the cape is orange-yellow with dark blue feather edges. The mantle is an iridescent green with feathers margined with black; the scapulars are crimson, the secondaries purple, the coverts brown and the flights dark brown. The long central tail feathers are brown with a fine tracery of black; the outer feathers are barred brown and black. The entire under surfaces are scarlet, and there is chestnut on the abdomen. The area of bare skin on the face is yellow. The bill is horn-coloured; the legs and feet are yellowish. The female is mainly warm brown with darker striations; she is smaller than the male and has a shorter tail.

Range Central China.

Status Uncertain, but almost certainly declining in those parts of the range which are closest to cultivated areas.

Avicultural rating Undoubtedly the most popular and widely-kept of all pheasants.

Feeding Diet M, with adequate supplies of grit.

Breeding They are very easy to breed, although many of the females are uncertain sitters. Clutch 6–12; incubation 23–24 days. If the hen continues to lay beyond the normal clutch size, the probability is that she will not sit; other individuals may prove excellent mothers, but careful assessment is needed. The chicks are easy to rear either naturally or artificially.

General management Straightforward. This species has the ability to survive in a restricted space, but fares much better in spacious pens. Provide natural cover in the pen, including trees or shrubs for essential shade. They are very hardy and virtually totally weather-resistant. They are good liberty birds, but may not stay. They are excellent flyers. Yellow mutations are available.

Chrysolophus amherstiae
Lady Amherst's Pheasant

50–68 in (127–172 cm), including a 30–46 inch (76–117 cm) tail. The crown is bronze-green; the stiff crest feathers are red. The cape is white, the feathers margined with blue; the throat, upper breast and mantle are metallic green, the feathers edged with black giving a scaly appearance. The lower back is black edged with orange-yellow, and the feathers of the rump are also edged with vermilion. The under surfaces are mainly white, the flanks with brown markings. The long tail feathers are white barred with black. The facial skin is blue. The bill is horn-coloured, the legs and feet blue-grey. The female is considerably smaller than the male, but larger than the female Golden Pheasant and with more pronounced streaking and spotting of darker brown plumage than the hen of that species. Crimson feathers in the white underparts identify hybrids.

Unfortunately, much casual hybridisation with other species (but mainly the Golden Pheasant) has led to a situation where a high percentage of Amherst stock is impure.

Range South-eastern Tibet, south-western areas of China, northern Burma.

Status Uncertain; probably declining in some areas but they may be holding on in the more inhospitable—and therefore inaccessible—parts of the range.

Avicultural rating This very popular bird vies with the Golden Pheasant as the most widely kept pheasant species.

Feeding/Breeding/General management Similar to Golden Pheasant.

Polyplectron bicalcaratum [PLATE 31] (II)
Grey Peacock Pheasant, Burmese Peacock Pheasant

PLATE 31 *Polyplectron bicalcaratum*, Grey Peacock Pheasant

28 in (71 cm) including a 16 in (41 cm) tail. The overall colour is grey but with ocellae on the mantle, wings, tail coverts and tail; those on the mantle and wings are blue with purple reflections. The ocellae on the coverts and tail are similar, but bigger and showing green and purple iridescence. The throat is white; there is a small, forward-curving crest. The bill, legs and feet are grey. The female is considerably smaller and of slighter build than the male; her ocellae are smaller and darker.

Range Central and southern Burma, Thailand, Laos, northern Vietnam, Hainan.

Status Almost certainly decreasing in most areas of the range, as habitat is utilised for crops etc.

Avicultural rating Now becoming more popular as the availability of captive-bred stock increases.

Feeding This pheasant needs small grain or seed. Diet D provides the basis; but they also take some insectivorous mixture (Diet G) and chopped hard-boiled egg.

Breeding They are fairly ready breeders once established; the females are usually good mothers, but eggs removed for artificial hatching usually produce double or treble clutching. Normal clutch 2; incubation 21–22 days. The chicks need care during the first few days; some live food is needed in the diet to begin with.

General management Wild birds inhabit damp (and sometimes chilly) forests on mountain slopes up to 5000 feet (1500 m). This may be the reason why they have adapted to temperate or damp climates better than many other pheasants. Well-drained ground is recommended; and some natural ground cover is required. Provide an open-fronted shed for shelter. They thrive in relatively small pens, and roost off the ground. They should be housed in pairs.

Polyplectron emphanum (I)
Palawan Peacock Pheasant, Napoleon Peacock Pheasant

---

20 in (51 cm) including a 10 in (25 cm) tail. The crown and upstanding, pointed crest are an iridescent blue-green; the cheeks and a stripe from the base of the bill, above the eye and to the sides of the neck are white. The head, neck, chin, throat and under surfaces are black; the mantle is black with a suffusion of blue. The back is

blackish mottled with chestnut. The long tail coverts are blackish with blue-green ocellae; the tail feathers also have ocellae plus white spots and edges to the feathers. The bill, legs and feet are black. The female is mainly dark brown with a smaller crest. She has a grey-white face, chin and throat; ocellae are present on her covert and tail feathers but are poorly defined.

Range Island of Palawan.

Status Scarce, but there may be reasonable populations within the relatively small territory.

Avicultural rating Never widely kept, mainly because of their scarcity and consequent high price. This species needs careful, sympathetic management.

Feeding Similar to Grey Peacock Pheasant.

Breeding This is an uncertain breeder. Clutch usually 2; incubation 19–20 days. The chicks are difficult starters if incubator or bantam hatched; they are frequently reluctant to feed. The females are usually good sitters but need quiet and seclusion. Provide small live food with the rearing diet for young chicks; very tiny larvae of housefly ('feeders') are excellent, provided that they are meticulously cleaned before feeding.

General management Similar to Grey Peacock Pheasant, but not as hardy. They are best shut into a well-lit shelter during severe weather; minimal heat in the shelter is desirable during prolonged low temperatures. Frostbite can be a problem; straw as a temporary floor covering may help.

Pavo cristatus [PLATE 32]
Common Peafowl, Indian Peafowl, Blue Peafowl

78–90 in (198–229 cm). There is a crest from the back of the head, with bare shafts and fan-shaped tips. The head, neck and breast are a metallic royal blue; the lower under surfaces are a shining green merging with black on the abdomen. The back feathers are scale-like and bronzy-green. The wing feathers are buff with black mottling and vermiculations; the flight feathers are chestnut. The feathers of the upper tail coverts are elongated to form a train which is raised in display; this is coppery-green with ocellae which are green, blue and copper with bronze-green 'eyes'. The bill and legs are horn-coloured. The upper surfaces of the female are brown, the head rufous; the breast is dark brown merging with pale buff lower underparts. There is some iridescent green on the lower neck; the crest is similar to that of the male. There are black-shouldered, white and variegated forms available.

Range India, Sri Lanka.

Status Declining in most areas, following habitat destruction; they are also hunted for food and following crop damage.

Avicultural rating Very straightforward at liberty; they have been popular subjects for centuries (King Solomon kept them).

Feeding Diet M, with an adequate supply of grit. They will find much additional food if at liberty.

Breeding They are free breeders. One cock to 4–5 hens is a good ratio; the hens are good mothers and may produce two (occasionally three) broods each year. Otherwise artificial incubation and rearing are successful. Clutch 3–8; incubation 28 days. The chicks are fairly slow developers and the cock's spectacular train does not appear until the third year.

General management If confined, they need very substantial aviaries indeed. They are best in a large garden or park setting; they rarely stray and are totally hardy. Pinioning is not necessary but young birds should be clipped annually until maturity. They roost high to avoid predators. In small gardens they are likely to wander and possibly damage neighbouring plots.

PLATE 32 *Pavo cristatus*, Common Peafowl

Pavo muticus [PLATE 33] (**II**)
Green Peafowl, Javanese Green Peafowl, Javan Peafowl

84–96 in (213–244 cm). The crown is iridescent green; and there is an upright crest. The head, throat and upper neck are blue-green; the mantle and breast are a scale-like green with blue centres. The remainder of the upper surfaces are a shining green-bronze; the lower underparts are dark green, with grey on the abdomen. The face is naked with a single line of black feathers from the base of the bill to the eye on a yellow and blue skin. The train is very similar to that of the Common Peafowl, but more golden. The bill, legs and feet are black. Like the preceding species, the cock has spurs on each leg. The female is similar but lacks the iridescent feathers, train and spurs.

Range Assam, Burma, Thailand, Vietnam, Laos, Malaysia, Java; the birds most frequently seen in aviculture are the subspecies *P.m. imperator* which ranges from south-eastern Assam to Thailand.

Status Declining in most parts of the range.

Avicultural rating Fairly popular, but its uncertain temperament makes for some difficulties. It is not completely hardy.

Feeding/Breeding Similar to Common Peafowl.

General management The temperament of the males renders semi-liberty virtually impossible for this species; it is very savage and will not hesitate to attack other birds, animals and humans, when its powerful spurs are brought into use. Green Peafowl need spacious aviaries or roofed paddocks; they are not weather-hardy and need protection (not heat) during severe weather. They are noisy but the call is slightly less strident than that of the Common Peafowl.

PLATE 33 *Pavo muticus*, Green Peafowl

PLATE 34 *Acryllium vulturinum*, Vulturine Guineafowl

Acryllium vulturinum [PLATE 34]
Vulturine Guineafowl

23 in (58 cm). The bare skin of the head and neck is blue; there is a ruff (or cowl) of short, plush-like brown feathers at the back of the head. The feathers of the lower neck are elongated and pointed, black with white shafts and blue edges. The upper surfaces are black, heavily spotted with white; the breast and abdomen are black with cobalt blue patches on the sides of the breast. The bill is black, the legs and feet dark grey. The female is similar but usually slightly smaller and lacks spurs.

Range Southern Somalia, eastern areas of Uganda and Kenya; north-east Tanzania.

Status Hunting has reduced numbers in some parts of the range, but the species manages to maintain numbers.

Avicultural rating Not widely kept, although it is less noisy than other members of the family. It dislikes damp conditions, and is not winter weather-resistant.

Feeding Diet M, with adequate grit. These birds forage well in a large paddock.

Breeding They should nest if in the right environment (a generous-sized paddock with tall grasses) and especially during prolonged warm, dry weather. Clutch 8–20; incubation variable, 24–28 days. The nest is usually placed in tall grass or similar vegetation; or it can be a scrape in the ground. The chicks' survival may depend as much on the weather as on correct feeding; start off with hard-boiled egg, pinhead oatmeal, and a fine-grade insectivorous mixture. Parent-rearing is best as the chicks are slow to become independent; or use a turkey-broody of modest size.

General management They must have a dry, well-drained paddock, and are best kept in groups. Provide winter shelter in the form of a dry, well-lit shed for occupation in severe weather, especially frost; slight heat is desirable. They greatly enjoy sunbathing, and like to roost high; but watch for predators and straying.

Numida meleagris
Helmeted Guineafowl

---

22 in (56 cm). The head and neck are bare; the casque or helmet is brown. The plumage is almost entirely lavender grey, heavily spotted with white; the upper mantle and breast lack spots. The wattles are bright red. The bill is dark red, the legs and feet grey-brown. There are no spurs. The female is similar but usually slightly smaller.

Range More than 20 subspecies are described, ranging over much of Africa where the habitat is suitable; mainly Cameroon through Central Africa and Chad to Sudan, Ethiopia; then south through Kenya, Tanzania, Zaire, Angola to South Africa. The subspecies *Numida meleagris galeata* (found from Senegal eastwards to Chad and the Central African Republic) is generally regarded as the bird from which domestic guineafowl are descended.

Status Abundant in many parts of the range.

Avicultural rating Not widely kept; like its descendants, it has an unlovely voice. These are hardy, and are also excellent 'watchdogs'.

Feeding/Breeding/General management Similar to Vulturine Guineafowl, but this species is generally much tougher and more adaptable. Shelter in severe weather is essential, however, if the birds are not to suffer from frostbite. The chicks are not difficult to rear, but, like their relatives, are slow to attain independence; if entrusted to a domestic broody she will almost certainly abandon them too early.

GRUIFORMES

Turnicidae (Button Quail)

There are more than a dozen species in this family, which are small quail-like birds. They are restricted to an Old World distribution. The main visual distinguishing characteristic that separates these birds from true quail is the absence of a hind toe. The females are bigger and more distinctively marked than the males; they are polyandrous and extremely pugnacious with each other. The female also has a loud booming call and is the dominant sex, but assists with nest-building. The males undertake incubation and rearing; the chicks are very small.

Turnix tanki
Indian Button Quail, Indian Bustard Quail, Yellow-legged Hemipode

8 in (20 cm). The top of the head is dark brown, and the feathers have darker margins; there is a chestnut half-collar on the rear of the neck. The upper surfaces are grey-brown with darker and lighter streaks and stipplings; the underparts are warm buff-brown, with a suffusion of chestnut shading to buff and white on the abdomen. The bill and upper mandible are brown, the lower mandible yellow; the legs and feet are yellow. The male is substantially smaller ($5\frac{1}{2}$ in, 14 cm) and lacks the female's more distinctive markings, including the half-collar on the rear of the neck.

Range India, Burma, parts of southern China; also Andaman and Nicobar Islands.

Status Their small size and retiring habits have helped to maintain populations; but habitat destruction in some areas has drastically reduced their numbers.

Avicultural rating Now rarely imported. They are fairly easy to manage and well worth keeping, if you can find them.

Feeding Diet B, with the addition of 20 per cent chick crumbs and some small live food. They also enjoy green food and seeding grass.

Breeding It is difficult to reproduce a controlled environment to permit the normal polyandrous cycle to be successful; they may be best kept on a 1:1 ratio. Provide good ground cover in the aviary (grass, rushes etc) or heaps of dried grasses pushed between wigwam-shaped twig frames. Well-drained soil is essential; damp, wet conditions are not tolerated. Clutch 3–6; incubation, by the male, 12–13 days. Very small live food is needed during the first few days, then chick crumbs etc.

General management My own birds were housed indoors during winter; they were not weather-resistant. Cold, damp conditions quickly debilitate them. They can be kept with most small passerines and do well in sheltered small aviaries in the summer months. Provide shelter from rain and a dust-bathing facility.

The only other species I have kept was the Barred Button Quail (*Turnix suscitator*), also from south-east Asia. It is smaller than the foregoing species, although size and plumage dimorphism is evident. It requires very similar management to the Indian Button Quail.

Gruidae (Cranes)

The crane family has a worldwide distribution, except for South America. There are 14 species, all in need of protection, with almost half seriously endangered. The main characteristics are a tall, graceful build, long legs and necks; they have few bright colours but there are occasional adornments on the head or throat (wattles). Spectacular dancing displays are often accompanied by equally extrovert resonant trumpeting. Both sexes participate in elaborate courtship displays: bowing, wing-spreading and aerial leaps. They are good avicultural subjects; hardy and adaptable but best in spacious accommodation and on well-drained ground. Several species breed, therefore aviculture can play an important role towards conservation.

Anthropoides virgo [PLATE 35] **(II)**
Demoiselle Crane

---

36 in (91 cm). This is the smallest of the family. The plumage is mainly blue-grey, but the head is much darker with white tufts extending from the cheeks down the sides of the neck; the neck is blackish with elongated feathers extending downwards over the breast. The bill is dark olive, lighter towards the tip; the legs and feet are black. The sexes are identical in appearance.

Range South-eastern Europe, north-east Africa eastwards to central Asia and parts of northern China.

Status Declining in some areas, despite protection.

Avicultural rating Excellent avicultural subjects if enough space is available; they are hardy and easily managed.

Feeding Diet R, with the addition of natural protein items which are likely to be picked up in a large planted area; or ensure that the recommended daily intake of locusts, earthworms etc is available. Offer whole maize later in the day.

PLATE 35 *Anthropoides virgo*, Demoiselle Crane

Breeding They are by no means free breeders, but an increased rate of success is now being achieved, albeit with an emphasis on artificial incubation and hand-rearing techniques at present. Modern sexing facilities allow pairs to be introduced with certainty; bear in mind that cranes are monogamous and usually pair for life, therefore alterations to established pairs should be made only as a final resort. They are ground nesters; the sexes share the incubation and rearing. Clutch usually 2; incubation 28–30 days. Eggs can be incubator-hatched and chicks hand-reared successfully; ensure that young chicks exercise after feeding, otherwise leg weakness may result. As methods, including diets, continue to improve the success ratio will increase for natural and artificial hatching and rearing.

General management The first requirement is space; do not try to keep cranes in confined areas. They are almost totally weather-resistant, but prolonged cold, wet conditions will debilitate them; a shelter with deep peat or straw may be necessary in winter to avoid frost damage to their toes. No winter heat is needed. Outside the breeding season they are very gregarious; and groups do well together. It is essential that the birds are pinioned (not clipped, which means an annual chase, and possible accidental loss); they can be kept at liberty on large lawned areas. They can be very long-lived: a Great White Crane (*Grus leucogeranus*) lived for 60 years in Washington Zoo, USA. There is a record of Demoiselle Cranes breeding in Osterley Park (Middlesex, England) 'menagerie' in the mid-sixteenth century, which suggests that we should by now be self-sufficient with the species and is a sad indictment of the inability of aviculturists and zoos to capitalise on this early success.

Anthropoides paradisea [PLATE 36] (**II**)
Stanley Crane, Blue Crane, Paradise Crane

---

40 in (102 cm). The plumage is mainly blue-grey in colour; the wing coverts are much darker and greatly elongated to produce feathers that almost sweep along the ground. The feathers of the head produce a curious swollen appearance. The bill is flesh-coloured, the legs and feet black. The sexes are alike; but the female is usually smaller.

PLATE 36 *Anthropoides paradisea*, Stanley Crane

Range South Africa.

Status Strict protection appears to have maintained the population at a viable level.

Avicultural rating Only occasionally available, which inhibits a wider interest in a very attractive species.

Feeding Diet R, with the addition of suitable extra animal protein. Oystershell grit of suitable size is also essential for these and other crane species.

Breeding Similar to Demoiselle Crane.

General management They need a large grass area; they are inclined to be nervous, but individual birds are very tame. Winter shelter is needed.

Balearica pavonina
Black-necked Crowned Crane, Sudan Crowned Crane

36–38 in (91–97 cm). This is the smaller of the two African crowned cranes. The upper and lower surfaces are mainly dark slate grey; the wings show considerable white and buff, and the flights are maroon. The forehead has a black plush-like area and there is a tuft of upstanding, straw-coloured feathers on the nape. The cheeks have conspicuously white and pink bare skin; the small wattle is red. The bill is black, the legs and feet dark grey. The sexes are alike.

Range Senegal to Chad, northern Zaire and eastwards to Uganda, Kenya and northern Tanzania.

Status Still fairly abundant in areas of suitable habitat, but numbers are declining across the range.

Avicultural rating See Demoiselle Crane.

Feeding/Breeding/General management Broadly similar to Demoiselle Crane; individual birds become very tame.

Balearica regulorum (II)
Grey-necked Crowned Crane, South African Crowned Crane, Blue-necked Crowned Crane

---

40–46 in (102–117 cm). Similar to the preceding species, but slightly bigger and with paler grey feathers on the neck and under surfaces.

Range Eastern and southern Africa; also central Africa where suitable habitat exists.

Status and avicultural rating See Black-necked Crowned Crane.

Feeding/Breeding/General management Similar to Black-necked Crowned Crane. This species is regarded as slightly more robust than *B. pavonina*.

Psophidae (Trumpeters)

This is a family of three rarely imported denizens of humid rain forest areas in northern South America, which are best known for their bugle-like call. This is often their downfall, as hunters imitate it and the birds come to investigate. They are the size of a domestic chicken, but with a longer neck and long legs, and are mainly dark-coloured; they are very poor flyers and therefore are much persecuted by hunters. Hand-reared birds have the reputation of making excellent pets; many imports are of this quality, which means, in other terms, that the majority of these easily domesticated birds shipped out of South America are 'imprints' and therefore of little value as potential breeding stock.

Psophia crepitans [PLATE 37]
Common Trumpeter, Grey-winged Trumpeter

---

22 in (56 cm). The plumage is mainly black with the feathers of the upper breast edged with iridescent blue-green; the feathers growing from the centre of the back are elongated and grey in colour. The bill is dark olive-yellow, the legs and feet dark grey. The sexes are alike.

Part II: The Birds

PLATE 37 *Psophia crepitans*, Common Trumpeter

Range Southern Venezuela, the Guianas, northeast Brazil, Ecuador, north-eastern Peru.

Status Excessive hunting and habitat destruction in some areas have reduced populations.

Avicultural rating Occasionally available and much sought after by collectors; this species needs care and is best in experienced hands.

Feeding Diets E and M will provide the basis of a suitable diet for trumpeters. They should also be given live food. Some authorities recommend minced raw beef in the diet but this is of doubtful value; a small quantity of processed (tinned) dog meat is preferable. Live food should include the occasional mouse, frog etc.

Breeding Little is recorded. Clutch believed to be 4–7; incubation 28 days. There are various reports of wild nests in tree holes, in branches of a tree, and at ground level; these are possibly contradictory. Incubation is believed to be carried out by the female; the nestlings are nidifugous.

General management They need a large enclosure and a sheltered situation; a tropical house environment is best in the winter but solid (concrete) floors cause foot problems. They are usually tame, but are not safe companions for smaller, weaker birds. They need to roost high. They are hardy enough for outdoor accommodation in the summer when well-drained ground is desirable; they are excellent swimmers.

Rallidae (Rails, Coots and Gallinules)

The Rallidae comprises nearly 130 species with worldwide range in suitable environments, which are reed-fringed waterways, marshes, lakes and rivers. They range in size from that of a sparrow to a domestic hen. They are mainly soberly plumaged; gallinules are brighter—many with purple and green plumage. Most are secretive and infrequently seen if there is plenty of cover in the aviary; they are aggressive and are not safe companions for weaker species, whether terrestrial or aquatic in habit.

Eulabeornis cajaneus
Grey-necked Wood Rail, Cayenne Wood Rail

15 in (38 cm). The head and neck are mid-grey; the mantle and wings are olive-green merging with the black of the lower back and tail. The breast and under surfaces are rufous, darker on the lower underpars and with a black area in the centre of the abdomen. The bill is yellow-orange, green at the tip; the legs and feet are red. The sexes are similar.

Range Southern Mexico, through Central America to parts of Panama; then southwards (where habitat is suitable) to Paraguay and northern Argentina.

Status Numbers have been reduced in many areas through loss of habitat, but there are good populations where they are undisturbed.

Avicultural rating Fairly specialist, they need

separate accommodation; they are hardy and not difficult to manage.

Feeding Relatively omnivorous. Use a mix of Diet G and Diet M, with the addition of some minced raw beef. Live food is also necessary and in a large naturally-planted aviary the birds will catch a good deal of animal food.

Breeding They are fairly ready breeders, in a good environment; they need plenty of ground cover (coarse grass, rushes, reeds etc) preferably near a pool or water source. The nest of vegetation is usually well concealed. Clutch 4–8; incubation 20–21 days. The chicks follow their parents when a few days old; provide fine-grade insectivorous mixture for them, plus extra finely-chopped hard-boiled egg, and small live food. They are very difficult subjects for hand-rearing. The chicks mature quickly and may need separate accommodation within 4–5 weeks of hatching; the cock is aggressive during the breeding period and may attack his own growing offspring.

General management They need a spacious aviary with preferably no companions or, in a very large area, only the bigger softbills such as touracos or Fairy Bluebirds. They become hardy but need shelter in severe weather. They are not strong flyers, but are expert climbers.

Other large rail species are occasionally available. Almost all will thrive on treatment similar to that outlined for the Grey-necked Wood Rail. They are not to be trusted with small or weaker birds and also fight among themselves, especially birds of the same sex.

Laterallus leucopyrrhus [PLATE 38]
White-breasted Rail, Red and White Crake

6½ in (17 cm). Most of the upper surfaces are chestnut; the lower back and tail are brown. The under surfaces are white; the flanks are finely barred with black. The bill is greenish-yellow, the legs and feet red. The sexes are alike.

PLATE 38 *Laterallus leucopyrrhus*, White-breasted Rail

Range Southern Brazil, Paraguay, Uruguay; occasionally northern Argentina.

Status Still fairly abundant in many parts of the range.

Avicultural rating Somewhat exacting in requirements, but an attractive and interesting species. It is not hardy.

Feeding Diet G, with a variety of small live food; seed may be offered but may not be taken.

Breeding They are now being bred more frequently. They make a fairly bulky nest, usually fairly high, but it may be in tall vegetation at or near ground-level. Clutch 4–6; incubation 24–26 days. The nestlings are nidifugous and follow their parents within hours of hatching. Provide plenty of small live food, fine-grade insectivorous mixture and grated hard-boiled egg. The best early rearing food is undoubtedly small aquatic insects; live *Daphnia*, *Tubifex* and glass worms are available from aquatic stockists, and should be offered in very shallow dishes of water with small pebbles in the bottom.

General management They mix well with other species and provide an unusual addition in a suitable aviary; problems occasionally arise when rails seek high roosting perches at dusk and disturb their companions. They thrive in outdoor aviaries in the summer but must be housed indoors (or in a tropical house) during late autumn and winter. Provide a small pond; they also enjoy cascades and waterfalls, as they are great splashers, and therefore need ample indoor accommodation to prevent a quagmire effect!

Porzana flavirostra
Black Crake

9 in (23 cm). The plumage is entirely black with a contrasting yellow-green bill and red legs and feet. The sexes are similar.

Range Wide distribution. Senegal eastwards to Sudan and then through central and eastern Africa to South Africa; wherever suitable habitat occurs.

Status Widespread and fairly abundant.

Avicultural rating Occasionally available; interesting and attractive. They are easy to manage but not completely hardy.

Feeding Similar to the White-breasted Rail.

Breeding They need good ground cover of coarse grass, reeds or rushes. The nest of aquatic vegetation is built in cover close to water; usually at or a little above ground-level. Clutch 3–5; incubation 20–22 days. The chicks are nidifugous but remain in the nest for up to 48 hours, then they follow their parents. Feed as for White-breasted Rail. They grow quickly; watch for parental aggression when they are self-supporting, and separate if necessary.

General management They need a secluded aviary with a pool. Related birds are not tolerated; the best companions (if any) are the larger softbills. They behave like jacana, and have long toes to walk on lily-pads etc. They need indoor quarters during the winter months.

Porphyrio porphyrio [PLATE 39]
Purple Swamphen, Purple Gallinule, Green-backed Gallinule, King Reedhen

PLATE 39 *Porphyrio porphyrio*, Purple Swamphen

18–20 in (46–51 cm). The upper surfaces are mainly green, the wings purple-blue; the face, chin and throat are bright blue but darker on the head and neck. The under surfaces are blue. The bill and a large shield on the forehead are red, the legs and feet orange-red. The sexes are similar.

Range Parts of south-west Europe, north-west Africa, then eastwards to eastern and southern Africa, Madagascar; also Iran eastwards across south-east Asia, Malaysia, Sumatra, Borneo, New Guinea, various western Pacific islands, local distribution in Australia.

Status Fairly abundant in some parts of the range; decreasing where habitat has been reduced.

Avicultural rating Easily managed and spectacular; they are aggressive with their companions. They are good climbers, and therefore are not paddock birds.

Feeding Similar to Grey-necked Wood Rail. They are omnivorous and need animal matter; they will steal eggs and nestlings of other species.

Breeding Established pairs are fairly reliable breeders. They need spacious quarters, an area of water and appropriate rushy, reed-type cover. The nest is built of aquatic vegetation, grass etc, either close to the water or (if facilities allow) as a raft-like structure on the water. Clutch 3–5; incubation 28 days. The chicks are nidifugous but may not leave the nest for 48 hours or more after hatching. Provide rearing food similar to that suggested for the Grey-necked Wood Rail, with the addition of a very small quantity of finely scraped raw beef.

General management This species needs a spacious aviary and is not a good mixer; it is certainly not safe associated with other ground-dwelling species and waterfowl. The birds enjoy reedy cover in the aviary, and need an area of water. They are hardy but need shelter in the winter, otherwise frostbite damages their feet.

Eurypygidae (Sun Bitterns)

The beautiful Sun Bittern is placed in a family by itself between the Kagu (Rhynochetidae) of New Caledonia and the Seriemas (Cariamidae) of South America. Three subspecies are described. When hunting its prey (in the wild Sun Bitterns feed on various large insects and small fish), the bird personifies slow motion, when its mainly brown, buff and chestnut plumage provides excellent camouflage in its habitat along river margins and flats. Seen in sunlight with its wings characteristically spread, the Sun Bittern is transformed into a bird of quiet but undeniable beauty. These birds also have an attractive, whistling call. Although they have, on occasion, been kept in private collections with good results, Sun Bitterns are rarely available.

Eurypyga helias
Sun Bittern

17–20 in (43–51 cm). The mantle and upper surfaces are red-brown with black barring; the underparts are buff with darker barring. The eye and throat stripes are white; there is a large area of orange-chestnut on the wings, mainly visible when the wings are expanded. The upper mandible is black, the lower orange-yellow; the legs and feet are yellow. The sexes are similar.

Range Southern Mexico, through Central America to Peru, Brazil, Bolivia, the Guianas; parts of Colombia and Ecuador.

Status Numbers have been reduced in many parts of the range.

Avicultural rating Rarely available. The Sun Bittern needs special accommodation and is recommended only for experienced aviculturists.

Feeding Diet G, with some raw, minced beef, and strips of raw, white fish; abundant live food (locusts etc). They will also take small frogs, newts etc.

Breeding A rare event under controlled conditions. Clutch 2; incubation 27 days. The nest is a well-constructed platform, usually on a horizontal limb a few feet above ground-level. The young remain in the nest for a varying period after hatching; they are fed by both parents.

General management They fare best in a tropical house environment; they like humid, well-planted conditions. Provide plenty of boughs and perches so that the birds can 'stalk' (slow, deliberate movements with wings half-spread). They enjoy basking.

Cariamidae (Seriemas)

These are almost the South American equivalent of the snake-hunting Secretary Bird (Sagittariidae) from Africa. But, although Seriemas do kill and eat snakes their main diet is composed of insects, small mammals and amphibians as well as some vegetable matter. They are good avicultural subjects and are sometimes kept at semi-liberty. Like the Sun Bittern these attractive birds tend to stalk live prey in an unhurried, dignified manner before pouncing.

Cariama cristata [PLATE 40]
Red-legged Seriema, Crested Seriema, Crested Cariama

PLATE 40 *Cariama cristata*, Red-legged Seriema

32 in (81 cm). These birds have mainly grey-brown plumage with lighter pencillings, and a bristly, upstanding frontal crest. The wings and tail are banded in black and white. The bill is red, the legs and feet red-brown. The sexes are similar.

Range North-eastern Argentina and Paraguay to southern and central Brazil.

Status Hunting and habitat destruction have reduced numbers in many areas.

Avicultural rating Infrequently available; they are not difficult to manage.

Feeding They are omnivorous. Diets G and M provide the basis, plus some minced raw beef and live food including locusts, and the occasional dead mouse or frog. Individual birds enjoy chopped fruit.

Breeding They are fairly ready breeders. They nest in trees up to 10 ft (3 m) above ground-level. They may use an open basket as a base for the nest. Clutch 2; incubation 25–26 days. The chicks may remain in the nest for several days before following their parents; provide them with fine-grade insectile mixture, some scraped raw beef, live food etc.

General management They need good-sized aviaries with ample ground cover. Their stately, aldermanic walk belies the speed with which smaller companions may be attacked and occasionally eaten. They need sheltered accommodation in the winter, and a tropical house environment suits them. Watch their feet on hard, man-made surfaces. They are similar in appearance to the Secretary Bird and seriemas also eat snakes. They have a loud, piercing call-note; they are good 'watchdogs' but may not be appreciated by neighbours.

Chunga burmeisteri
Burmeister's Seriema, Black-legged Seriema

28 in (71 cm). This species is smaller and more grey in colour than the Red-legged Seriema. There is a white stripe above the eye and the crest is shorter. The bill, legs and feet are black. The sexes are similar.

Range North-west Argentina, parts of south-western Paraguay.

Status See Red-legged Seriema; this species is protected in some areas.

Avicultural rating See Red-legged Seriema.

Breeding/Feeding/General management Similar to Red-legged Seriema, these birds have also bred successfully in aviaries.

CHARADRIIFORMES

Jacanidae (Jacanas)

Jacanas are also called lily-trotters and lotus birds, because of their ability to traverse lily pads and floating aquatic plants. There are eight species; their distribution includes Africa south of the Sahara, Madagascar, India, Malaysia, Borneo, the Philippines and Australia. There is one New World species. Occasionally available to aviculturists, they need a tropical house environment with shallow pools which have a good surface area. In the correct environment they are not difficult to manage. The females are usually bigger than the males.

Actophilornis africana [PLATE 41]
African Jacana, Lily-trotter

PLATE 41 *Actophilornis africana*, African Jacana

9–11 in (23–28 cm). Most of the plumage is bright chestnut, except the black crown and hind neck; the white chin, throat and upper breast are conspicuous, and are separated from the chestnut under surfaces by a band of yellow. The bill and frontal shield are bluish-white; the legs and feet are grey-green. The sexes are alike but the hen is usually slightly bigger.

Range Widely distributed in Africa: Senegal to Sudan and south to Cape Province.

Status Still fairly abundant where appropriate habitat exists.

Avicultural rating This species needs careful treatment, so is suitable for experienced aviculturists only. It is rarely available.

Feeding Diet G, plus some finely scraped raw beef, abundant live food (including aquatic live food).

Breeding Success is unlikely except in a near recreation of the bird's natural breeding environment. Clutch 4; incubation 23–24 days. In the wild the eggs are often part submerged but waterproofing prevents damage. The chicks are nidifugous and follow their parents within hours of hatching. Abundant small live food and aquatic insects would be necessary for successful rearing.

General management Tropical house accommodation is essential for success. Ensure that there is a resilient natural floor covering to safeguard their delicate toes. Warm, humid conditions suit these birds.

Jacana spinosa
Common Jacana, Northern Jacana, American Jacana

10 in (25 cm). The whole of the head, neck, upper back and underparts are glossy black, the tail reddish-brown. The upper surfaces are chestnut. The bill and frontal shield are yellow, the legs and feet grey-green. In another race the bill and frontal shield may be yellow and red. A black (melanistic) form of the Common Jacana occurs fairly frequently and examples often live alongside normal chestnut and black forms.

Range Southern Mexico, central America, parts of Colombia, Venezuela, the Guianas; western Ecuador, eastern Peru, northern Argentina.

Status Still fairly abundant in many parts of the range where suitable habitat occurs.

Avicultural rating See African Jacana.

Feeding/Breeding/General management Similar to African Jacana.

Glareolidae (Coursers and Pratincoles)

Although a number of the 16 species of coursers and pratincoles have been available to aviculturists from time to time, their unusual behaviour renders them dubious subjects for controlled conditions. Advances in avicultural husbandry may make this opinion obsolete in due course, but at present the ability to maintain individual species for self-gratification or exhibition only cannot justify keeping them. One species is seen more regularly in aviaries and tropical houses and seems more likely to offer reasonable success in the long term.

Pluvianus aegyptius
Egyptian Plover, Crocodile Bird

9 in (23 cm). The crown, face and mantle are black; there are white stripes from the nostrils above the eye and meeting on the nape. The rest of the upper surfaces are grey; the underparts are tawny-buff with a conspicuous black chest band. The bill is black, the legs and feet blue-grey. The sexes are alike.

Range Senegal and Gambia eastwards to Chad and Central African Republic; south to Zaire, northern Angola. It is also found in Egypt. It frequents river margins and sandbanks.

Status Still fairly abundant in many areas.

Avicultural rating Only occasionally available; it has some special needs but is not difficult in the correct captive environment.

Feeding Diet G, with an abundance of small live food. It is best if small aquatic live food can be supplied; scraped meat or mealworms are frequent substitutes but are not recommended. The need for the appropriate live diet cannot be overstressed.

Breeding This is likely to prove extremely difficult. Clutch 2–3; incubation is not recorded. The eggs are known to be buried in sand when, according to some authorities, the adult incubates only after dark when the temperature falls, for 19–21 days; during the daytime the adult appears to be more concerned to provide shade over the spot where the eggs are buried 2–3 in (5–8 cm) down. It is also reported that the adults carry water and regurgitate it over the 'nest' area to keep the eggs moist. The chicks are nidifugous; they squat when danger threatens, and the parents temporarily bury them in sand. From this, some idea of the difficulties facing potential breeding attempts may be ascertained.

General management They can be housed outside during the hottest months of the year. They prefer a dry, sandy or semi-sandy environment and a shallow pool with a good surface area. Provide a well-lit shelter with a similar but smaller-scale living area. In the winter months warm tropical house-type accommodation is needed. The species moves very rapidly, both running and in flight; the potential for injury is obvious in small accommodation which allows enough take-off or acceleration space but no braking area.

Charadriidae (Plovers)

The plover family has a worldwide distribution, except Antarctica. Some species are gregarious in the non-breeding season. The birds inhabit flat land areas close to water, and some species are highly migratory. A small number of species is available to aviculturists from time to time; most do well in outdoor flights during the summer months but the warm-climate species need appropriate winter accommodation. Breeding results have been increasing over recent years and there is no reason why they should not continue to do so.

Anitibyx armatus [PLATE 42]
Blacksmith Plover

PLATE 42 *Anitibyx armatus*, Blacksmith Plover

11 in (28 cm). The forehead, crown, rear neck, rump and lower under surfaces are white. The mantle and wings are grey; the back of the head, cheeks, chin, throat, lower under surfaces, upper back and flight feathers are black. The bill, legs and feet are black. The sexes are alike.

Range Suitable areas of central Africa and Kenya southwards to Angola; parts of South Africa.

Status Still fairly abundant in areas where suitable habitat exists.

Avicultural rating This species needs fairly carefully planned accommodation, but is a good avicultural subject in experienced hands.

Feeding Diet G, with abundant live food; provide occasional scraped raw beef. Small invertebrates are valuable additions to the diet.

Breeding Established and compatible pairs may breed. Clutch 4; incubation 26–28 days. The parents share the incubation and rearing. The nest is a scrape among vegetation. Fine insectile food and plenty of small live food are essential for rearing. They fare best if they are the sole occupants of an aviary.

General management The lack of sexual dimorphism is a problem when establishing pairs; surgical sexing techniques are helpful but compatibility may be a problem even on a 1 : 1 basis, and conflict can be savage. Provide a grassed flight in summer, and remove to temperate accommodation in winter. Warm tropical house winter accommodation is not recommended; the basic requirements are a large grassy enclosure, an area of pebbles or shingle, water and some degree of protection from low temperatures and frost. Breeding is more likely if pairs are housed in an environment of this kind throughout the year.

Vanellus spinosus
Spur-winged Plover

12 in (30 cm). The top of the head, chin, throat,

breast and tail (except the base) are black; there is a conspicuous area of white on the cheeks and the sides of the neck. The upper surfaces are grey-brown; the abdomen and base of the tail are white. The bill, legs and feet are black. The sexes are alike.

Range Middle East, central and East Africa.

Status Still fairly abundant in some parts of the range.

Avicultural rating See Blacksmith Plover.

Feeding/Breeding/General management Similar to Blacksmith Plover.

Vanellus coronatus
Crowned Lapwing

---

11 in (28 cm). The top of the head is black with a white ring on the crown. The upper surfaces are grey-brown; the chin is white merging with the fawn-brown breast. The lower underparts are white. The bill is red with a black tip; the legs and feet are red. The sexes are alike.

Range Central and East Africa, southwards to Angola and South Africa.

Status Abundant in many parts of the range.

Avicultural rating See Blacksmith Plover.

Feeding/Breeding/General management Similar to Blacksmith Plover, but a drier, well-drained aviary is desirable. It is a fairly ready breeder, but nervous, so it needs seclusion, and it is best if it is the sole occupant of the accommodation.

Charadrius pecuarius [PLATE 43]
Kittlitz's Plover, Kittlitz's Sand Plover

---

6 in (15 cm). The forehead, sides of the neck and under surfaces are white; the upper parts are fawn-grey. The forehead is white bordered with a narrow black line; a second black line extends from the base of the bill, through the eye and down the sides of the neck, then forms a collar across the nape. The bill, legs and feet are black. The sexes are alike.

PLATE 43 *Charadrius pecuarius*, Kittlitz's Plover

Range Senegal eastwards to Chad, Central African Republic; Sudan and East Africa south to Cape Province; Madagascar.

Status Wide distribution. Abundant in many areas of suitable habitat.

Avicultural rating Occasionally available. See Blacksmith Plover.

Feeding Similar to Blacksmith Plover.

Breeding Established pairs can prove fairly free breeders. A sandy area is desirable in the flight which should be kept dry. Clutch 2; incubation 20–21 days. The female frequently buries the eggs by kicking sand over them when leaving the nest scrape.

General management Similar to Blacksmith Plover, but with an area of sand and pebbles in the aviary. This species is usually winter hardy.

Thinocoridae (Seedsnipe)

The Thinocoridae comprises four species which closely resemble quail in appearance and were once classified with them. They are confined to western and southern parts of South America. Essentially ground-dwelling, they can fly with a rapid zig-zag flight. They are only rarely available to aviculturists.

Thinocorus rumicivorus [PLATE 44]
Least Seedsnipe

PLATE 44 *Thinocorus rumicivorus*, Least Seedsnipe

6 in (15 cm). The upper surfaces are brown with light and dark mottling and streaking; the underparts are grey merging to white on the abdomen. There is a black area on the throat, then a vertical black line to the lower breast producing a 'necktie' effect. The bill is horn-coloured; the legs and feet yellow. The females are similar, but lack the necktie.

Range Western areas of South America from south-west Ecuador to southern Argentina; Tierra del Fuego, Falkland Islands.

Status Fairly common in many parts of the range.

Avicultural rating Rarely available, but the indications are that they may be interesting and potentially successful avicultural subjects.

Feeding Diet O, with the addition of chick crumbs and some pinhead oatmeal; they also take small live food and green food, although I found mine reluctant at first.

Breeding Little information is recorded. Clutch usually 4; incubation not recorded. There seems to be no reason why breeding should not take place in a suitable aviary environment. They prefer a dry, pebbly floor, with some natural ground cover such as heathers and low shrubs. Seclusion is probably important. In the wild, they breed from late summer to early winter.

General management They fare best in aviaries, indoor or outdoor. They are hardy birds, but should be enclosed during severe weather (frost, snow). They seem to prefer a dry environment and take frequent dust-baths. No aggression was seen when they were housed with small softbills and finches.

Laridae (Gulls, Terns)

More than 80 species with near world-wide distribution. Although rare in aviculture, gulls are generally easy to manage. Terns, with the exception of the described species, are not good avicultural subjects.

Larosterna inca [PLATE 45]
Inca Tern

16 in (41 cm). Most of the plumage is blue-grey, lighter on the throat and upper breast. There is a prominent moustachial streak from the base of the bill to below the eye; the feathers curl outwards from the cheeks. The bill, legs and feet are red. The sexes are similar.

PLATE 45 *Larosterna inca*, Inca Tern

Range Coasts of Peru and Chile.

Status Very abundant in many parts of the range.

Avicultural rating Non-standard accommodation is required; for experienced aviculturists only.

Feeding They are essentially fish-eaters. Give strips of raw fish or (better) whole small fish such as whitebait; frozen whitebait is readily available and excellent for these birds, but expensive.

Breeding This is the only tern to breed in burrows or rock crevices. Clutch 2; incubation 29 days. Although it is difficult to reproduce an environment conducive to successful nesting, it is more feasible nowadays; artificial glass-fibre cliffs etc are available that go some way towards achieving the desired objective. Thriving colonies have been established in zoos.

General management They need a large aviary with an area of rocks or pebbles and water. Running water (a small cascade powered by an electric pump is acceptable) is an advantage. They have a fluttering flight and are usually tame and confiding. They can cope with cold but are not resistant to frost or prolonged cold and wet conditions.

COLUMBIFORMES

Pteroclididae (Sandgrouse)

Sandgrouse are the size of small pigeons; they have short legs which make their progress on the ground no more than an ungainly walk. They are distributed mainly in Africa and Asia and there are 16 species. They are much persecuted by hunters who wait at water holes for vast flocks to arrive at dusk and dawn. One species (Pallas' Sandgrouse) from Central Asia undergoes periodic population 'explosions' when irruptive migrations take the species far from its normal range; in 1908 one such movement brought birds thousands of miles westwards to Europe, including Britain. Sandgrouse have unique drinking habits; the crop can hold more than 1 fl. oz (28 ml) of water which is transported back to the young birds. Round trips of 60–70 miles for water are not unusual; grain and seeds are the main diet.

Pterocles orientalis
Black-bellied Sandgrouse

14 in (36 cm). The head and upper surfaces are sandy-grey with some orange spotting. The throat is chestnut with a black area below; the breast is pinkish-fawn. The lower underparts are black. The bill, legs and feet are blue-grey; the legs are feathered. The female has some barring on the wings; the upper parts are slightly lighter in colour and the head and neck are spotted. The throat is yellow.

Range Spain, North Africa eastwards through Middle East to southern Russia, northern Afghanistan, India.

Status Still vast numbers in some areas, despite heavy predation.

Avicultural rating An interesting species, but recommended for experienced aviculturists only; it does well in the correct environment.

Feeding Diet O, with an adequate supply of grit. Some chick starter crumbs can also be offered; green food is sometimes taken but individual birds vary in their interest.

Breeding This and other sandgrouse species have been bred several times in aviaries. They need a dry, sandy environment and shelter from winds and driving rain. These birds do not tolerate damp conditions at all. Clutch 2–3; incubation 26–28 days, usually shared on a day/night basis. The chicks leave the nest soon after hatching and are fed by regurgitation.

General management Provide a roomy flight attached to an equally spacious shelter; the latter should be well-lit so that the birds can be confined during severe winter weather. Most species are hardy but unable to tolerate damp or wet. Utilise a glass-substitute material to protect the flight but admit maximum light.

Other species of Sandgrouse which are occasionally imported include the Asiatic Pallas' (*Syrrhaptes paradoxus*) (V), Pin-tailed (*Pterocles alchata*) (V) from southern Europe, and Chestnut-bellied (*P. exustus*) (V) which ranges over much of Africa, the Middle East and India. All do well if correctly fed and housed, are usually long-lived and will breed.

Columbidae (Doves, Pigeons)

The Columbidae comprises nearly 400 species which can conveniently, if unscientifically, be broken down into three groups: fruit pigeons (Africa, Asia, East Indies and Australia); seed-eating pigeons and doves of both the Old and New World; and the much bigger crowned pigeons from New Guinea. Including various subspecies, there are well over 300 forms of fruit pigeon and nearly 500 seed-eating pigeons, but only seven crowned pigeons. The Rock Dove (*Columba livia*) is the ancestor of all 'street' pigeons and most of the strains used for sport, exhibition and more utilitarian purposes. Many kinds of pigeons and doves are long-established avicultural favourites. Seed-eating species are most widely kept; most are hardy, easy to feed and can be housed with other birds. Despite the fact that the dove is a universal symbol of peace, they do not tolerate their own kind at all; one pair to each unit of accommodation is the rule. Fruit pigeons are more troublesome and cannot be regarded as hardy. Many species of pigeons and doves breed freely in controlled conditions.

Columba guinea (III)
Triangular-spotted Pigeon, Speckled Pigeon, Guinea Pigeon

15 in (38 cm). The plumage is mainly blue-grey, with a vinaceous tinge on the neck. The mantle and wing coverts are chestnut, the latter heavily spotted with white; there is a broad black band on the tail. The conspicuous orbital patch is red. The bill is black; the legs and feet red. The sexes are similar.

Range Senegal eastwards across Africa to Ethiopia, Tanzania, Kenya, then south and south-west to Namibia and South Africa.

Status Common in many parts of the range where savannah exists; also now found in farming and urban areas.

Avicultural rating Occasionally available; they are hardy and easily managed, but their large size makes selection of companions difficult.

Feeding Diet O, with the addition of some maize. Grit is essential. They may take green food.

Breeding They are fairly willing breeders if a true pair is obtained. Like others in the family, they are not good nest-builders, but plastic trays (used by packers to hold small amounts of choice fresh fruit) can serve as a base for the nest. Alternatively, make a rudimentary 'tray' of welded mesh and fix it in place, then add a few twigs to encourage the birds to use it. Clutch 2; incubation 16–17 days. They need seclusion and will abandon the nest with minimum provocation.

General management They are hardy and can remain outside throughout the year. Provide a roomy aviary and comfortable shelter. Choose companions carefully; these birds are robust and can be aggressive. They have a fairly strident call which is repeated at frequent intervals.

Streptopelia senegalensis (III)
Senegal Dove, Laughing Dove, Palm Dove

11 in (28 cm). The upper surfaces are soft vinous-brown; the lower back, wings and rump are grey. The under surfaces are pale vinous, whiter towards the abdomen. The foreneck is a deeper vinous-red with conspicuous black spots. The bill is grey-black, the legs and feet red. The sexes are similar.

Range Widespread throughout Africa, wherever suitable habitat occurs; Morocco, Algeria, Libya, Egypt, then locally distributed southwards through East Africa to Cape Province; also Iran, Afghanistan, India.

Status Variable, but very common indeed in some areas of the range.

Avicultural rating Very easily managed; popular, hardy and inexpensive.

Feeding Diet O, with some green food. Adequate grit is essential.

Breeding They are very free breeders. The only inhibiting factor may be whether a 'pair' is true. Clutch 2; incubation 16–17 days. The young should be removed from the aviary as soon as they are independent; as the parents start on a further brood the cock may injure or kill the older chicks.

General management Easily managed, and totally weather-resistant. They need an outdoor aviary with a shelter. They are normally good mixers, but can be a problem when breeding. They have an attractive cooing call (hence the synonym) which has little nuisance value.

Streptopelia chinensis
Spotted Dove, Chinese Necklaced Dove, Chinese Spotted Dove, Spotted Chinese Turtle Dove

---

12 in (30 cm). The upper surfaces are pale brown; the underparts pale vinous. The nape ring is black, heavily spotted with white. The bill is brown; the legs and feet red. The sexes are similar.

Range India, Sri Lanka; eastwards to Burma, eastern China; Taiwan, Hainan, Greater Sundas, Palawan, Sumatra, Borneo. Also introduced into the USA, Hawaii, Mauritius, New Zealand, Australia etc.

Status Very common in many parts of their natural range; numbers are increasing in several areas where they have been introduced.

Avicultural rating See Senegal Dove.

Feeding/Breeding/General management Similar to Senegal Dove.

Macropygia unchall
Barred Cuckoo-Dove

---

15 in (38 cm). The forehead and crown are grey-brown with some darker spotting; the nape and upper back have iridescent purple-green feathers. The rest of the upper surfaces, including the long, graduated tail, are rufous with distinctive black barring; of the underparts the throat is buff, the breast vinous, and the lower under surfaces buff (often with some light barring on the upper breast). The bill is black, the legs and feet red. The female is similar, but her underparts are dark buff with some fine barring.

Range Himalayas eastwards to northern Burma, parts of west and south-east China, northern Laos and Vietnam, Hainan Island, Malaysia, Sumatra, Java, Lombok Island.

Status Common in some parts of the range.

Avicultural rating An attractive and interesting member of the family, which is easy to manage.

Feeding Diet O, with adequate grit. Also offer some chick crumbs. Individual birds will take berries, especially hawthorn.

Breeding This species prefers natural cover in the aviary. Clutch 1–2; incubation 16–18 days. They will use a basket or tray, and often only one squab is reared.

General management This is not difficult, but they need some protection from severe winter weather; a well-lit shelter to 45°F (7°C) is best. They are occasionally aggressive to some companions, but are generally good mixers.

Turtur abyssinicus (III)
Black-billed Wood Dove

---

8 in (20 cm). The forehead and crown are grey; the mantle grey-brown. The remainder of the upper surfaces are earth-brown, with two darker bands across the rump. The underparts are vinous, lighter on the abdomen. There are iridescent violet-blue spots on the wings. The bill is black, the legs and feet red. The sexes are similar.

Range Senegal and Mauritania east to Chad,

Central African Republic, Sudan, Ethiopia, northern Uganda, north-west Kenya.

Status Common in some parts of the range.

Avicultural rating An attractive and easily managed species which is frequently available.

Feeding Diet A, with adequate grit. Individuals may enjoy green food but most birds ignore it.

Breeding Sexing is difficult, but true pairs are usually willing breeders. Nests are built in wire or plastic trays or boxes, or in shrubby vegetation, bunches of spruce or heather etc. Clutch 2; incubation 12–14 days.

General management Excellent occupants of a garden aviary; they mix well with small passerines and quail but not with related species. They are hardy but need a comfortable shelter; slight heat is desirable during severe winter weather. Like other doves and pigeons they are prone to night fright; precautions against injury can include a fabric net stretched above the roosting perch.

Turtur afer (III)
Blue-spotted Wood Dove

8 in (20 cm). Very similar to Black-billed Wood Dove, but the mantle is browner and the bill is red with a yellow tip. The sexes are similar.

Range Senegal eastwards to Central African Republic, parts of Zaire, Ethiopia then south to Tanzania, Malawi, northern Transvaal.

Status Common in areas of suitable habitat.

Avicultural rating See Black-billed Wood Dove.

Feeding/Breeding/General management Similar to Black-billed Wood Dove.

Turtur chalcospilos
Emerald-spotted Wood Dove

8 in (20 cm). Similar to Blue-spotted Wood Dove, but the upper surfaces are lighter, the wing spots iridescent green and the bill red with a black tip. The sexes are similar.

Range South-west and south-east Africa; Cape Province.

Status Common in many parts of the range.

Avicultural rating See Black-billed Wood Dove.

Feeding/Breeding/General management Similar to Black-billed Wood Dove.

Turtur tympanistria [PLATE 46] (III)
Tambourine Dove

PLATE 46 *Turtur tympanistria*, Tambourine Dove

8 in (20 cm). The upper surfaces are earth-brown with dark-blue spots on the wings; the forehead,

eye stripe and underparts are white. The bill, legs and feet are red-brown. The female has a grey forehead, eye stripe and underparts. Her wing-spots are not as iridescent as those of the cock.

Range Wide distribution in Africa south of a line between Sierra Leone and southern Ethiopia.

Status Common in many parts of the range.

Avicultural rating An easily managed and popular species which is imported fairly infrequently.

Feeding Similar to Black-billed Wood Dove. They will also take live food.

Breeding Ease of sexing this species makes breeding a more realistic prospect. They reproduce fairly freely in some collections, but are less willing in others. They need a secluded aviary with abundant natural cover. Clutch 2; incubation 13–14 days. Provide wire or plastic trays or shallow boxes fixed into place among cover.

General management They are good aviary birds but not completely hardy and are best kept indoors during the winter. They have a quiet disposition (except when with other doves and pigeons) and settle quickly. They are usually confident birds.

Oena capensis [PLATE 47]
Namaqua Dove, Cape Dove, Masked Dove

9 in (23 cm). The upper surfaces are grey-brown; the forehead, chin, throat and breast are black. The lower underparts are grey-white. There are dark bands across the rump and metallic blue spots on the wing coverts. The rufous patches on the wings are conspicuous in flight. The bill, legs and feet are purple. The female lacks the black area. This is a very slenderly built dove with a long, graduated tail.

Range Mauritania and Senegal eastwards through Mali, Niger to Chad, Central African Republic;

PLATE 47 *Oena capensis*, Namaqua Dove

Sudan southwards to Zambezi; Cape Province; Madagascar.

Status A common resident in hot, arid parts of the range; this species prefers semi-desert bush, dry woodland.

Avicultural rating Attractive, frequently imported and usually fairly inexpensive; it is delicate in the normal European climate.

Feeding Diet A, with adequate grit.

Breeding These doves frequently take a long time to settle and breed (two to three years), then may produce several broods over succeeding seasons. Clutch 2; incubation 12–14 days. Provide suitable nesting receptacles (trays or baskets) fixed securely in position among natural or artificial cover. The site is usually within a few feet of the ground.

General management They need a sheltered, well-lit aviary; at their best during spells of really hot weather, they are then very active, and look like giant butterflies in flight. They do not tolerate damp conditions and must be wintered inside in moderate temperature (45–50°F/7–10°C). They also need extra artificial light during short winter days. They are gentle companions for other small birds except their own family. In aviculture the cocks outnumber the hens by a big margin; many imported 'hens' prove to be juvenile males.

Chalcophaps indica [PLATE 48]
Emerald Dove, Indian Green-winged Dove, Green-winged Pigeon

PLATE 48 *Chalcophaps indica*, Emerald Dove

10 in (25 cm). The forehead and eyebrow are white, merging with the blue-grey nape and crown; the hind neck is vinous, the mantle and wings dark metallic green. The lower back is black with two grey-white bars; the under surfaces are vinaceous red. The bill, legs and feet are red. The female is similar but lacks the white eyebrow; her underparts are duller.

Range Much of south-east Asia; India and Sri Lanka eastwards to Philippine Islands, Moluccas, Greater and Lesser Sunda Islands, New Guinea, northern Australia.

Status Generally fairly abundant, but some of more than a dozen subspecies are becoming rare.

Avicultural rating An excellent avicultural subject; it is a colourful bird which settles well to aviary life and is completely hardy when acclimatised.

Feeding Diet O, with adequate grit. Individual birds accept green food; most will take berries and some live food.

Breeding Compatible pairs are fairly reliable breeders. They usually build high among vegetation, occasionally on a shelf or on top of a nest box. Clutch 2; incubation 14–15 days. The nest is more substantial than the normal flimsy platform favoured by most doves and pigeons but is loosely constructed.

General management They fare best in a large aviary, although they alternate between periods of activity and lethargy. They spend much time at ground level; therefore other birds of terrestrial habit (quail etc) may not be suitable companions, although doves normally have a peaceful disposition with unrelated species. They roost high and may disturb smaller passerine companions at dusk.

Ocyphaps lophotes
Australian Crested Pigeon, Crested Pigeon, Crested Bronze-winged Pigeon

13 in (33 cm). The upper surfaces are fawn-brown, the wing coverts iridescent green, the secondaries violet-blue (the feathers have light-coloured margins). The upright crest is black, and the underparts silvery-grey. The bill is lead-grey, the legs and feet red. The sexes are similar.

Range Mainly eastern and southern areas of Australia; parts of Queensland, western Victoria, South Australia, Western Australia. This species is slowly extending its range.

Status Common in many areas.

Avicultural rating Limited captive-bred stock is available. It is hardy and a fairly free breeder, but does not thrive in cold, wet conditions.

Feeding Diet O, with adequate grit. Green food is enjoyed by many birds, and live food is appreciated.

Breeding Compatible true pairs are fairly reliable breeders. Clutch 2; incubation 14–15 days. The nest is a frail platform of twigs; but they will use a tray or basket. The cock defends his territory (which may be the entire aviary) vigorously; therefore choose companions with care.

General management They are very easily managed. They do best in a spacious outdoor aviary. Although this species does not appreciate damp, it is completely weather-resistant if a comfortable unheated shelter is available. Unlike other members of the Columbidae, it is active and busy in the aviary.

Geopelia cuneata [PLATE 49]
Diamond Dove, Red-eyed Dove

7 in (18 cm). The head, neck and breast are pale blue-grey; the nape, back and wings grey-brown. The lower under surfaces are white, and the wings are spotted with white. The orbital patch is red. The bill is dark grey, the legs and feet flesh-coloured. The sexes are similar but the female is often smaller, of slighter build and slightly more brown in colour.

Range Northern and parts of central Australia.

Status Variable; common in some areas.

Avicultural rating This is the most popular and

PLATE 49 *Geopelia cuneata*, Diamond Dove

widely kept of all foreign doves and pigeons. Now virtually domesticated, it is easily managed and hardy; the Diamond Dove should be in every mixed collection of seedeating species.

Feeding Diet A, with adequate grit. Individual birds may take insectivorous food (not an essential part of the diet) and greenstuff.

Breeding True pairs are usually free breeders at any time of the year. Clutch 2; incubation 14–15 days. They will use a shallow basket or tray, adding twigs etc. As with many other small dove species, the young leave the nest early, usually after less than two weeks; remove them when independent and feeding.

General management These doves are peaceful occupants of an aviary, and mix well with the smallest finches. They are hardy but need dry winter quarters. Do not keep more than one pair per aviary, or with other dove species. A silver mutation has been available for some years; now other colours including red and yellow are available.

Geopelia striata
Zebra Dove, Barred Ground Dove

---

8 in (20 cm). The forehead, cheeks and throat are pale grey; the back of the head and nape are suffused with pinkish-brown. The rest of the upper surfaces are light brown, with the mantle and wings barred black. The under surfaces are buff-pink shading to white on the abdomen; the sides of the neck, breast and flanks have black and white barring. The bill is grey, the legs and feet pale red. The sexes are similar, but the hen is usually slightly smaller than the cock and has paler underparts.

Range Southern Burma, Thailand, Malaysia, Borneo, southern New Guinea to northern, central and parts of Western Australia (where it is called the Peaceful Dove). Introduced to Hawaii, Madagascar.

Status Fairly common in most parts of the range but becoming rare in some areas; generally an adaptable and successful species.

Avicultural rating Despite its similarity to the Diamond Dove, this species is much less widely kept. They are hardy and good breeders.

Feeding/Breeding/General management Similar to the Diamond Dove. They can be prolific.

Columbina talpacoti [PLATE 50]
Ruddy Ground Dove, Talpacoti Dove, Pygmy Ground Dove
One of several small South American doves described in the bird trade as 'Pygmy Ground Doves' or 'Pygmy Doves'.

---

6 in (15 cm). The head and neck are blue-grey; the upper surfaces vinaceous-pink. The chin and throat are grey-white, the rest of the underparts cinnamon-pink. There are some black spots on the wings. The bill is horn-coloured, the legs and feet flesh-coloured. The female is mainly grey-brown above with paler underparts.

Range Mexico, Central America to northern Colombia, Venezuela, the Guianas then south to Brazil, Bolivia, Paraguay, central Argentina.

Status Fairly common in many parts of the range.

Avicultural rating This is one of the best-known South American doves; it is easily managed but needs comfortable winter quarters.

Feeding Diet A, with adequate grit. Offer chick crumbs and some small live food.

Breeding A fairly reliable breeder but not as free as Diamond or Zebra Doves. It uses a shallow tray or basket or an open box; it occasionally nests in vegetation close to the ground. Clutch 2; incubation 14–15 days. The chicks grow quickly and should be removed from the breeding aviary when independent (at about 4 weeks old).

PLATE 50 *Columbina talpacoti*, Ruddy Ground Dove

General management This and related species are best kept indoors and in slight heat (45°F/7°C) during the winter; they may survive prolonged cold or wet weather but are likely to be debilitated by the experience. Despite their small size, they are aggressive to other members of the family; they may also attack passerine species when breeding. They are ground-dwellers.

Columbina minuta
Plain-breasted Ground Dove, Pygmy Ground Dove, Pygmy Dove

6 in (15 cm). The upper surfaces are grey-brown, lighter on the head and neck with a pronounced blue suffusion; the wings are grey-brown. The underparts are fawn-pink, lighter towards the abdomen. The bill is grey, the legs and feet red. The female is browner than the male: the blue wash on the head and neck are replaced by very pale chestnut.

Range Southern Mexico, Central America, northern Colombia, Venezuela, the Guianas. Southwest and central Brazil, Bolivia, southern Peru, Paraguay.

Status Variable, abundant in a few areas, but mainly small, sporadic populations.

Avicultural rating Occasionally imported. See Ruddy Ground Dove.

Feeding/Breeding/General management Similar to Ruddy Ground Dove. This species thrives in small aviaries and tolerates passerine companions better than the Ruddy Ground Dove.

Columbina cruziana
Gold-billed Ground Dove, Peruvian Ground Dove, Croaking Ground Dove, Pygmy Ground Dove

6 in (15 cm). The head and neck are bluish-grey; the upper surfaces are mainly grey-brown. The wings have black spots and a dark red-brown bar. The underparts are pink-brown. The bill is orange with a black tip; the legs and feet are pink. The female is smaller and has a yellow bill. Her head is browner, and her other shades slightly more muted.

Range North-west Ecuador, western and eastern areas of Peru, western Chile.

Status Fairly common in some parts of the range; now in many cultivated areas.

Avicultural rating See Ruddy Ground Dove.

Feeding/Breeding/General management Similar to Ruddy Ground Dove. Occasionally there are 3 eggs; established pairs can be prolific. They are extremely aggressive to their companions when breeding, so are best kept in a separate small aviary if possible.

Columbina passerina
Passerine Ground Dove, Common Ground Dove, Scaly-breasted Ground Dove

6 in (15 cm). The upper surfaces are grey-brown; the wings are slightly darker and have some black spots. The underparts are pink-brown; the feathers have darker edgings, giving a scaly effect. The bill is black, the legs and feet pink. The female is slightly smaller with greyer underparts.

Range Parts of south-eastern USA, Mexico, Central America to Colombia, Venezuela, northern Brazil, the Guianas, Antilles, Bahamas, Bermuda.

Status Common in many parts of the range.

Avicultural rating See Ruddy Ground Dove.

Feeding/Breeding/General management Similar to Ruddy Ground Dove.

Geotrygon versicolor
Mountain Witch Dove, Crested Quail Dove, Blue Dove

12 in (30 cm). This dove has a hood-like crest. The forehead is black; the crown and nape grey-brown with the hind neck an iridescent bronze. The mantle and wing coverts are metallic purple-maroon, the primaries chestnut. The under surfaces are grey, with chestnut on the abdomen; the rump and tail are dark green. The bill is black, the legs and feet pink. The sexes are similar.

Range Jamaica.

Status Vulnerable because of its limited range.

Avicultural rating A highly-prized species which is not difficult to manage, but expensive to purchase.

Feeding Diet O, with the addition of maize; they also enjoy live food. Grit is essential.

Breeding They are not free breeders. Clutch 2; incubation 16–18 days. They will use an open box or tray, but they need seclusion. The nesting receptacles are best concealed among natural or artificial cover.

General management They are hardy, but best kept on well-drained ground. They need a comfortable, well-lit shelter in the winter months.

Caloenas nicobarica [PLATE 51] (I)
Nicobar Pigeon

16 in (41 cm). The head, neck and breast are slate-blue; there are elongated hackle-feathers which

PLATE 51 *Caloenas nicobarica*, Nicobar Pigeon

protrude from the sides and back of the neck. The rest of the plumage is green with some copper and blue reflections on the under surfaces. The short tail is white. The grey-black bill is hooked and there is a small knob behind the nostrils. The legs and feet are dark red. The female is smaller and slighter.

Range Andaman and Nicobar Islands; Luzon, Sundas, New Guinea, Solomon Islands.

Status Rare in most parts of the range. There are only one or two large colonies.

Avicultural rating Rarely available but much sought after.

Feeding Diet O, with adequate grit; they also enjoy diced fruit, berries and some larger grain such as maize.

Breeding Spacious, lofty aviaries with abundant cover are required; use baskets fixed high among the vegetation. Clutch 1; incubation 28 days. The nestling period can be up to 12 weeks.

General management They are not normally aggressive with other species, but they may attempt to kill and eat small passerines sharing their terrestrial territory. They feed at ground level and are accustomed to the forest floor; therefore they are happier in well-grown (even gloomy) aviaries or a tropical house environment. When not foraging at ground level, the birds occupy a horizontal branch among cover in a characteristically alert attitude.

Gallicolumba luzonica (II)
Luzon Bleeding-heart Pigeon

10 in (25 cm). The forehead and crown are grey; the rest of the upper parts are grey-brown with a suffusion of metallic green-purple. The cheeks, chin, throat and breast are white; an area in the centre of the breast is bright red. The lower under

surfaces are salmon-yellow. The bill is black, the legs and feet red. The female is smaller and the 'bleeding heart' is usually smaller and paler than that of the cock.

Range Luzon and Polillo Islands.

Status Numbers are decreasing.

Avicultural rating A much-prized species which is a reasonably good breeder, and easy to manage.

Feeding Diet O, with adequate grit. They also enjoy chick crumbs and green food. Provide a small quantity of Diet G from time to time, plus grated cheese, berries and live food.

Breeding They fare best in an aviary with abundant cover, including vegetation at ground level. Take care if introducing a pair; the cock's aggressive behaviour towards the hen sometimes leads to tragedy, hence the need for cover. The cock will display, then chase the hen; a good level of compatibility is usual after the initial introduction period. Clutch 2; incubation 16–17 days. They will use a tray, basket or box off the ground; occasionally a shelf will be used. Watch the cock for aggressive intent as the chicks grow.

General management They mix with other birds, but better breeding results will be achieved if they are in separate accommodation. They are hardy but need slight heat in the winter, and are best enclosed during severe weather because of the possibility of frostbite. They often become very confiding and will feed from their owner's hand.

Gallicolumba criniger
Bartlett's Bleeding-heart Pigeon

11½ in (29 cm). Similar to Luzon Bleeding-heart, but this pigeon is bigger and darker in colour.

Range Mindanao, Leyte, Samar and Basilan Islands.

Status Rare throughout the range.

Avicultural rating Less often seen than the Luzon Bleeding-heart Pigeon, and much in demand.

Feeding/Breeding/General management Similar to Luzon Bleeding-heart Pigeon. Clutch only a single egg. Like the previous species, this pigeon enjoys both rain and dust bathing.

Goura cristata (II)
Blue Crowned Pigeon

---

30 in (76 cm). Most of the plumage is blue-grey; the wings and tail are darker. The wing coverts and a wide band across the back are chestnut; there is a black mask from the base of the bill through the eyes to the cheeks. The chin is black; there is some white on the wing coverts. There is a distinctive fan-shaped crest. The bill is black, the legs and feet red. The sexes are alike.

Range North-west New Guinea and adjoining islands.

Status Uncertain. The population has been depleted as a result of excessive hunting; the European and North American millinery trade is responsible for popularising 'goura' (head) feathers. Numbers may now be recovering, helped by difficult (inaccessible) terrain near some of their areas of territory.

Avicultural rating Rarely available and expensive. This bird's rarity demands that it should be acquired by experienced aviculturists only.

Feeding Try a mix of Diets G, O and P, with some larger grain such as maize. Additional items which can be offered include trout pellets (with boiled rice), berries and live food.

Breeding Success is infrequent, but it was first accomplished in the mid-eighteenth century. They will sometimes use a large box or basket fixed a few feet above ground level; and are best housed amid natural or artificial cover. Clutch 1; incubation 28–30 days. The chick leaves the nest after 4–5 weeks and is fed by the parents for a further 8 weeks.

General management They do best in moist tropical house conditions, but high humidity is not recommended; they are also excellent in a large grassed enclosure during the summer. They spend most of the hours of daylight at ground level and then roost high for the night. Normally gentle, they mix well with other birds but may be aggressive when nesting. They are not hardy.

Treron waalia (III)
Bruce's Green Pigeon, Yellow-bellied Green Pigeon, Waalia Fruit Pigeon

---

12 in (30 cm). The head, neck and upper breast are grey-green; the rest of the upper surfaces are olive. The lower breast and underparts are yellow. There is a mauve shoulder patch. The bill is whitish, lilac at the base; the legs and feet are orange. The sexes are alike.

Range Senegal eastwards to southern Chad, Central African Republic, Sudan, Ethiopia, northern Uganda, northern Kenya; also Gabon, Zaire; parts of southern Arabia.

Status Common in suitable habitat in many parts of the range.

Avicultural rating This pigeon needs care when first imported but established birds are usually hardy. A mainly fruit diet is required.

Feeding Diet P, with recommended additions. Provide as varied a diet as possible within these guide-lines.

Breeding They are frequent breeders but the chicks have proved difficult to rear. Provide baskets or trays fixed securely into place among high natural or artificial cover. It is best if the breeding pair are the sole occupants of an aviary. Clutch 1–2; incubation 16–18 days.

General management They live well in an outdoor aviary with a roomy, frost-free shelter for winter occupation. A planted flight is best, but bear in mind the copious nature of their droppings and the need for constant attention if the plants are not to succumb. Their voice is much harsher than the normal pigeon or dove call notes.

Treron curvirostra
Thick-billed Green Pigeon

---

10 in (25 cm). The forehead, crown and neck are grey; the mantle and lesser wing coverts are maroon. The rump and upper tail coverts are olive, the central tail feathers yellow-olive; there is a broad black band across the tail. The underparts are yellow-olive. The bill is pale green, red at the base; the legs and feet are red-brown. The female is similar but lacks the maroon on her mantle and coverts.

Range Western Nepal; Thailand, Laos, Vietnam, Hainan Island, Malaysia, Sumatra, Borneo, Simalur, Nias, Sipora, Siberut, Batu and Enggano Islands.

Status Widespread and common in many parts of the range.

Avicultural rating See Bruce's Green Pigeon.

Feeding/Breeding/General management Similar to Bruce's Green Pigeon. Clutch 2; incubation 14–16 days. They have bred fairly frequently, but there have been few successful rearings. In the UK the first young were raised in the mid-1960s.

Several other Fruit Pigeons appear on the market from time to time. Their basic requirements are similar to those outlined for Bruce's Fruit Pigeon. Many species (including some no longer available) have successfully hatched and reared young in aviaries. There is clearly potential for interested aviculturists to establish these beautiful birds in controlled conditions.

PSITTACIFORMES

This order contains more than 300 species, mainly from tropical and sub-tropical regions. They generally have gaudy plumage, a strong, hooked bill, and two front and two hind toes. Many have a strident voice and some are good mimics of familiar sounds, including the human voice. They are generally birds of forests. Their main foods are seeds, nuts, fruit and berries; some are specialist feeders, including nectar drinkers. Most nest in hollow trees; some in burrows and termite nests. With few exceptions they are hardy and easily managed, although accommodation for many species must be much more strongly built than that for other aviary birds. Many of the larger species are popular pets, but often housed in entirely unsuitable small cages; feather-plucking and other stress-related problems arise as a direct result.

Loriidae (Lories and Lorikeets)

Eos bornea [PLATE 52] (II)
Red Lory, Moluccan Lory

12 in (30 cm). This bird is mainly red; the secondaries and flight feathers are black. The wing and under tail coverts are blue. The bill is orange, the legs and feet dark grey. The sexes are alike.

Range Islands of Amboina, Saparua, Ceram, Buru and Kei.

Status Common in some parts of the range.

PLATE 52 *Eos bornea*, Red Lory

Avicultural rating One of the most popular and widely-kept of lories; it has a strident voice.

Feeding Diet K, plus fruit (either cubed or put through a liquidiser). Some individuals enjoy green food, spinach, chickweed etc. Also offer berries (cultivated and wild) and small live food; seed is best limited to small amounts of plain canary and millet, offered soaked.

Breeding They are fairly frequently bred. Clutch 2; incubation 24 days. Provide a stout nest-box with absorbent material (1–2 in, 13–26 mm) in the base (peat, wood shavings etc). The chicks remain in the nest for 7–8 weeks. It is essential that these and related species are the sole occupants of their accommodation. This species has bred in cages.

General management They fare best in outdoor aviaries comprising a stoutly-built flight and sleeping shelter (although most roost in a nest box). Concrete floors are necessary to maintain cleanliness; wash perches, shelves and utensils regularly. They are hardy once properly acclimatised.

Eos reticulata [PLATE 53] (II)
Blue-streaked Lory

PLATE 53 *Eos reticulata*, Blue-streaked Lory

12 in (30 cm). The plumage is mainly red but there are conspicuous violet-blue streaks on the neck and mantle; a purple-blue band extends from the eyes down the sides of the neck. The bill is orange, the legs and feet grey. The sexes are alike. This species has a longer tail than other *Eos* lories.

Range Tanimbar Islands; introduced to adjacent islands.

Status Believed to be still fairly abundant.

Avicultural rating A popular species, first bred in the USA in 1939. It is hardy, and has been more readily available from the 1970s.

Feeding/Breeding/General management Similar to Red Lory. The nestling period may be as long as 12 weeks.

Pseudeos fuscata [PLATE 54] (II)
Dusky Lory

10 in (25 cm). The forehead, cheeks and hind neck are olive-brown with the feathers of the cheeks and neck edged with dull yellow; the crown is yellow. The upper surfaces are mainly very dark brown; the rump is buff-white. The yellow collar runs between the brown of the chin and throat and the brown of the upper breast; there are further bands of yellow and orange-brown; the abdomen is also dull yellow-orange. The bill is yellow-orange and an area of bare skin at the sides of the lower mandible is the same colour; the legs and feet are black. The sexes are similar but the adult plumage is extremely variable, frequently giving the impression of sexual dimorphism.

Range New Guinea; also the islands of Salwati and Japen.

Status Widespread and abundant in some parts of the range.

Avicultural rating Good avicultural subjects;

PLATE 54 *Pseudeos fuscata*, Dusky Lory

Trichoglossus haematodus [PLATE 55] (II)
Green-naped Lorikeet

PLATE 55 *Trichoglossus haematodus*, Green-naped Lorikeet

they are hardy, easily managed, and fairly consistent breeders.

Feeding Diet K, with various soft fruits; offer plain canary seed or millet (dry and soaked), millet sprays, berries and green food.

Breeding This species is a ready breeder. Clutch 2; incubation 24–26 days. The young remain in the nest for up to 10 weeks. The adults are usually aggressive when breeding.

General management Easily managed. Provide a flight with shelter and a box for roosting. They are active and intelligent, and have shrill voices.

11 in (29 cm). The head and chin are purple-black, darker on the throat; the nape is greenish-yellow. The upper surfaces are green. The feathers of the lower throat and breast are red with dark purple edges giving a barred effect. The centre of the abdomen is green; the rest of the underparts are yellow—most of the feathers being edged with green. The bill is red; the legs and feet grey. The sexes are alike.

Range Western New Guinea and adjoining islands; Buru, Amboina, Ceram, Watubela, Goram and Ceramlaut.

Note The following are some of the subspecies most likely to be encountered in aviculture. More than 20 are described in all.

Mitchell's Lorikeet (*T. h. mitchellii*): islands of Bali and Lombok.
Forsten's Lorikeet (*T. h. forsteni*): island of Sumbawa.
Weber's Lorikeet (*T. h. weberi*): island of Flores.
Swainson's Lorikeet (*T. h. moluccanus*): eastern Australia from Cape York southwards; Tasmania.
Rainbow Lorikeet, Blue Mountain Lorikeet, Red-collared Lorikeet (*T. h. rubritorquis*): northern Australia.
Edwards' Lorikeet (*T. h. capistratus*): island of Timor.
Coconut Lorikeet (*T. h. massena*): Bismark Archipelago, Solomon Islands, New Hebrides.

Status Abundant in many parts of the range; there is evidence of crop-damage in some fruit-growing areas which leads to persecution of local populations.

Avicultural rating Very popular avicultural subjects, despite their somewhat discordant voices. They are hardy and easily managed.

Feeding Diet K, plus fruit; also offer berries, green food, live food, plain canary and millet seed (dry or soaked), and millet sprays.

Breeding Most of the subspecies breed, and some pairs are extremely prolific. Clutch 2; incubation 24–28 days. The nestling period is fairly long (8–10 weeks). It is usual to keep just a single breeding pair in an aviary. Successful colony breeding has also been recorded on a number of occasions. Indiscriminate matings between subspecies should be avoided; the resulting fertile progeny could lead to a loss of the pure subspecies through mixed blood infusion, as well as making the already difficult task of correct identification almost impossible.

General management Provide a roomy aviary with a flight and a shelter. Adults are playful and active and appreciate space. Provide a nest box for roosting. Cleanliness is essential. They are not suitable companions for other species.

Trichoglossus flavoviridis meyeri (II)
Meyer's Lorikeet

$7\frac{1}{2}$ in (19 cm). The top of the head is brown with a suffusion of gold on the forehead. The upper surfaces are green, the underparts yellow-green, the feathers with dark green margins. The cheeks are dull green, the ear coverts yellow. The bill is orange, the legs and feet dark grey. The female is similar to the male but usually has ear coverts which are less yellow.

Range Sulawesi.

Status Uncertain.

Avicultural rating An attractive small lorikeet, but this subspecies is not readily available.

Feeding Diet K with fruit. Some birds will take sweetened boiled rice and milk. Some birds enjoy mealworms.

Breeding Individual pairs may prove to be fairly free breeders. They have only been available to aviculturists within recent decades and have already proved fairly prolific, although there were some nestling losses initially. Clutch 2; incubation 23–24 days. Some pairs will attempt to nest throughout the year.

General management They thrive in small aviaries and require a flight and small shelter with a nest box for roosting and breeding. It is best if they are the sole occupants, especially in small units. Maintain cleanliness; cleaning operations should be carried out from outside if a pair is nesting, especially in limited space accommodation.

Trichoglossus goldiei (II)
Goldie's Lorikeet

---

7 in (18 cm). The forehead and crown are red; the cheeks and ear-coverts are mauve-pink with some darker blue streaking. The upper surfaces are green, the underparts lighter green flecked with

yellow. The bill is black, the legs and feet brown. The sexes are similar.

Range Much of central New Guinea; Weland Mountains eastwards to south-east Papua.

Status Available information suggests that there are only modest populations throughout the range.

Avicultural rating This is likely to be one of the most popular small brush-tongued species, if captive-bred populations can be built up; they are attractive, easily managed and hardy.

Feeding Diet K, with items of soft fruit; pears and grapes are especially favoured. Plain canary, millet and millet sprays may be offered; my own pair are not interested in eating the latter, but enjoy stripping the sprays.

Breeding Although virtually unknown in aviculture until 1977, this species has proved a willing breeder; home-bred stock is increasingly available. Clutch 2; incubation 23–25 days. The male has an amusing display. Pairs will breed in small aviary units.

General management Acclimatised birds are perfectly hardy; they will winter outdoors. They do best in a small shelter and flight unit. The shelter should be well lit if the birds are to be enclosed in severe weather; they seldom roost in nest boxes. Despite their small size it is best if they are the sole occupants of the aviary.

Lorius garrulus (II)
Chattering Lory

12 in (30 cm). Most of the plumage is scarlet; the wings and thighs are green. There is some yellow on the bend of the wing, and also under the wing coverts. The bill is orange-brown, the legs and feet black. The sexes are similar.

Range Islands of Halmahera and Weda, Moluccas.

Status Fairly common in some parts of the range.

Avicultural rating One of the best known of the genus, and quite widely kept, they are easily managed and hardy.

Feeding Diet K, with items of soft fruit; the latter is enjoyed sectioned or cubed in preference to puréed. Offer soaked seed (plain canary or millet), millet sprays and live food.

Breeding They are frequently bred. Clutch 2; incubation 26 days. The nestlings may remain in the box for 10 weeks. The adults are aggressive when young are present. Watch the male parent as the offspring move towards independence when he may attack them.

General management Acclimatised birds are hardy. Provide a flight and a shelter; also a box for roosting. They are invariably tame and amusing; they are also very active and fairly noisy (hence the name). They are not suitable companions for other species. They spend much time climbing, including on the aviary wire so watch for predator problems, and also ensure the cleanliness of the mesh if it is much used.

Lorius garrulus flavopalliatus (II)
Yellow-backed Lory

12 in (30 cm). Very similar to the nominate subspecies, except that this species has an area of yellow on the mantle.

Range Islands of Batjan, Obi and Morotai (*L. g. morotaianus*), Moluccas.

Status There are still good populations. The subspecies *L. g. morotaianus* was once said to be the most numerous parrot species on the island of Morotai.

Avicultural rating See Chattering Lory.

Feeding/Breeding/General management Similar to Chattering Lory. This bird has also proved a free breeder.

Lorius lory (II)
Black-capped Lory

12 in (30 cm). The forehead, crown, nape and lores are black. The mantle, lower breast and underparts are a dark purplish-blue and the wings are dark green washed with bronze. There is a vivid band of yellow across the undersides of the flight feathers. The abdomen is bright blue whilst the under tail coverts are paler. The remainder of the plumage is red except for a blue band on the hind neck. The bill is orange, the legs and feet dark grey. The sexes are similar.

Range New Guinea and adjacent islands.

Status It is widely distributed, and there are believed to be substantial populations in some parts of the range, but loss of habitat will eventually affect its present fairly secure status.

Avicultural rating These are excellent avicultural subjects, but noisy.

Feeding/Breeding/General management Similar to Chattering Lory. True pairs are fairly willing breeders. Hand-reared birds often prove excellent mimics. Such birds are also invariably tame and confiding, but nesting pairs often display great hostility to humans.

Lorius domicellus (II)
Purple-capped Lory, Purple-naped Lory

11 in (28 cm). The forehead and crown are black; there are violet or purple feathers on nape. The wings are green; the bend of the wing is white and blue. The rest of the plumage is scarlet except for some yellow across the upper breast and violet-blue thighs. The bill is orange-red; the legs and feet are dark grey. The sexes are similar.

Range Islands of Ceram and Amboina; also Buru, where it has been introduced.

Status Uncertain but apparently not abundant.

Avicultural rating Popular and has been frequently bred; stock is difficult to obtain in some countries, including the UK.

Feeding Diet K, with items of fruit. Offer plain canary seed or millet, green food and occasional live food.

Breeding This species is a frequent breeder in collections in Europe and the USA. Clutch 2; incubation 24–26 days. The nestling period is very variable (7–14 weeks).

General management Provide a flight and a shelter; also a box for roosting. These birds are hardy, and have a strident voice. They have a reputation of being very intelligent, and often forming close attachments with their owner.

Cacatuidae (Cockatoos)

Probosciger aterrimus [PLATE 56] (II)
Palm Cockatoo, Great Palm Cockatoo

25–31 in (64–79 cm). The entire plumage is grey-black; the naked cheek patches are red, but the shade varies according to the bird's mood. There is a very large crest of narrow, elongated feathers. The bill is huge, and the thighs bare. The bill is black; the legs and feet are dark grey. The female is of smaller and slighter build; the bill is less massive than that of the male.

Range Cape York Peninsula (northern Australia), New Guinea, Aru Islands, Misol Island.

Status Relatively common in some parts of the main range; now exterminated in developed areas.

Avicultural rating Extremely rare and expensive; these birds have a good disposition, despite the cock's fearsome appearance, and are an excellent avicultural subject.

PLATE 56 *Probosciger aterrimus*, Palm Cockatoo

Feeding Diet H, with additional unshelled nuts. They are especially fond of almonds. Use of one of the modern pellet foods (see p. 27) is recommended. Also offer fruit and various raw vegetables; trial and error is often the sole way of establishing the varied diet necessary for these and other large parrots with highly individual dietary tastes. Some enjoy pea-pods; others develop tastes for exotic fruits. Cheese is a valuable food for those birds which will take it. Grit is essential, of course. Both mineralised and oystershell should be supplied in separate dishes.

Breeding This species is bred only infrequently. A large hollow log is probably the best nesting receptacle; provide a good depth of wood chippings in the base. The log should be 4–6 ft (1.2–1.8 m) high, minimum 15–18 in (38–46 cm) diameter and with a good-sized nest-chamber. Clutch 1; incubation 31–35 days. Compatibility is one of the first obstacles to overcome before breeding is possible; if a new pair is being introduced watch the male carefully for aggressive intent. Even when pairs have eggs or chicks there is still potential for problems; desertion is fairly common and the owner must be ready to hand-rear valuable youngsters. It is useful to provide plenty of suitable natural timber; adult birds can destroy this, and with luck leave the nest log relatively intact.

General management They must have strong, concrete-floored aviaries with strong metal framing, and heavy welded mesh. Concrete slabs are best for the construction of shelters. Many prefer to roost in the flight; but it is desirable to provide a shelter for sleeping during severe weather. They are usually very hardy, but provide some measure of protection in open flights by use of rigid plastic windbreaks or screens. They have various call notes, including a penetrating screech.

Eolophus roseicapillus (II)
Roseate Cockatoo, Galah

14 in (36 cm). The forehead to nape (which includes the crest) is white with a tinge of pink; the rest of the upper surfaces are grey. The face, neck, chin, throat and most of the underparts are rose-pink; the lower abdomen is pale grey. The bill is light horn-coloured; the legs and feet are grey. The sexes are similar but the male's iris is dark-brown and red or red-brown in the case of the female.

Range Most of Australia except coastal areas; Tasmania.

Status Abundant, even increasing, in some areas; but it is regarded as a pest by farmers and growers who lose no opportunity to destroy the birds, sometimes in substantial numbers. This is a ludicrous waste when viewed in the context of Australia's present ban on exports of its fauna.

Avicultural rating Once among the most popular and widely kept of parrots; now it is only rarely available. Adaptable, hardy and easy to manage, it is also expensive.

Feeding Diet H, with an equal amount of pellet food (see p. 27), with recommended additives, fruit, green food etc. Grit is essential.

Breeding A classic example of a species once so inexpensive and freely available that few aviculturists would expend time and facilities to breed from their birds. Now, with individual examples realising nearly-four-figure sums, there is regret that aviary-bred strains are not already established from the substantial numbers once available. Clutch 2–4; incubation 25 days. A stout barrel seems to be the favourite nest receptacle. Individual adults may be aggressive when breeding. Birds will line the nest chamber with twigs etc. The young fledge at 7–8 weeks.

General management Similar to Palm Cockatoo. It is completely hardy. It can create noise problems (but is not the worst offender among cockatoos). Provide ample natural wood for chewing.

Cacatua leadbeateri [PLATE 57] (II)
Leadbeater's Cockatoo, Major Mitchell's Cockatoo

PLATE 57 *Cacatua leadbeateri*, Leadbeater's Cockatoo

14 in (36 cm). The crown, including the forward-pointing crest, is white with a suffusion of salmon pink (the extended crest is tipped white and has an orange area in the centre of a pink band). The upper surfaces, tail and abdomen are white; the face, neck, breast and underparts (except the white abdomen) are salmon-pink. The bill is white; the legs and feet are grey. The sexes are similar except that the iris is dark brown in the male, reddish-pink in the female.

Range Western and central areas of Australia; not found in the north or east.

Status Declining in most areas, especially where urban development has taken place. It is still fairly common but in a diminishing number of areas.

Avicultural rating A prized avicultural subject, and now, sadly, very expensive. Hardy and adaptable, it is possibly the most beautiful of all cockatoos.

Feeding Diet H, with equal amounts of pellet food (see p. 27), plus fruit and green food. Grit is essential.

Breeding A fairly infrequent breeder, despite a substantial number of individual pairs in many collections. The best successes seem to be achieved by hand-rearing. Provide a barrel or hollow log for the nest. Clutch 3; incubation 27–30 days. Adults are often very aggressive when breeding. The fledgling period is 6–8 weeks. Watch for parental aggression when the young fledge; the cock is the usual instigator.

General management Similar to Palm Cockatoo. A long aviary shows off the birds in flight to better advantage. They are very able destroyers of timber and wire-netting, so ensure that the accommodation is strongly built and on a concrete base.

Cacatua sulphurea (II)
Lesser Sulphur-crested Cockatoo

13 in (33 cm). The plumage is mainly white; the

ear-coverts and expandable crest are bright yellow. There is some yellow on the undersides of the flight and tail feathers. The bill, legs and feet are dark grey. The area of naked skin around the eye is white. The sexes are similar but the male's iris is dark brown, whereas it is reddish-brown in the female.

Range Celebes and island of Buton.

Note One of the more frequently available subspecies is the Citron-crested Cockatoo (*C. s. citrinocristata*) from the island of Sumba. It is distinguished from the nominate subspecies by the more orange crest and ear-coverts.

Status Variable; abundant in some parts of the range but scarce elsewhere.

Avicultural rating A long-established favourite, despite its discordant voice. Unfortunately most imported birds are sold as pets; they are unsatisfactory for this role unless hand-reared.

Feeding Diet H, with roughly equal quantities of pellet foods (see p. 27), plus fruit, green food etc. Seeding grasses and soaked seed (plain canary seed or millet) are enjoyed by many of these birds. Grit is essential.

Breeding Infrequently bred by comparison with the numbers imported. Various nesting receptacles may be used, such as the hollow log, barrel or grandfather clock type; all must be strong and substantially built, but the birds may still reduce them to matchwood. Clutch 2–3; incubation 28 days. The chicks are in the nest for 10–12 weeks. Watch for parental aggression when fledged; remove the young as soon as they are independent. The adults are often savage to humans when nesting; it is best if feeding etc can be carried out from outside the aviary.

General management Provide a flight and a shelter of metal or welded mesh construction, with a concrete base. They are hardy, and their calls are penetrating. They are not safe with other aviary occupants. They are fond of rain bathing.

Cacatua galerita [PLATE 58] **(II)**
Greater Sulphur-crested Cockatoo, Sulphur-crested Cockatoo

PLATE 58 *Cacatua galerita*, Greater Sulphur-crested Cockatoo

20 in (51 cm). This is a near-identical but larger version of the Lesser Sulphur-crested Cockatoo. The sexes are similar but the male has a dark brown iris and the female's is red-brown.

Range Eastern and south-eastern Australia, Tasmania. Introduced into New Zealand.

Note The two subspecies most often encountered are the following. The Triton Cockatoo (*C. g. triton*) is from New Guinea and adjoining western islands; it has been introduced to Ceramlaut and Goramlaut islands (Indonesia) and Palau Island (Pacific). It is distinguished by its broader and more rounded crest feathers. The area of bare skin around the eye is blue. Eleonora's Cockatoo (*C. g. eleonora*) is from the Aru Islands and differs from *C. g. triton* only in having a smaller bill. The nominate subspecies is now extremely scarce in aviculture since imports from Australia ceased in 1960.

Status Abundant in many parts of the range; they are regarded as pests in crop-growing areas, and farmers wage a constant battle against them. They have been known to dismember timber-built houses in some Australian suburbs.

Avicultural rating Formerly extremely popular, especially as pets. Reports of individual birds exceeding a 100-year life span in captivity are not unusual.

Feeding/Breeding/General management Similar to Lesser Sulphur-crested Cockatoo. Breeding successes with the three subspecies referred to have been infrequent. Clutch 2–3; incubation 28 days. The young leave the nest after 10–12 weeks. They are often very destructive.

Cacatua moluccensis (II)
Salmon-crested Cockatoo, Moluccan Cockatoo

20 in (51 cm). Most of the plumage is pale salmon-pink. There is an impressive backward-sweeping crest with long, broad feathers of a deep salmon-pink shade. The undersides of the flight feathers are suffused with deep salmon-pink; the undersides of the tail feathers are washed with salmon-pink and pale orange. The bill is black; the legs and feet are dark grey. An area of bare skin around the eye is white with a bluish tinge. The sexes are similar but the male's iris is black; the female's iris is dark brown.

Range Islands of Ceram, Saparua and Haruku, southern Moluccas. Introduced to Amboina.

Status Abundant in some parts of the range; this cockatoo is regarded as a pest in cultivated areas.

Avicultural rating Individual hand-reared young birds are often gentle and confiding, therefore they are in great demand as pets, although imported adults are unsuitable. They have extremely loud and discordant voices.

Feeding/Breeding/General management Similar to Lesser Sulphur-crested Cockatoo. Clutch 2; incubation 28–30 days. Accommodation for these birds needs to be immensely strong; they are capable of dismantling all but purpose-built steel and concrete structures.

Cacatua goffini [PLATE 59] (II)
Goffin's Cockatoo

PLATE 59 *Cacatua goffini*, Goffin's Cockatoo

12 in (30 cm). The plumage is mainly white; there is some salmon-pink between the base of the bill and the eye, also in the feathers of the short crest. The undersides of the flight and tail feathers are suffused with yellow. The bill is greyish-white, the legs and feet dark grey. The area of bare skin around the eye is white. The sexes are similar but the iris is dark brown in the male and reddish-brown in the female.

Range Tanimbar Islands, Indonesia; also introduced to Tual in the Kai Islands.

Status Extremely precarious following the near-total destruction of their forest habitat in the early 1970s. This deforestation coincided with the large-scale export of the species which was previously rare in aviculture.

Avicultural rating The low price (by comparison with other cockatoo species) ensured Goffin's Cockatoo an important role in the pet trade. The species is generally unsuited to the close confines of a cage and often does not settle down.

Feeding Diet H, with some fruit and green food. Pellets (see p. 27) should be offered, but many birds refuse them.

Breeding Successes are fairly infrequent at this stage, but efforts are being maintained by many enthusiasts who realise that the species is facing major problems in its wild habitat. Clutch 3; incubation 28–30 days. The young are in the nest for 10–12 weeks. Pairs will use a box or a small hollow log.

General management They are extremely destructive, so ensure that aviaries are built to appropriate standards using heavy metal or wire etc. They are noisy, but not excessively so. They are completely hardy and will often roost out in severe weather.

Nymphicinae (Cockatiel)

This is an aberrant member of the cockatoo family, and, together with the Budgerigar, is in my opinion the species best suited to the role of 'pet' parrot. The original wild form is grey; it is now available in a bewildering range of mutations and patterns. It is an excellent cage or aviary subject. Young birds (preferably hand-reared) make delightful tame pets; most are good mimics and some learn to imitate the human voice.

Nymphicus hollandicus
Cockatiel, Cockatoo Parrot

12 in (30 cm). The plumage is mainly grey, paler on the under surfaces. The head, crest, cheeks, chin and throat are yellow, the ear-coverts orange. There is a conspicuous area of white on the wing coverts; the tail is dark grey with a black underside. The bill, legs and feet are grey. The female is similar but lacks the bright head colours which are present but muted; the underside of her tail is barred grey and yellow. A growing number of colour mutations are available.

Range Most of Australia, except coastal areas.

Status Abundant.

Avicultural rating Very popular; apart from the Budgerigar, this is the most widely kept of the parrot order. It has all the virtues and no vices, is easy to house and feed, a free breeder, mixes with other birds, has an attractive appearance, becomes tame and makes an excellent pet.

Feeding Diet J is recommended for Cockatiels; in addition they can have millet sprays, various items of green food and regular slices of apple. Grit is essential.

Breeding They are very free breeders, either as single pairs or on a colony system. An aviary 9 × 6 × 3 ft (2.7 × 1.8 × 0.9 m) will house a single pair. They can be bred in cages, but this is not recommended. Clutch 4–7; incubation, by both sexes, 19–21 days. The young are usually reared without trouble. Sponge cake or wholemeal bread soaked in milk makes an excellent rearing food, but ensure that it does not sour during very hot weather.

General management They are very easily managed. They thrive in a typical shelter and flight arrangement and are hardy. They mix well with other non-aggressive birds, even small finches. They are not destructive, and have a pleasant, whistling call.

Psittacidae (Parrots)

Psittaculirostris desmarestii [PLATE 60] (II)
Desmarest's Fig Parrot, Golden-headed Fig Parrot

PLATE 60 *Psittaculirostris desmarestii*, Desmarest's Fig Parrot

7½ in (19 cm). The forehead is red merging with the orange-yellow crown and nape; the patch below the eye is bright blue. The cheeks and throat are light green, streaked with yellow-gold and bordered on the throat with blue; a brownish-red band borders the blue. The upper surfaces are dark green, the underparts a lighter green. The inner wing coverts have an orange edge; there is a yellow band across the underside of the flights. The bill is black, the legs and feet dark grey. The sexes are alike.

Range Western and southern New Guinea; also the islands off western New Guinea.

Note Five subspecies are described in addition to the nominate. All exhibit plumage differences and occupy territory in the range noted above.

Status Relatively uncommon. There is no evidence of abundant populations but they are more numerous in some areas than in others.

Avicultural rating An avicultural rarity, and best in the hands of experienced people; they are occasionally imported and are usually expensive.

Feeding Essentially seed and fruit; use Diet A, with a selection (daily) from figs (fresh, or thoroughly soaked dried ones), apples, grapes and various berries. Bananas are recommended by some and are probably widely used, but I have been anti-bananas (in relation to diets for birds) for many years. Nectar, of the kind supplied to lories and lorikeets or smaller species such as sunbirds and honeycreepers, is sometimes fed to these birds; it may be better eliminated from the diet, except as a pick-me-up for a stressed or sick individual (as some birds consume it to the exclusion of other items). Grit should be offered.

Breeding This species has bred only on rare occasions. Little is recorded about its nesting activity, except that a cavity in a tree is used in the wild. Probably a nest box or small log would be acceptable in the aviary. Clutch believed to be 2–3; incubation probably 23–25 days.

General management Despite their small size, they are surprisingly hardy when acclimatised. They fare best in a shelter and flight unit; the shelter should be well lit and capable of semi-permanent occupation during really severe weather. They roost in a nest box, but beware of the possibility of winter egg-binding as a consequence.

Opopsitta diophthalma [PLATE 61] (II)
Double-eyed Fig Parrot, Dwarf Fig Parrot

6 in (15 cm). The main colour is green, darker on the mantle, back and wings, grass-green on the under surfaces. The forehead and crown are scarlet with an orange-yellow patch on the back of the head; the stripe from the base of the bill to above the eye is blue. The cheeks, chin and throat are red; a patch of violet-cobalt borders the bottom of the cheeks. The flanks and sides of the breast are yellow. The bill is grey at the base, becoming nearly black at the tip; the legs and feet are

PLATE 61 *Opopsitta diophthalma*, Double-eyed Fig Parrot

have been recorded. Use a log or nest-box. Clutch 2; incubation 19–20 days. Once established, this is a good avicultural subject. It has lived in collections for several years and may prove as hardy as Desmarest's Fig Parrot, although it requires courage on the part of an owner to expose rare and valuable exotics such as these to the rigours of a hard winter.

Eclectus roratus [PLATE 62] (II)
Eclectus Parrot

PLATE 62 *Eclectus roratus*, Eclectus Parrot (female)

greenish-grey. The female is similar, but has salmon-buff cheeks.

Range New Guinea, including islands off the western coast; Aru Islands; north-eastern Australia.

Note Eight subspecies in all throughout the range; all show plumage differences.

Status Relatively common in some parts of New Guinea; believed rare in other parts of the range, but it may be difficult to observe in some areas.

Avicultural rating See Desmarest's Fig Parrot. This species is rarely available.

Feeding/Breeding/General management Similar to Desmarest's Fig Parrot. Few successful breedings

13–14 in (33–36 cm). Despite the most marked sexual dimorphism of any parrot species, *Eclectus* present considerable identification problems. A total of 10 subspecies are described; the males of many of these are difficult to distinguish and there is considerable confusion in separating several of them. The nominate subspecies, *E. r. roratus*, is known as the Grand Eclectus Parrot. Another name familiar to aviculturists is the Red-sided Eclectus (*E. r. polychloros*), which is slightly bigger than the Grand Eclectus and with sufficient

[160]

plumage differences to make positive identification feasible.

The Grand Eclectus (*E. r. roratus*) male is mainly green with a slight yellow tinge on the head; there is blue on the bend of the wing, the outer webs of the primaries are mauve-blue, and the underwing coverts red. The areas on the sides of the body are red. The under tail coverts are yellow-green and the central tail feathers are tipped with buff-white. The lateral feathers are green, with a blue suffusion towards the tips and edged with buff-white. The underside of the tail is grey-black tipped with buff-white. The upper mandible is yellow-orange with a yellow tip, the lower mandible black. The upper mandible is notched. The legs and feet are dark grey.

The Grand Eclectus (*E. r. roratus*) female has mainly red plumage, darker on the back and wings. The lower breast, abdomen and a band across the upper mantle are purple. There is mauve on the bend of the wing and the outer webs of the primaries. The under wing coverts are dull purple, the under tail coverts red, and there are yellowish tips to the longer feathers. The upper surfaces of the tail are red tipped with orange-yellow; below they are dusky orange tipped with yellow-orange. The bill is black; the legs and feet dark grey.

The Red-sided Eclectus (*E. r. polychloros*) male is similar to *E. r. roratus* but larger. The green areas of plumage have a slight yellow tinge. The central tail feathers are green with pale yellow tips; the lateral feathers are blue tipped with pale yellow, suffused on the outer webs with green. There is a little green on the outermost tail feathers.

The Red-sided Eclectus (*E. r. polychloros*) female is similar to *E. r. roratus*, but the red is brighter on the head and upper breast, and more dull and brownish on the back and wings. The lower breast and the band across the upper mantle are blue; the under tail coverts are red, the tail red tipped with orange. There is a narrow blue ring around the eye.

Range Distribution of all 10 subspecies is spread over the following locations: New Guinea and offshore islands; Moluccas, Tenimber, Kei, Admiralty, Aru and Solomon groups; islands of Sumba and Biak; north-eastern Australia. In some cases individual subspecies are confined to a single island; others are more wide-ranging.

Status They are reported as common in some parts of the range, but hunting is clearly reducing numbers in many areas. Some observers report a preponderance of males to females.

Avicultural rating Few are imported nowadays. They are often difficult to establish during the immediate post-import period. This is a bird for the connoisseur, but it is easily managed and hardy after acclimatisation.

Feeding Prospective owners should fix in mind the need for a consistent input of Vitamin A. Fresh vegetables are vital: carrot, corn cobs, tomatoes, pea-pods etc. Ensure that the supply is maintained and add wild green food such as chickweed, dandelion and groundsel. Diet H can provide the main seed intake, but with the addition of plain canary seed or millet (both dry and soaked). Various berries are appreciated. Grapes and pomegranates are relished, mainly for the pips or seeds.

Breeding Eclectus are now being more widely bred, but there is some way to go before captive stocks are sufficient to ensure avicultural security. In some zoos Eclectus have been kept and bred successfully on a colony system; but aviaries for such ventures must be large and are probably beyond the scope of most private aviculturists. One of the biggest advantages of this system is the compatibility of pairs following self-selection. Individual pairs may breed after a period of careful introduction. They need a 4 ft (1.2 m) box or a log. Clutch 2; incubation 28–30 days. The chicks leave the nest after 10–11 weeks.

General management It is best if they are housed in spacious aviaries; a 15 ft (4.5 m) length is recommended. A flight and shelter arrangement is recommended but construction must be strong. The birds are hardy after acclimatisation.

Polytelis swainsonii [PLATE 63] (**II**)
Barraband Parrakeet, Superb Parrot, Green Leek Parrot

PLATE 63 *Polytelis swainsonii*, Barraband Parrakeet

16 in (41 cm). This species is mainly green in colour, lighter on the underparts; the forehead, cheeks and throat are yellow. A band across the lower throat to the sides of the neck is scarlet. The hind crown is tinged blue; the outer webs of the primaries are blue. The tail is green with glossy black below. The bill is red, the legs and feet grey. The female is duller and lacks the areas of yellow on the forehead, cheeks and throat.

Range A limited range in northern Victoria and the interior of New South Wales.

Status Relatively abundant, even in areas utilised for agriculture.

Avicultural rating Elegant and colourful; now widely seen in aviculture. They are hardy, and good breeders.

Feeding Diet J, with adequate grit. Also offer a selection of greenstuff, fruit and seeding grasses.

Breeding Usually reliable breeders, many pairs prefer a natural log to a nest box. Clutch 4–6; incubation 19 days. They can be bred on a colony system, but spacious accommodation is necessary; individual cocks are sometimes aggressive to newly-fledged young males.

General management Provide spacious accommodation; a flight length of 15–20 ft (4.5–6 m) is recommended, plus a shelter. They are hardy and weather-resistant.

Polytelis anthopeplus (**II**)
Rock Pebblar, Rock Pepplar, Regent Parrot

16 in (41 cm). The forehead, crown and nape are yellow-green; the mantle and back are dark olive-green. The under surfaces, shoulders and rump are bright yellow; a band across the inner wing coverts is red. The tail is blue-black, the bill coral, the legs and feet grey. The female is generally duller; the lateral tail feathers have rose-pink margins and tips.

Range This species occupies two widely separated territories: south-western Australia; and north-west Victoria to eastern South Australia.

Status Abundant in the south-western area of the range; much less so in the south-east.

Avicultural rating It has been bred in Europe since 1880, but is still not widely available and is therefore expensive.

Feeding Diet J, with adequate grit. Also offer the usual range of green food, fruit etc.

Breeding Generally not as free-breeding as related species. Provide a box or hollow log. Clutch 4–6; incubation 19 days.

General management Similar to Barraband Parrakeet.

Polytelis alexandrae (II)
Princess of Wales Parrakeet, Princess Parrakeet, Princess Alexandra's Parrakeet

18 in (46 cm). The crown and nape are light pastel blue; the rest of the upper surfaces are pale olive, except for the violet-blue rump, and shades of light green and dull sky blue in the secondaries and primaries. The chin, throat, sides of the neck and thighs are pale pink; the rest of the underparts are grey blue with a suffusion of yellow-green. The long tail is olive-green with some blue and pink. The bill is red; the legs and feet dark grey. The female's crown is more grey, her rump and wing coverts duller and her tail shorter.

Range Interior of central and western Australia.

Status Rare in the wild, but it occupies somewhat inhospitable territory where disturbance is fairly minimal.

Avicultural rating Popular and fairly widely kept, they are hardy, good breeders and easily managed.

Feeding Diet J, with adequate grit. Provide green food such as spinach, cress and chickweed. Seeding grass heads are appreciated in season, as is some fruit.

Breeding Over the years some individual pairs have proved prolific whereas others are much less consistent. Colony breeding is possible, but all birds must be introduced simultaneously. Clutch 4–6; incubation 19 days. They will use a nest-box or an upright hollow log. Lutino and blue mutations have been produced.

General management Provide shelter and flight accommodation. The flight should be 15–20 ft (4.5–6 m) in length. They are normally confident and friendly birds. They have a very penetrating call, uttered frequently. They are weather-resistant.

Barnardius barnardi (II)
Barnard's Parrakeet, Mallee Ringneck Parrot

13 in (33 cm). The forehead is red, the crown and nape green. There is an area of grey-brown from the back of the eyes across the neck, bordered yellow. The mantle and back are dark blue-green; the rump is green. The under surfaces are turquoise, the cheeks, shoulders and wing margins light blue. There is a yellow band across the lower breast. The tail is green and blue. The bill is grey-white, the legs and feet grey. The female is similar but duller than the male; the mantle and back are dark grey-green.

Range Interior of south-eastern Australia.

Status Declining in many areas.

Avicultural rating Widespread, but not common, this is a very desirable avicultural subject.

Feeding Diet J, with adequate grit. Also offer a wide variety of green food, fruit etc. These birds are fond of seeding grasses, and individual birds consume some live food.

Breeding Provide a hollow log or a tall (4-ft/1.2-m) box. Clutch 4–6; incubation 19 days. Provide some soaked seeds for rearing.

General management Provide a shelter and flight. The recommended minimum length of the flight is 15 ft (4.5 m). The species is very destructive to woodwork. Keep single pairs to each unit as they can be belligerent, and some are aggressive to humans. They are hardy.

Platycercus elegans (II)
Pennant's Parrakeet, Crimson Rosella

14 in (36 cm). Most of the plumage is crimson. The cheek patches are violet-blue; the nape, back and wings are black with feathers bordered with crimson to give a laced effect. The bend of the wing, the under coverts and the outer webs of the flights are blue. The central tail feathers are blue with a slight green suffusion; the lateral feathers are blue, tipped white. The bill is grey-white, the legs and feet grey. The sexes are alike.

Range Eastern and south-eastern Australia; introduced to New Zealand and Norfolk Island.

Status Fairly plentiful in some areas of the range (Australia); the Norfolk Island population is reported to be flourishing, but the species is much less common in New Zealand.

Avicultural rating An excellent avicultural subject. It is easy to manage, hardy and a willing breeder.

Feeding Diet J, with adequate grit. Also offer a selection of green food, fruit, berries etc.

Breeding Normally free breeders. Provide a tall (4-ft/1.2-m) box or hollow log. Clutch 4–8; incubation 19–20 days. Although eggs are produced readily by many pairs, some hens have a tendency to abandon or destroy their clutches.

General management They fare best in lengthy shelter and flight accommodation, similar to the recommendation for Barnard's Parrakeet. They are hardy, but aggressive and not suitable for communal housing. They have a pleasant melodious call.

Platycercus icterotis (II)
Stanley Parrakeet, Western Rosella

10 in (25 cm). The entire head and under surfaces are red except for yellow cheek patches. The mantle, back and wings are black, the feathers edged with dark green; there are a few red feathers in individual birds. The shoulders are black; the outer webs of the flights and the under wing coverts are blue. The tail is green and blue. The bill, legs and feet are grey. The female's head, face and upper breast are dull green, faintly marked with red and yellow; her forehead is red and cheek patch dull yellow.

Range South-western Australia.

Status Reasonable numbers, but not plentiful.

Avicultural rating A popular avicultural subject; their small size and ease of sexing are advantages over related species. They are hardy and free breeders.

Feeding Diet J, with adequate grit; also offer green food and slices of apple.

Breeding Provide a hollow log or a box with a good depth of decayed wood chips. Clutch 4–8; incubation 19–20 days. They are usually single-brooded.

General management Their smaller size enables the utilisation of shorter flights; a 10 ft (3 m) minimum length is recommended. Individual males are occasionally over-aggressive with their females; watch their behaviour carefully when the birds approach breeding condition.

Platycercus eximius ceciliae [PLATE 64] (II)
Golden-mantled Rosella

12 in (30 cm). The head, neck and upper breast are red; the cheek patches are white. The mantle and back are black, the feathers margined with yellow; the under wing coverts are blue. The lower breast is yellow; the rump and abdomen pale green. The tail feathers are green and blue, the vent and under tail coverts red. The bill is grey-white, the legs and feet grey. The females are similar, but the areas of red on the head and breast are much duller.

PLATE 64 *Platycercus eximius ceciliae*, Golden-mantled Rosella

Range South-eastern Australia; Tasmania. Introduced into New Zealand.

Status There are good numbers in the Australian part of the range, and the New Zealand population is increasing. They are uncommon in Tasmania.

Avicultural rating Probably the most widely kept of the larger Australian parrakeets, this subspecies is hardy, easily managed and a good breeder.

Feeding Diet J with adequate grit. Also offer a selection of green food items; apple is much enjoyed and various berries may also be taken.

Breeding They are usually free breeders; there are sometimes two clutches in a season. Clutch 4–7; incubation 19–21 days. As with other members of the genus, introduction of the pairs is often difficult; watch the cock carefully and remove the hen if the male's behaviour is very savage. Some breeding pairs appreciate sweetened wholemeal bread and milk when rearing. Supply a box or log similar to that suggested for Barnard's Parrakeet.

General management Provide a lengthy shelter and flight similar to the recommendation for Barnard's Parrakeet. These birds are often aggressive to their neighbours; so double-wiring is essential in adjoining aviaries. They have a penetrating (but not discordant) call.

Psephotus haematonotus (II)
Red-rumped Parrakeet

10 in (25 cm). The plumage is mainly green, brighter on the head and breast; the mantle has a suffusion of blue. The lower under surfaces are yellow; the rump is red. There is some yellow on the shoulder of the wing; the tail is blue and green. The bill is black, the legs and feet grey. The female is mainly grey-green in colour with lighter, yellow-green underparts. A yellow mutation is widely available.

Range Mainly the interior of south-eastern Australia.

Status Abundant and extending its range.

Avicultural rating A long-standing avicultural favourite, this species is easily managed, hardy and a free breeder, as well as being inexpensive.

Feeding Diet J, with adequate grit. Also offer a selection of green food, seeding grasses, apple etc.

Breeding They are willing breeders; provide a hollow log or a box. Clutch 4–6; incubation 19–20 days. Cocks often resent their newly-fledged male offspring; the former may need to be removed until the chicks are self-supporting. Two clutches are produced by many pairs.

General management Provide a lengthy shelter and flight; 10 feet (3 m) is the minimum length recommended. They are generally belligerent and not safe with other birds. They have a pleasant call. This is an excellent beginner's parrakeet.

Psephotus varius (II)
Many-coloured Parrakeet, Mulga Parrot

---

11 in (28 cm). This species is like a multi-coloured Red-rumped Parrakeet. Much of the plumage is green, darker on the back, paler on the lower breast. The forehead is yellow, the crown red. The lower abdomen and thighs are yellow, usually with orange-red markings; the upper tail coverts and rump are light blue-green banded with green, yellow-green and with some red on the upper coverts. The shoulder of the wing is yellow; the under wing coverts and the outer webs of the primaries are blue. The bill is grey, the legs and feet grey-brown. The female is mainly olive-brown with duller markings on her forehead, crown, wings etc.

Range Interior of central and southern Australia.

Status Abundant in many parts of the range.

Avicultural rating This parrakeet has a generally poor reputation in UK aviculture, but seems to do better in Continental Europe and the USA.

Feeding Diet J, with adequate grit. Also offer a selection of wild and cultivated green food, slices of apple, carrot etc. These birds are fond of millet sprays and seeding grasses.

Breeding This is an uncertain breeder. It usually takes a long time to settle in new quarters and does not react well to change. Provide a hollow log or a box, and offer a choice of sites. Clutch 4–7; incubation 19–20 days. Provide a good variety of suitable rearing foods: bread and milk, soaked seeds, spray millet, extra green food etc. They are usually among the earliest nesters.

General management Newly acquired birds are often difficult to establish and subject to various diseases which may be stress-related. Provide shelter and flight accommodation similar to that specified for the Red-rumped Parrakeet, but longer if possible. A concrete base and scrupulous cleanliness are essential because of their susceptibility to parasitic worms. They are generally hardy, and have a pleasant soft call.

Cyanoramphus novaezelandiae (I)
Red-fronted Kakariki, Red-fronted Parrakeet

---

11 in (28 cm). The plumage is mainly green in colour, lighter on the underparts. The forehead, crown, a line through the eye and small cheek patch are red; there are patches of red on the sides of the rump. The outer webs of the flight feathers are violet-blue. The bill is blue-grey, the legs and feet grey-brown. The sexes are similar.

Range New Zealand and adjoining islands.

Status Declining rapidly in New Zealand, but populations on small offshore islands appear to be secure.

Avicultural rating Occasional avicultural subjects for more than a century, they were then widely kept from the 1970s onwards. They are easily managed and free breeders.

Feeding Diet J, with grit and a plentiful supply of green food; individual birds appreciate fruit and berries. Live food is enjoyed by most.

Breeding Usually willing breeders. Provide a box 10 in (25 cm) square and 12 in (30 cm) high; remove this in autumn to prevent year-round breeding. Clutch 3–10; incubation 19–20 days. Extra rearing food is important with large clutches; bread and milk, soaked seed, millet sprays, seeding grasses and proprietary budgerigar rearing foods are recommended. Watch for aggression from the cock towards young male offspring when the latter are independent.

General management Provide a shelter and a flight

at least 12 ft (3.6 m) long. Concrete floors aid hygiene, but provide fresh turves from time to time for the birds to pick through. This species spends much time at ground level. It is hardy, and has a fairly shrill call. It is not long-lived, but is prolific.

Cyanoramphus auriceps (**II**, ssp. *forbesi* **I**)
Yellow-fronted Kakariki, Yellow-fronted Parrakeet

10 in (25 cm). The plumage is mainly green, slightly more yellowish on the under surfaces. The forehead and a streak to the front of the eye are red; the crown is yellow. A patch on each side of the rump is red; the outer webs of the flight feathers are violet-blue. The bill is blue-grey, the legs and feet grey-brown. The sexes are similar.

Range New Zealand and adjoining islands.

Status Recovering in some areas after a decline. There are good populations on some offshore islands.

Avicultural rating See Red-fronted Kakariki.

Feeding/Breeding/General management Similar to Red-fronted Kakariki. The two species will hybridise freely in aviaries, but this should be avoided to ensure the purity of each. The call is slightly less shrill than that of the foregoing species. *C. auriceps* spends less time at ground level in aviaries.

Neophema bourkii (**II**)
Bourke's Parrakeet

8 in (20 cm). The upper surfaces are earth-brown, the wing coverts with lighter margins; the underparts are lighter brown suffused with pink, with a deeper pink on the lower breast and abdomen. The forehead, under tail coverts and sides of the rump are blue; the wing margins are violet-blue. The bill is greyish horn-coloured, the legs and feet brown. The female is similar, but usually shows less blue on the forehead, is whiter about the face, and the pink shades are more muted.

Range Interior areas of south-western and central Australia.

Status After suffering a severe decline (the species was added to a listing of endangered species in 1958), it is now making a good recovery in many parts of its range.

Avicultural rating A long-established avicultural favourite. It is easily managed, hardy and a good breeder.

Feeding Diet J, with grit. Offer a selection of green food daily; some birds enjoy a slice of apple.

Breeding They are usually willing breeders. Nest boxes should be 12 in (30 cm) deep and about 9 in (23 cm) square. Clutch 4–6; incubation 18–19 days. There are usually two clutches. The young are very nervous immediately after fledging, so avoid disturbance as much as possible; transfer them to separate accommodation when independent.

General management Provide a shelter and flight. The flight should be a minimum of 8 ft (2.4 m) long and the shelter capable of accommodating birds in periods of severe weather. Although breeding pairs need individual accommodation, this species can be mixed with other small aviary birds. It is not destructive to plants. The birds are usually very active at dusk. They have a pleasant quiet song with no discordant call-notes.

Neophema chrysostoma (**II**)
Blue-winged Grass Parrakeet

8 in (20 cm). The upper surfaces are olive-green; the throat and breast are pale green with the lower underparts yellow. The forehead is deep blue edged with pale blue; the wing coverts are dark blue. The tail is blue-grey. The bill is blue-grey, the legs and feet grey-brown. The sexes are similar but the female is usually duller.

Range South-eastern Australia; Tasmania.

Status There are good populations in many parts of the range; they are especially abundant in Tasmania.

Avicultural rating They are not as widely kept as other members of the genus but are hardy and easily managed.

Feeding/Breeding/General management Similar to Bourke's Parrakeet, and totally inoffensive. The young are usually less nervous than related grass parrakeets. They can be housed in a planted aviary.

Neophema elegans (II)
Elegant Grass Parrakeet

9 in (23 cm). The upper surfaces are olive-green; the underparts lighter and suffused with yellow which is more pronounced on the lower breast, belly and abdomen. The forehead is dark blue edged with light blue; the cheeks (from the base of the bill to the back of the eye) are yellow. The wing coverts are blue, the under wing coverts dark blue and the tail blue and olive. The bill is black, the legs grey-brown. The female is similar but duller; some cocks have a small orange patch on the abdomen, which the hen lacks, but this is not a certain guide.

Range South-western and south-eastern Australia; Kangaroo Island.

Status Abundant in many parts of the range and increasing in numbers in areas where land has been cleared for agricultural use.

Avicultural rating Popular and widely kept.

Feeding/Breeding/General management Similar to Bourke's Parrakeet. They are usually of a quiet disposition and not as active as others of the genus.

Neophema pulchella [PLATE 65] (II)
Turquoise Grass Parrakeet

PLATE 65 *Neophema pulchella*, Turquoise Grass Parrakeet

8 in (20 cm). The top of the head and upper parts are bright green; the forehead, face and cheeks are turquoise-blue. The under surfaces are yellow. A band on the inner wing coverts is red; the upper wing coverts are light blue with the outer webs of the flights dark blue. The tail is green, yellow on the underside. The bill is horn-coloured, the legs and feet grey. The female is duller and lacks the red wing band.

A variety now well established in aviculture has a large orange patch on the abdomen; both males and females carry this colour. It is generally believed to be the result of careful selection in captive breeding.

Range South-eastern Australia to northern Victoria.

Status Now uncommon in most parts of the range.

Avicultural rating Popular and widely kept, they are hardy, easily managed and good breeders.

Feeding/Breeding/General management Similar to Bourke's Parrakeet. Juveniles are nervous when newly fledged and prone to injury through panic-flight. Watch the adult cock for signs of aggression while the young male offspring are in the same aviary; remove the latter as soon as they are independent. They are usually double-brooded; do not allow more than two clutches per season. This is one of the more aggressive members of the genus so do not house pairs in adjoining flights.

Neophema splendida (II)
Splendid Grass Parrakeet, Scarlet-chested Grass Parrakeet

---

8 in (20 cm). Most of the upper surfaces are grass-green. The forehead, crown, cheeks, chin and throat are bright blue, deep cobalt on the face and throat; the breast is bright red and the remainder of the underparts yellow. The wing coverts are pale blue; the under wing coverts and outer webs of the flights are dark blue. The tail is blue and green. The bill is grey-black, the legs and feet grey-brown. The female lacks the red breast and is generally duller in colour.

Range Interior of southern Australia.

Status Local distribution, and generally uncommon throughout the range.

Avicultural rating The most colourful of the grass parrakeets, it is popular and widely kept.

Feeding/Breeding/General management Similar to Bourke's Parrakeet. Generally very confiding, they are hardy birds but do not tolerate damp conditions.

Melopsittacus undulatus [PLATE 66]
Budgerigar, Warbling Grass Parrakeet, Shell Parrakeet

PLATE 66 *Melopsittacus undulatus*, Budgerigar

7 in (18 cm). The forehead, forecrown, face, chin and throat are yellow; the nape, mantle, upper back and wings are yellow-brown with darker barring. The underparts, rump and tail coverts are grass-green; the tail is blue-green. There is a small patch of violet-blue on the lower cheeks and a necklace of black spots across the throat. The bill is yellow, the legs and feet blue-grey. The female is similar, but is distinguished from the male during the breeding season by her brown cere; at other times both sexes have a blue or whitish-blue cere.

This species is now available in a very substantial range of colours and colour combinations. The main basic colours include various shades of green and blue (plus mauve and violet), albino, lutino etc. There are many plumage patterns within that range: clearwing, opaline, whitewing and rainbow, to mention but a few.

Range Australia.

Status Variable; they are abundant in a few places but are generally fairly scarce.

Avicultural rating Unquestionably the most widely-kept member of the parrot family; huge numbers are kept as pets as well as studs maintained by serious breeders.

Feeding Diet A, with millet sprays, seeding grasses and green food. Grit is essential.

Breeding They are free breeders in a cage or aviary; they can be bred on a colony system in a spacious aviary, but bickering (occasionally more serious fighting) between pairs usually reduces the breeding potential. Most breeders employ indoor cages or units. Provide nest boxes with suitable entrance-holes. Clutch 3–8; incubation 17–18 days. Little additional food is needed for rearing. Many pairs would breed throughout the year if permitted; discourage this by removing the nest boxes or splitting breeding pairs.

General management Hardy and easily-managed little birds, they are good occupants of an outdoor shelter and flight but need dry, draughtproof roosting quarters. They mix reasonably well with more robust exotic species such as Java Sparrows, weavers and whydahs, but do not house them with small waxbill-type birds. Many of these birds can be trained to mimic the human voice; it is essential that a young bird (6 weeks old) is selected for training. Patience is needed and the first essential is to win the bird's confidence; hens can learn to talk just as effectively as cocks. Free-flying budgerigar flocks have been established by many aviculturists; but this is not likely to be successful in urban areas as they need a quiet situation with mature trees near to the aviary.

Psittacus erithacus (**II**, ssp. *princeps* **I**)
Grey Parrot

13 in (33 cm). The plumage is mainly grey, slightly darker on mantle and wings; the feathers of the face, head and neck are edged with pale grey. The rump is grey-white, the tail red. The primaries are grey-black; the area of bare skin around the eye is white. The bill is black, the legs and feet grey. The sexes are alike.

Range West Africa eastwards through Central African Republic to western Kenya, northern Angola.

Status Abundant in many parts of the range, although the numbers are much reduced around developed areas where their habitat has been lost.

Avicultural rating Widely kept as a pet for generations, it is now coming into its own as an avicultural subject. It is hardy, and easy to manage.

Feeding Diet H, plus adequate grit. Also offer plain canary seed, millet (including millet sprays), green food, fruit, berries etc. Individual birds enjoy peas (in the pod), corn-on-the-cob and other vegetables.

Breeding This is by no means the easiest species to breed; compatible pairs may take years to settle down. It is best if several individuals are housed in aviaries to pair on a self-selection basis, but this can be expensive at present-day valuations of large parrotlike species. These birds do best in a secluded aviary; arrange feeding, watering etc via hatches so that there is no need to enter. Provide a hollow log or a box (12 in or 30 cm square and 24 in or 60 cm high) with a good depth of wood chippings inside. Clutch 3–4; incubation 28–30 days. The chicks leave the nest after about 12 weeks. Provide extra rearing foods such as bread and milk, sponge cake in honey water, soft fruits and boiled sweetened rice. The adults are often savage when nesting.

General management Provide a strong metal-framed and welded mesh aviary. Shelters are best prefabricated of concrete blocks or similar. Hand-reared birds make amusing pets as they are highly intelligent and good mimics. If caged, ensure that they have the maximum possible space and frequent 'outings'; close permanent confinement leads to stress-related problems including feather-plucking. The majority of birds offered for sale are imports and need careful acclimatisation, especially if they are acquired in the autumn or winter. Dealers sometimes offer birds described as 'growlers', which are usually wild adults or birds with a fear of human beings; these rarely become tame and do not make good pets.

Psittacus erithacus timneh (II)
Grey Parrot, Timneh Grey Parrot

12 in (30 cm). The plumage is very dark grey and the tail dark maroon-red. The upper mandible is reddish with a black tip, the lower mandible black and the legs and feet grey.

Range Southern Guinea, Sierra Leone, Liberia, western areas of Ivory Coast.

Status There are still good populations in many parts of the range.

Avicultural rating This distinctive subspecies is frequently offered by dealers as 'Timor', 'Timnor' or 'Timnar' Grey Parrots. They are usually offered at lower prices than the nominate subspecies; they have a poor reputation as pets and talkers. The majority of imports appear to be mature birds; young, hand-reared examples can make excellent mimics.

Poicephalus senegalus [PLATE 67] (II)
Senegal Parrot

9 in (23 cm). The head, face and nape are grey, paler on the cheeks. The upper surfaces and a band across the breast are green; the lower underparts are yellow, and the thighs bright green. The primaries and tail feathers are dark brown,

PLATE 67 *Poicephalus senegalus*, Senegal Parrot

suffused with green. The bill is grey, the legs and feet grey-brown. The sexes are alike.

Range Senegal, Gambia, Guinea, southern Mali.

Note The Orange-bellied Senegal Parrot (*P. s. mesotypus*) of eastern and north-eastern Nigeria, south-west Chad and northern Cameroon differs from the nominate subspecies in having an orange abdomen; the green of the upper surfaces and breast is also paler. The Scarlet-bellied Senegal Parrot (*P. s. versteri*) of the Ivory Coast westwards to Ghana and Nigeria differs from the nominate subspecies in having a red abdomen.

Status Variable. There are good numbers in some parts of the range but they are much reduced elsewhere, especially in areas of development or cultivation.

Avicultural rating Newly imported birds need sympathetic and careful handling but they eventually become hardy. The low price and typical small parrot appearance ensure their popularity in the pet trade, but most imports are adult and totally unsuited to caging which will impose great stress or even cause death. Young, hand-reared birds can make good pets.

Feeding Diet J, with grit. Millet sprays and seeding grasses are also appreciated. Patience is needed to persuade the birds to take a varied diet, but it is important that apple and other fruit, green food and sprouted seeds are included.

Breeding Considering the vast numbers which are imported annually, these birds are relatively infrequent breeders. Provide a stout nest box or a hollow log; the box should be approximately 12 in (30 cm) square and 24–30 in (60–75 cm) high. Clutch 3–4; incubation 28 days. Provide extra rearing food such as sprouted plain canary seed or millet, bread and milk etc.

General management Recently imported adult Senegal Parrots need careful handling as stress-related problems surface quickly if not checked by a secluded and quiet environment, coupled with gentle and patient care. Provide a shelter and flight; a minimum length of 10 ft (3 m) is recommended. Strong construction is essential.

Poicephalus meyeri (II)
Meyer's Parrot

---

9 in (23 cm). The upper surfaces are grey-brown; the area on the crown is yellow. The rump and underparts are blue-green; the bend of the wing, thighs and under wing coverts are yellow. The bill, legs and feet are dark grey. The sexes are similar.

Range Southern Chad and north-eastern Cameroon eastwards through Central Africa to southern Sudan; western areas of Ethiopia.

Status Abundant in many parts of the range where suitable habitat exists.

Avicultural rating Imported more regularly since the 1960s, this is a popular and easily managed species.

Feeding/Breeding/General management Similar to Senegal Parrot. This species has proved slightly more free-breeding than the previous species; the fledglings are said to be extremely nervous after leaving the nest.

Agapornis cana (II)
Madagascar Lovebird, Grey-headed Lovebird

---

6 in (15 cm). The whole of the head, neck and breast are light grey; the rest of the upper surfaces are dark green. The underparts are yellow-green; the tail is green with some yellow in the lateral feathers, and the subterminal band is black. The bill, legs and feet are pale grey. The female is entirely green.

Range Mainly coastal areas of Madagascar. Introduced to Zanzibar, Mauritius, Rodrigues, Comoro and Seychelles.

Status Abundant in many areas of the natural range. The introductions have been less successful, except on Comoro where a thriving population exists in some parts.

Avicultural rating They are less widely kept than others in the group. Imports are intermittent and there have been fairly limited breeding successes. These birds are susceptible to respiratory problems and not as hardy as other lovebird species.

Feeding Diet J, plus grit. Offer a variety of green food (groundsel, chickweed, cress, spinach) and regular slices of apple or carrot.

Breeding Provide a small nest box with rotted wood chips in the base. Clutch 4–7; incubation 23 days. Some birds strip willow or apple twigs to add to the nest, but others ignore other materials. Provide sprouted seeds (sunflower, plain canary seed or millet) plus millet sprays, as well as bread and milk.

General management They fare best in outdoor aviaries with comfortable sleeping quarters, except in winter when indoor flights are safer. Seclusion is important for this species. They do not enjoy prolonged damp or foggy conditions. They are initially nervous after accommodation transfers, so handle them carefully.

Agapornis pullaria (II)
Red-faced Lovebird

5½ in (14 cm). The plumage is mainly green in colour, the upper surfaces slightly darker than the yellow-green underparts. The forehead and face are tomato-red, the rump blue. The tail is green with yellow and red in the laterals and a subterminal black band. The bill is red, the legs and feet grey. The female is similar but the forehead and face are more orange.

Range Guinea westwards to Congo, northern Zaire, southern Sudan, Ethiopia, north-west Tanzania.

Status Variable, but believed to be abundant in certain areas of the range.

Avicultural rating Now rarely available. They are difficult to establish and not completely hardy even after acclimatisation. They are infrequently bred.

Feeding Diet J, with appropriate grit. Also provide millet sprays, sprouted seeds, green food, apple etc.

Breeding They have been very rarely bred in the UK and the USA. Breeders in Continental Europe and also in South Africa have enjoyed rather more—but still infrequent—successes. Wild birds nest in termite mounds, and peat bales, peat-filled boxes or barrels, and cork blocks have all been used in captive breeding. The birds excavate a tunnel or chamber in the selected site. Clutch usually 4–6; incubation 22–24 days. A secluded outdoor aviary may suit in the summer, but there are records of successful breedings later in the year when indoor flights were utilised. Provide willow or apple twigs for stripping.

General management This is probably the most difficult of lovebirds to establish after importation. They are nervous and highly strung, so house them in a quiet situation. Colony groups in suitable aviaries are feasible. They are delicate birds and do best in slight heat (4°C, 40°F) in winter.

Agapornis roseicollis [PLATE 68] (II)
Peach-faced Lovebird

PLATE 68 *Agapornis roseicollis*, Peach-faced Lovebird

6 in (15 cm). The plumage is mainly green in colour, lighter and more yellowish on the underparts; the forehead, face, chin, throat and upper breast are rose-pink. The rump is blue; the tail is green above and blue below with some orange colouring in the laterals and subterminally barred black. The bill is horn-coloured, the legs and feet grey. The sexes are alike. The species is also available in many colour mutations.

Range South-west Africa.

Status Abundant in many parts of the range dependent on the availability of water sources.

Avicultural rating One of the most popular and widely-kept members of the genus, it is now nearly domesticated. It is hardy and easily managed.

Feeding Diet J, with adequate grit. Also offer millet sprays, seeding grasses, varied green food, and slices of apple and carrot.

Breeding This species is usually a free breeder. Provide a strong ply nestbox 7 in (18 cm) square and 12 in (30 cm) high. Provide ample willow twigs for stripping; both sexes carry bark slivers to the nest by threading them through the rump and back feathers. Clutch 4–6; incubation 23 days. They will attempt to breed throughout the year; allow a maximum of three broods, then either remove the nest-boxes or split up the pairs. If the boxes are removed the birds should be kept indoors during bad weather.

General management Not suitable for colony-system housing. It is best to keep single pairs in small aviaries: 6 × 6 × 3 ft (1.8 × 1.8 × 0.9 m) is suitable, although excellent results can be achieved in aviaries of half these dimensions. They are hardy birds and do best in unheated indoor accommodation when not breeding in the winter. They are spiteful with other birds, and have noisy, penetrating calls.

Agapornis taranta (II)
Abyssinian Lovebird, Black-winged Lovebird

6½ in (17 cm). The plumage is mainly green in colour, slightly lighter and more yellowish on the under surfaces. The forehead and eye-ring are red, the flights and under wing coverts black. The tail is green with some yellow in the lateral feathers and subterminally barred black. The bill is red, the legs and feet grey. The female lacks the red and her under wing coverts are green.

Range Highlands of Ethiopia.

Status Variable, but good numbers in areas of suitable habitat.

Avicultural rating A popular avicultural species, but not as widely kept as Peach-faced, Masked and Fischer's Lovebirds. They are hardy, and reasonably free breeders.

Feeding Diet J, with grit. Also offer millet sprays, green food, figs, apple etc. Wild birds consume juniper berries which have a high content of B-group vitamins. There is some evidence that the species has a particular need for B vitamins, and deficiency symptoms arise if these are not available.

Breeding Provide a nest box (similar to the type suggested for the Peach-faced Lovebird) with a good depth of decayed chippings in the base; also provide some willow twigs for stripping, but the finished nest is not elaborate. Clutch 3–4; incubation 25–26 days.

General management Similar to Peach-faced Lovebird. This species is very hardy, and quieter than most other lovebirds.

Agapornis fischeri [PLATE 69] (II)
Fischer's Lovebird

PLATE 69 *Agapornis fischeri*, Fischer's Lovebird

5½ in (14 cm). The body plumage is mainly green, darker on the back, wings and tail, lighter and more yellowish on the underparts. The forehead, face and throat are orange-red; the collar and upper breast are yellow to yellow-orange. The under wing coverts are blue-green; the rump and upper tail coverts are blue. The tail is green with some yellow in the laterals, and subterminal black and terminal blue bands. The eye-ring is white. The bill is red, the legs and feet grey. The sexes are alike.

Range Northern Tanzania.

Status Abundant in a few areas, but under pressure in crop-growing districts.

Avicultural rating A popular and widely-kept avicultural subject, they are hardy and easily managed and free breeders.

Feeding/Breeding/General management Similar to Peach-faced Lovebird. The adults have a tendency to pluck chicks in the nest.

Agapornis personata (II)
Masked Lovebird, Yellow-collared Lovebird, Black-masked Lovebird

---

6 in (15 cm). The plumage is mainly green in colour; the neck, collar and upper breast are yellow. The whole of the head, cheeks, chin and throat are sooty-black; the eye ring is white. The under wing coverts are green to grey-blue. The tail is green with yellow-orange in the laterals, and black bands across the base and tip. The bill is red, the legs and feet grey. The sexes are similar.

Range North-eastern Tanzania; introduced into parts of Kenya.

Status Abundant in some areas, but regarded as a crop pest and under pressure where cultivation is established.

Avicultural rating See Fischer's Lovebird.

Feeding/Breeding/General management Similar to Peach-faced Lovebird. Sexing is difficult. Various mutations have been produced and are now becoming more readily available.

Loriculus vernalis (II)
Vernal Hanging Parrot

---

5 in (13 cm). The plumage is generally bright green in colour, paler on the under surfaces. A patch on the throat is blue; the rump and upper tail coverts are red. The tail is green, bluish below. The bill is orange-red with a yellow tip; the legs and feet are yellow-orange. The sexes are similar but the adult female has a brown iris whereas the male's is white.

Range South-west India, eastern India and eastern Himalayas, Burma, Thailand, southern Laos, Cambodia, southern Vietnam; Andaman Islands and islands off Mergui (western Burma).

Status Abundant in many parts of the range; scarce in Himalayas, Burma.

Avicultural rating These are popular and interesting small psittacines which do best in tropical house type accommodation. A specialised diet is needed.

Feeding Diet K used to soak sponge cake and Diet E with additional berries as available; also offer plain canary seed, millet and a small quantity of sunflower. Live food is taken by most hanging parrots and mealworms are especially enjoyed; offer 3–4 per bird each day, more if breeding. Ensure that nectar foods are not in open dishes or the birds may transfer the sticky liquid to their face or head when feeding. Many newly-exported birds appear bald after quarantine as result of poor feeding arrangements. Offer nectar in a tubular, clip-on drinking fountain. Provide grit.

Breeding Infrequently bred. A hollow log or a tall box are the best sites. Bark strips and leaves are carried to the nest chamber, tucked under the breast and rump feathers. Clutch 2–3; incubation 20–22 days. Provide a varied diet as recommended above, plus extra live food.

General management Properly acclimatised birds will winter outside; damp conditions are not tolerated and slight heat in the shelter is needed during prolonged cold spells. A secluded planted flight is ideal; otherwise use tropical house type accommodation, although consistent high temperatures are not recommended. They sleep hanging upside down like bats; but do not allow roosting in an unprotected outdoor flight as birds suspended from the roof netting will attract predators. They do not like to come to the ground for food, so place food and water vessels close to a perch. If housed in cages, these must be spacious and kept scrupulously clean (especially the wire fronts, as birds expel liquid droppings onto the bars). They can be housed with other small birds. It is believed that extra protein in the diet may stimulate breeding condition as well as being an important rearing food. These birds are delicate and need sympathetic, gentle handling until established after importation. Their call notes are not strident.

PLATE 70 *Loriculus galgulus*, Blue-crowned Hanging Parrot

Loriculus galgulus [PLATE 70] (II)
Blue-crowned Hanging Parrot

5 in (13 cm). Most of the upper surfaces are bright green, the underparts paler. A triangular-shaped area on the mantle is golden-buff, the lower back yellow. The throat, rump and upper tail coverts are red; a patch on the crown is blue. The tail is blue-green. The bill is black, the legs and feet brown. The female lacks the red throat patch and area of yellow on the lower back; she is generally duller with a poorly defined area of blue on the crown and a golden-buff mantle.

Range Malaysia, Borneo, Sumatra; also Singapore and some small outlying islands in the region; Thailand.

Status Abundant in many parts of the range.

Avicultural rating See Vernal Hanging Parrot.

Feeding/Breeding/General management Similar to Vernal Hanging Parrot.

Loriculus philippensis (II)
Philippine Hanging Parrot, Luzon Hanging Parrot

6 in (15 cm). Much of the plumage is bright green, but paler and with a yellowish tinge on the under surfaces. The forehead and fore crown are red with a narrow yellow border behind; the rear crown and nape are golden-orange. There is an area on the throat and upper breast which is orange-red; the rump and upper tail coverts are red with blue on each side of the rump. The bill is red, the legs and feet orange. The female lacks the area of red on the throat and breast.

Range Philippine Islands, including Sula Archipelago.

Note The following 11 subspecies are described, all exhibiting slight differences in plumage.
    Loriculus p. philippensis: islands of Luzon, Polillo, Marinduque, Catanduanes and Banton.
    L. p. mindorensis: island of Mindoro.
    L. p. bournsi: islands of Tablas, Romblon and Sibuyan.
    L. p. panayensis: islands of Ticao, Masbate and Panay.
    L. p. regulus: Guimaras and Negros islands.

L. p. chrysonotus: formerly Cebu Island; now probably extinct.
L. p. worcesteri: islands of Samar, Leyte and Bohol.
L. p. siquijorensis: formerly Siquijor Island; now probably extinct.
L. p. apicalis: islands of Mindanao, Dinagat, Siargao and Bazol.
L. p. dohertyi: Basilan Island.
L. p. bonapartei: larger islands of Sula Archipelago.

Status Abundant throughout much of the range, although some subspecies have been affected as primary forest is cleared, and this process is likely to have further adverse effects.

Avicultural rating Occasionally available; see Vernal Hanging Parrot.

Feeding/Breeding/General management Similar to Vernal Hanging Parrot. Clutch 3–4; incubation 20–21 days. Few breeding successes have been recorded.

Psittacula eupatria (II)
Alexandrine Parrakeet

20–23 in (51–58 cm). The plumage is mainly green, brighter on the forehead and crown; the cheeks and back of the head are suffused with grey-blue. There is a black stripe over the lower cheek patches; the collar on the hind neck is pink. There is a large patch of red on the wing coverts; the long tail is blue-green. The bill is red, the legs and feet grey. The female is similar but duller and has no black stripe on her cheeks, and lacks the pink neck collar.

Range Eastern Afghanistan westwards to Pakistan, India, Burma, northern Thailand, Laos, Vietnam; Sri Lanka and Andaman Islands.

Status Wide distribution; abundant in many areas.

Avicultural rating A long-standing and popular avicultural subject, they are good breeders, hardy and easily managed.

Feeding Diet J, with adequate grit. Various items of green food (chickweed, groundsel, spinach etc), fruit and berries are appreciated. Timber for gnawing is important, as is a good, balanced diet; provide branches of fruit trees and other non-poisonous types daily.

Breeding When some effort has been made to persuade them to nest, Alexandrines are usually willing breeders; the problem is that, like many African and Asiatic seedeating and parrotlike species, they have been so readily available and inexpensive that few aviculturists have bothered to try. Provide a stoutly-built nest-box, 15 in (38 cm) square and 30 in (76 cm) high. Clutch 2–4; incubation 28–29 days. Bread and milk are taken as a rearing food; various infant cereals are also acceptable.

General management The size of the bill gives some indication of the destructive powers of this species, therefore their accommodation must be of substantial construction. Provide a shelter and flight, the latter a minimum of 15 ft (4.5 m) in length. The shelter should be capable of accommodating the birds in periods of severe weather. They are hardy but susceptible to frostbite. Adult birds which are offered by the trade as 'pets' do not settle down; caging imposes great stress and is cruel. Young birds which have been hand-reared prove intelligent companions, but even these should not be permanently housed in a cage; some freedom is vital for this quality of avian mind. Their voices are discordant.

Psittacula krameri manillensis (II)
Indian Ring-necked Parrakeet, Rose-ringed Parrakeet

---

16 in (41 cm). The plumage is mainly apple-green in colour with a yellowish suffusion on the lower underparts. A line from the cere to the eyes is black; the chin and the stripes across the lower cheeks are black. The collar on the hind neck is

pink; the back of the head and nape are pale lavender-blue. The long tail is blue-green. The bill is dark red, black towards the tip, and the lower mandible is black; the legs and feet are grey-green. The female is almost entirely green apart from a black line from cere to eye. There are now many colour mutations.

Range Eastern Pakistan, India, Nepal, Burma, Sri Lanka.

Status Abundant in many parts of the range.

Avicultural rating Popular and widely kept. It is very hardy, although, like related species, susceptible to frostbite.

Feeding Diet J, with adequate grit. Also offer various items of fruit and green food.

Breeding They are very free breeders. Provide a nest box 10 in (25 cm) square and 18–24 in (45–60 cm) tall; fix it securely at a good height in the flight. Clutch 3–6; incubation 23–24 days. This species can be bred on a colony system.

General management Slightly less destructive than the Alexandrine Parrakeet, but it needs substantially-built aviaries. Provide a shelter and flight as recommended for that species, but a flight length of 12 ft (3.6 m) is acceptable. These birds are easily managed and often long-lived. They are not recommended for keeping as pets.

Psittacula krameri (II)
African Ring-necked Parrakeet

15 in (38 cm). This bird is very similar to the more frequently seen Indian subspecies *P. borealis* (W. Pakistan, northern India, east to Central Burma), but has a smaller, slighter build; the bill is also smaller and very dark red.

Range Guinea eastwards to Uganda, Sudan and Ethiopia; introduced into Mauritius, Zanzibar, Egypt, Aden, Kuwait, Iraq, Iran, Hong Kong, Macao and Singapore.

Status Abundant in many parts of the range, but under pressure in areas of cultivation.

Avicultural rating Less widely kept than the Indian subspecies, but popular. It is easily managed.

Feeding/Breeding/General management Similar to Indian Ring-necked Parrakeet.

Psittacula cyanocephala (II)
Plum-headed Parrakeet

13 in (33 cm). The head is deep red suffused with purple-blue, giving a plum-like 'bloom'. The plumage is mainly green in colour, brighter on the mantle and with a yellowish tinge on the under surfaces. The chin, the stripe across the lower cheeks and the narrow neck collar are black; there is an area of bright green below the collar. There is a dark red patch on the wing coverts. The long tail is blue and yellow-green. The upper mandible is orange-yellow, the lower mandible black; the legs and feet are grey-brown. The female is similar but lacks the 'bloom' on the head, which is lavender-blue; the black collar is replaced by yellow.

Range Eastern areas of West Pakistan, Nepal, India, Sri Lanka.

Status Abundant in many parts of the range.

Avicultural rating A popular cage and aviary subject for many centuries, it is easily managed and hardy. The females are usually in short supply; many 'hens' prove to be immature cocks in import shipments.

Feeding/Breeding/General management Similar to Indian Ring-necked Parrakeet.

Psittacula alexandri fasciata (II)
Moustache Parrakeet, Banded Parrakeet

13–15 in (33–38 cm). The upper surfaces are

green, brighter on the hind neck; the head is grey with a bluish suffusion and a tinge of green round the eyes. There is a black stripe from the cere to the eyes and broad black 'moustache' markings from the base of the lower mandible, below the cheeks to the sides of the neck. The whole of the breast is rose-pink; the lower underparts are green. There is a yellow patch on the wing coverts; the long tail is blue and yellow-green. The upper mandible is red, the lower mandible black. The female is similar but the head is more blue, and the yellow on the coverts is more indistinct. The bill is completely black.

Range Nepal and northern India eastwards through Assam and Burma to southern China, northern Laos, Vietnam; Hainan and Andaman Islands, Java, Bali and some offshore islands. A population in southern Borneo is probably the result of introductions by man.

Note Although 8 subspecies are described, *Psittacula alexandri fasciata*, from northern India, Burma, eastwards to southern China, is the one usually encountered in aviculture. All subspecies vary slightly in plumage and, in some cases, in size.

Status Abundant in many parts of the range, and occasionally seen in very large flocks. They cause considerable damage to crops in areas of cultivation.

Avicultural rating Imported birds are somewhat unpredictable and often difficult to acclimatise, but they are hardy when established. They are infrequent breeders.

Feeding Diet J, with adequate grit. Also offer a variety of green food, apple, carrot etc. Provide fresh non-poisonous branches for gnawing.

Breeding They are infrequently bred, considering the scale of imports. Clutch 3–4; incubation 23–24 days. Provide bread and milk or sponge cake and honey water when the young are being fed.

General management Similar to the recommendations made for the Indian Ring-necked Parakeet. These birds are very noisy, and only hand-reared birds are likely to make pets; prospective owners should bear in mind their destructive (gnawing) capabilities. Permanent caging of these handsome, active birds is definitely not recommended.

Ara ararauna [PLATE 71] (II)
Blue and Yellow Macaw, Blue and Gold Macaw

32–34 in (81–86 cm). The forehead and front of the crown are green, the remainder of the upper surfaces bright blue. The throat is black, the rest of the underparts golden-yellow. The cheeks are pink-white with lines of tiny green-black feathers. The primaries and tail are a darker blue; the under tail coverts are sky-blue, and the underside of the tail dull yellow. The bill is black, the legs and feet dark grey. The sexes are alike.

Range Panama south to Colombia and eastwards to southern Venezuela, the Guianas, Trinidad; Brazil, Bolivia, Paraguay.

Status There are still good populations in some parts of the range, but numbers are declining mainly through habitat destruction.

Avicultural rating One of the most beautiful and impressive-looking of the parrot order. It is now being captive-bred with increasing frequency; surgical sexing techniques have helped, plus the realisation that despite their size they will reproduce in modest-sized quarters. They are also good liberty birds.

Feeding Diet H, with equal quantities of pellet food. Provide plenty of fruit, green food, vegetables; apple, carrot and corn-on-the-cob are among many items especially appreciated by these birds. Pine nuts and unshelled nuts (hazels, Brazils, almonds, etc) should be regular features of their diet. A widely-varied menu is very important for macaws; regard the recommendations here merely as a basis and make sensible trial-and-

PLATE 71 *Ara ararauna*, Blue and Yellow Macaw

error investigations to ascertain the likes and dislikes of individual birds or pairs. Appropriate sized grit is essential.

Breeding The relative ease of determining the sex of individual birds has now improved the frequency of breeding successes. Established, compatible pairs will nest in accommodation of fairly modest dimensions (a 12 × 6 × 6 ft, 3.6 × 1.8 × 1.8 m, flight is acceptable to some). Provide a large, strong wooden barrel for nesting; plastic and metal are not recommended because of the danger of internal high temperatures and condensation. In addition, the gnawing instincts of the sitting female are unsatisfied during incubation in a plastic or metal 'nest'. Ensure that the nest is sited in a position where some shade from strong sunlight is available. Clutch 3–4; incubation 26–28 days. The chicks leave the nest after 12–14 weeks. Large macaws can be extremely aggressive when breeding; very strong and with powerful bills, they should be treated with respect and care. Remember that bird-to-bird introductions (even with birds of the opposite sex) can be fraught with danger and fierce fights sometimes ensue. Some pairs prefer a nest box or a barrel at ground level. Provide plenty of additional blocks of wood inside, so that the hen can gnaw them to her preferred size or type of chippings or shavings. Bread and milk can be offered as a rearing food supplement; ensure that it is absolutely fresh at all times.

General management Hand-reared macaws are intelligent pets, but need caring owners who will devote plenty of time to them throughout their long lives. Such birds should not be acquired by people away from home on a regular daily basis. Many macaws are extremely destructive and use heavy mandibles rapidly to reduce stout timber to matchwood; therefore aviary accommodation needs to be of the appropriate strength of metal, concrete and welded mesh construction. The birds' calls are best described as raucous, and are rarely appreciated by close-proximity neighbours. Individual macaws of certain species are good liberty subjects; they fare best in sparsely-populated rural areas, and are certainly not recommended for release in an urban setting. The method usually employed to provide some 'guarantee' that liberty birds will stay is to wing-clip and release the bird(s) into a convenient tree or similar perch. Each night the birds are enclosed in the aviary again. By the time the moult renews their feathers the birds are imprinted with their surroundings and unlikely to abscond. The delay also enables their wing muscles to strengthen through regular exercise; closely confined macaws usually lose the power of strong flight fairly quickly and need time to recover it.

Ara macao [PLATE 72] (I)
Scarlet Macaw, Red and Yellow Macaw

33–35 in (84–89 cm). The plumage is mainly scarlet in colour; the wing coverts are yellow, with

[180]

Psittaciformes

PLATE 72 *Ara macao*, Scarlet Macaw

some feathers edged or tipped with green. The flight feathers are blue; the lower back, rump, upper and under tail coverts light blue. The tail is scarlet with a blue tip, the outer feathers of the tail blue. The area of white skin on the cheeks has lines of tiny red feathers. The upper mandible is white with a black tip, the lower mandible black; the legs and feet are dark grey. The sexes are alike.

Range Mexico and Central America, central and south-eastern Colombia, the Guianas, Trinidad, western and central Brazil, eastern Peru and part of northern Bolivia.

Status Numbers appear to be declining in many areas of the range; habitat destruction and excessive trapping are the probable causes.

Avicultural rating They are widely kept and, like other macaw species, now being bred in increased numbers.

Feeding/Breeding/General management Similar to Blue and Yellow Macaw. Scarlet Macaws are generally regarded as less docile than the previous species and are usually particularly aggressive when nesting.

Ara chloroptera (II)
Green-winged Macaw, Red and Blue Macaw

---

35 in (89 cm). Similar to Scarlet Macaw, but this species is a slightly darker shade of red, and lacks any yellow in the wings. There is some green in the lesser wing coverts. The upper mandible is ivory with black on the lower base, the lower mandible is black. The sexes are alike.

Range Eastern Panama, northern areas of Colombia eastwards to Venezuela, Guianas; northern and eastern Bolivia, Paraguay, northern Argentina and all but the most southerly and easterly areas of Brazil.

Status There are reasonable numbers in some areas, but, like others of the family, they are suffering a decline through loss of habitat and trapping.

Avicultural rating Popular, but not as widely kept as Blue and Yellow Macaw or Scarlet Macaw.

Feeding/Breeding/General management Similar to Blue and Yellow Macaw. This is the second largest of macaws; it is very powerfully built.

Ara severa (II)
Severe Macaw, Chestnut-fronted Macaw

---

18 in (45 cm). The plumage is mainly green in colour; the crown is tinged with blue. The forehead and chin are chestnut-brown; the bend of the wing and the under wing coverts are red. The primary coverts and flights are blue; the tail is

red-brown above, the underside brick-red. The area of bare skin on the cheeks is white with lines of tiny black feathers. The bill is black; the legs and feet are dark grey. The sexes are similar.

Range Panama and an extensive range in northern South America: Colombia, Venezuela, the Guianas, north-western areas of Brazil and south-east to Bahia.

Status Formerly abundant in many parts of the range, numbers are now reduced following habitat loss and associated pressures.

Avicultural rating Popular, but never as widely kept as the larger and more colourful species. They are hardy and easily managed.

Feeding Diet H, with adequate grit. Also provide pellet food with a variety of fruits, vegetables and green food. Offer some plain canary seed and millet.

Breeding This macaw has bred infrequently. Provide a stoutly-constructed nest box (approximately 12 × 12 × 24 in, 30 × 30 × 60 cm high); with chippings and larger pieces of wood in the box for the female to shred. Clutch 2–3; incubation 28 days.

General management Pairs thrive in modest-sized shelter and flight accommodation; the flight should be 12 × 6 × 6 ft (3.6 × 1.8 × 1.8 m) minimum. They are hardy but probably better enclosed in a roomy, well-lit shelter during prolonged cold or frosty weather.

Ara maracana (II)
Illiger's Macaw, Blue-winged Macaw

16½ in (42 cm). The forehead is red, the crown and nape blue; most of the upper surfaces, breast and underparts are green. The lower back and an area on the abdomen are red; the primaries, coverts and secondaries are blue. The tail is reddish-brown at the base, merging with blue; the underside is a drab yellow-olive. A triangular-shaped area of bare skin on the cheeks is yellow-buff. The bill is black, the legs and feet brown. The sexes are alike.

Range Eastern Brazil, Paraguay, north-east Argentina.

Status Abundant in a few parts of the range, but the loss of forest habitat in many areas is reducing the population.

Avicultural rating Although never popular with many aviculturists, the high price of the bigger macaw species has resulted in increased demand for smaller, inexpensive (by comparison) kinds.

Feeding/Breeding/General management Similar to Severe Macaw. Breeding aviaries should be secluded and there is a theory that the proximity of trees simulating a forest habitat may encourage nesting. Clutch 2–3; incubation 26–27 days. Provide high-protein rearing foods.

Ara nobilis cumanensis (II)
Noble Macaw

---

13 in (33 cm). The forehead and forecrown are blue, the rest of the upper parts green. The under surfaces are yellowish-green. The bend of the wing and under wing coverts are red. The tail is green; the underside of the flights and tail are dull yellow. The area of bare skin on the cheeks is white. The upper mandible is horn-coloured, the lower mandible grey; the legs and feet are dark grey. The sexes are alike.

Range Central areas of Brazil.

Note The nominate subspecies, *Ara n. nobilis*, is now known as Hahn's Macaw. It is slightly smaller and distinguished from *A. n. cumanensis* by a darker (blackish) bill. It has a wider range in north-eastern South America from the Guianas and eastern Venezuela to central, eastern and southern Brazil.

Status Variable; still abundant in some areas but deforestation is reducing the population.

Avicultural rating See Illiger's Macaw.

Feeding/Breeding/General management Similar to Illiger's Macaw. Some pairs are prolific and have been bred on a colony system. Their small size allows housing and even breeding in very small outdoor flights: 8 × 6 × 6 ft (2.4 × 1.8 × 1.8 m) minimum is recommended. Provide stoutly-constructed nest boxes 10 in (25 cm) square, and 24 in (60 cm) high. Clutch usually 4; incubation 25–26 days.

Aratinga guarouba (I)
Queen of Bavaria's Conure, Golden Conure

---

14 in (36 cm). Apart from the primaries, secondaries and outer wing coverts, which are dark green, the entire plumage is golden-yellow. The heavy bill is horn-coloured, the legs and feet flesh-coloured. The sexes are alike.

Range North-eastern Brazil.

Status Numbers are decreasing as their habitat is destroyed.

Avicultural rating A beautiful and much-coveted species, but it is very expensive. The birds are easily managed and true pairs are usually willing breeders.

Feeding Diet J, with adequate grit. Provide a wide variety of fruit; also offer vegetables and green food.

Breeding Visual sexing is difficult, but surgical techniques enable true pairs of these and other non-dimorphic psittacines to be introduced with certainty. Provide a stoutly-contructed nest-box with some wood chips. Clutch 2–3; incubation 26–28 days. Adults are often aggressive when breeding. Provide a high-protein rearing diet.

General management These conures are highly intelligent and need opportunities to occupy their inventive outlook. They are very prone to feather-plucking in an unsuitable environment. They do best in a reasonably spacious aviary furnished with plenty of natural timber (for gnawing). They are hardy, but it is best if they are persuaded to roost in a shelter. They have harsh calls, and are efficient destroyers of unprotected timber.

Aratinga jandaya [PLATE 73] (II)
Jendaya Conure, Yellow-headed Conure

PLATE 73 *Aratinga jandaya*, Jendaya Conure

12 in (30 cm). The head and neck are yellow with some orange on the forehead, cheeks and throat; the breast and lower underparts are orange-red. The back and wings are green; the flights and secondaries blue. The lower back is orange-red, the rump green. The tail is olive-green with a blue

tip, the underside dark grey. The bill is grey-black, the legs and feet grey. The sexes are alike.

Range North-eastern Brazil.

Status There are reduced numbers in some parts of the range following habitat loss.

Avicultural rating Formerly freely available, they are now much less so. They are hardy and fairly free breeders.

Feeding Diet J, with adequate grit. Offer fruit, vegetables and green food.

Breeding Provide a stoutly constructed nest box. Clutch 3–4; incubation 26 days.

General management Like related species, they will rapidly destroy exposed woodwork; so accommodation must be of robust construction using metal, concrete and welded mesh. They are hardy, and have discordant voices.

Aratinga solstitialis [PLATE 74] **(II)**
Sun Conure

12 in (30 cm). The greater wing coverts are green, with some feathers tipped with yellow, the primaries dark blue and the tail olive green; the remainder of the plumage is yellow with an orange suffusion on the forehead, cheeks, breast etc. The bill is dark horn-coloured, the legs and feet grey-brown. The sexes are alike.

Range North-eastern Venezuela, the Guianas and north-eastern Brazil

Status This conure appears to be thinly distributed throughout its range; nowhere is it abundant.

Avicultural rating Now a popular avicultural species, but limited in availability. It is hardy and a good breeder.

Feeding/Breeding/General management Similar to

PLATE 74 *Aratinga solstitialis*, Sun Conure

Jendaya Conure. Some pairs are very prolific.

Nandayus nenday **(II)**
Nanday Conure, Black-headed Conure

12 in (30 cm). The plumage is mainly green in colour; the underparts are paler and suffused with yellow. The head and forecheeks are sooty-black, the thighs red. The throat and upper breast are tinged with blue; the flight feathers are blue. The tail is olive-green, the underside grey. The bill is black, the legs and feet grey-brown. The sexes are alike.

Range South-eastern Bolivia, northern Argentina, Paraguay, parts of southern Brazil.

Status Still abundant in some parts of the range, but habitat loss is affecting some populations.

Avicultural rating A popular avicultural subject, but noisy. It is hardy, easily managed and a good breeder, either in aviaries or at liberty; a flourishing colony (initiated by escaped birds) is reported in California.

Feeding Diet J, with adequate grit. Provide abundant supplies of fruit, vegetables etc. Some birds enjoy bread and milk, which is also a useful rearing food.

Breeding They are fairly prolific. They will use a stout box or a log, but occasionally reduce the nesting receptacle to matchwood before settling down to breed. Clutch 3–5; incubation 26 days.

General management Ensure that the aviary accommodation is of suitably robust construction, i.e. metal-framed, concrete base and heavy gauge mesh. They are hardy and easily managed; and have raucous calls.

Pyrrhura frontalis (II)
Red-bellied Conure, Maroon-bellied Conure

---

10 in (25 cm). The plumage is mainly dark green, paler and more yellowish on the under surfaces. The narrow band on the forehead is chestnut; a patch on the ear-coverts buff-brown. The throat, breast and sides of the neck are olive-grey, the feathers margined with yellow and tipped with drab-olive to give a scale-like effect. The centre of the abdomen and a patch on the lower back are red-brown. The flight feathers are blue, the tail green at the base, then reddish. The bill is grey-brown, the legs and feet dark grey. Sexes alike.

Range South-eastern Brazil, Argentina, Paraguay and Uruguay.

Status Abundant in some areas but deforestation is reducing numbers.

Avicultural rating One of the most widely kept members of the *Pyrrhura* genus, they are easily managed, hardy and fairly reliable breeders. They have an inquisitive disposition and a reputation for escaping through small apertures, breaks in netting etc.

Feeding Diet J, with adequate grit. Provide a good selection of fruit, vegetables and green food on a regular basis.

Breeding They have been bred on a colony system, but most successes have resulted from single-pair occupancy of small aviaries. Clutch 4–6; incubation 26 days. Provide a stoutly-constructed nest box slighty bigger than a budgerigar box and up to 24 in (60 cm) high. Offer soaked seeds (including millet sprays), high-protein rearing foods and sponge cake with honey water when rearing.

General management They are usually confident, inquisitive birds, and often develop a good rapport with their owners. They are hardy and thrive in a typical shelter and flight unit. Provide plenty of stout branches for their climbing activity and replace as necessary. They are noisy, but their voices are not as strident as those of *Aratinga* and *Nandayus* species.

Myiopsitta monachus (II)
Quaker Parrakeet, Monk Parrakeet, Grey-breasted Parrakeet

---

11 in (28 cm). The forehead, cheeks and throat are grey; the grey-brown feathers of the breast are edged with grey-white. The upper surfaces are green. A broad band across the upper abdomen is a drab yellow-green; the lower abdomen, thighs and under tail coverts are brighter yellowish-green. The flight feathers are blue; the tail is green, blue towards the tip and with a blue-green under surface. The bill is yellow-brown, the legs and feet grey-brown. The sexes are alike.

Range Central Bolivia eastwards to southern Brazil; northern areas of Argentina.

Status Still abundant in some parts of the range but, like other South American species, decreasing as tracts of habitat are destroyed.

Avicultural rating An inexpensive and widely kept species, they are hardy, easily managed and good breeders. They are excellent liberty birds.

Feeding Diet J, with adequate grit. Also provide the usual fruit, vegetables and green food.

Breeding They can be very prolific. They will utilise nest boxes, or, if housed in a spacious aviary, construct a large natural nest in a tree using supplied twigs, branches etc. Clutch 4–7; incubation 28–30 days. Offer sprouted seeds, seeding grass, bread and milk when they are rearing young.

General management The major disadvantage with these otherwise admirable beginner's parrot-like birds is a particularly piercing voice; they are also destructive to timber and wire-netting accommodation. They are hardy and easily managed; and free breeders. A colony in a large aviary will often utilise a heap of brushwood on a wood and wire base as a foundation for a communal nest. Aviaries for these birds should be strongly built.

Forpus coelestis (II)
Celestial Parrotlet, Pacific Parrotlet

5½ in (14 cm). The forehead, crown, cheeks and chin are bright apple-green; the rest of the upper surfaces are darker green. The underparts are glaucous-green; the lower back, rump and wing coverts are cobalt-blue. There is an area of pale blue behind the eye. The bill is horn-coloured; the legs and feet are flesh-brown. The female is similar but lacks the areas of blue.

Range Western Ecuador, north-western areas of Peru.

Status Fairly abundant in some areas.

Avicultural rating Very popular, although it is now less freely available. It is easily managed, hardy, and a relatively free breeder.

Feeding Diet J, with adequate small grit. Fruit is usually ignored; green food is taken mainly when rearing young, but offer it regularly in an effort to persuade the birds to sample it. Also offer seeding grasses, and millet sprays.

Breeding True pairs are often prolific. Provide nest boxes about half the size of a standard budgerigar box with a layer of chippings or peat inside. Clutch 3–8; incubation 18–20 days. This species can be bred on a colony system in a large aviary. Despite their small size the birds can be aggressive; watch for attacks by the parents on newly-fledged chicks.

General management They fare best in small shelter and flight accommodation, although they have been bred successfully in roomy cages; aggression between pairs is more likely in a limited space. Although they are diminutive they are not safe companions for other birds. There is an urgent need to build up aviary-bred stocks as import restrictions reduce the availability of wild birds.

Forpus xanthops (II)
Yellow-faced Parrotlet

---

6 in (15 cm). The crown, cheeks and throat are yellow; most of the upper surfaces are grey-green. The underparts are yellow-green; the lower back, rump, upper tail and wing coverts are cobalt-blue. The bill is horn-coloured, with dark grey at the base of the upper mandible; the legs and feet are brown. The female is similar but the lower back and rump are pale blue; the secondaries and primaries have a blue tinge.

Range They are known only from a limited area in north-western Peru.

Status Not known.

Avicultural rating Since their introduction into aviculture in the early 1980s they have proved as adaptable and free-breeding as Celestial Parrotlets.

Feeding/Breeding/General management Similar to the Celestial Parrotlet. Birds breed successfully as individual pairs in small aviaries, or in larger groups where flight space allows. Watch for disruptive behaviour of single, unmated birds in colonies.

Brotogeris versicolurus (**II**)
Canary-winged Parrakeet

---

9 in (23 cm). The plumage is mainly dull green, more olive on the back; there is a faint blue-grey wash around the forehead and eyes. The secondaries are white with yellow markings, the primaries dark blue. The tail is green above, blue-green below. The bill is horn-coloured, the legs and feet flesh-brown. The sexes are alike.

Range Eastern Ecuador, north-eastern Peru, southern Bolivia eastwards across Brazil; French Guiana.

Note The subspecies *B. v. chiriri* is also listed in some works as the Canary-winged Parrakeet. It is distinguished from the nominate subspecies by its lighter green plumage, fully feathered lores and slightly larger size. Its range is from eastern and northern Bolivia, Paraguay, northern Argentina to southern and eastern Brazil.

Status Abundant in many parts of the range; it is also established in other areas (from escaped pet birds) including Lima (Peru)and Puerto Rico.

Avicultural rating A very popular species for many years, mainly as a pet, although the piercing voice is an obvious disadvantage. It is hardy and easily managed, but has a poor captive breeding record.

Feeding Diet J, with adequate grit. Provide a selection of fruit, green food and vegetables on a regular basis.

Breeding There have been relatively few successes, considering the vast numbers imported over the years. Provide a strongly built nest box about 10 in (25 cm) square and up to 24 in (60 cm) tall. Clutch 3–5; incubation 26–28 days. Provide bread and milk, sponge cake and honey water etc for rearing.

General management They thrive in suitably robust shelter and flight accommodation. Watch for birds which are distressed during severe (cold) weather; house them inside if necessary. Provide a box or log for roosting purposes.

Pionus menstruus (**II**)
Blue-headed Parrot, Blue-headed Pionus, Red-vented Parrot

---

11 in (28 cm). The head, neck and upper breast are blue with some red on the throat; the lower breast is green, the feathers edged with blue. The undertail coverts are red, the feathers tipped with blue-green. The rest of the upper surfaces are green; a patch on the shoulders of the wings is bronze-green. The tail is green, red at the bases of the feathers and tipped with blue. The bill is dark grey with red at the bases of the mandibles; the legs and feet are grey-green. The sexes are alike.

Range Costa Rica south to Amazonia, the Guianas, eastern Peru, northern Bolivia.

Status Variable; abundant in some parts of the range. It is mainly a forest dweller, so populations are likely to be affected by habitat loss.

Avicultural rating They were formerly fairly freely available, but never among the most popular avicultural subjects. There are some behavioural problems which may be stress-related. They are hardy, but provide frost-free shelters in which the birds can be enclosed during severe (cold) weather.

Feeding Diet J, with adequate grit. Also offer fruit, vegetables, sprouted seed, pine nuts etc.

Breeding Only an occasional breeder, except in

specialist collections. Clutch 3–5; incubation 26 days. They do best in a secluded setting provided by a leafy background to the breeding aviaries.

General management Provide spacious shelter and flight accommodation. Take care when introducing pairs; aggressive behaviour is not unusual. Otherwise they are much less active than many parrotlike species, but have fairly strident calls.

Amazona aestiva (II)
Blue-fronted Amazon Parrot

14–15 in (36–38 cm). The plumage is mainly green, the feathers of the neck and mantle edged with black; the underparts paler green. The forehead and lores are pale blue; there is a variable amount of yellow on the crown, sides of the head, chin and throat. The shoulder of the wing and the outer webs of the secondaries are red; the primaries are violet-blue towards the tips. The bill is black, the legs and feet grey. The sexes are alike.

Range Eastern Brazil, northern and eastern Bolivia. Paraguay, northern Argentina.

Status Still abundant in many parts of the range, but habitat loss affects them in several areas.

Avicultural rating These have been popular 'pet' parrots for generations, but are now less readily available and expensive. They are hardy, easily managed, and have occasionally bred.

Feeding Diet H, with an approximately equal part of pellet food (see p. 27). Offer suggested additions to the basic menu such as fresh fruit, vegetables etc. Bones with meat on them are greatly relished by these and many other large psittacine species and can be fed to advantage from time to time. Grit is essential.

Breeding The majority of Amazon Parrots imported are destined for the pet trade, which may be one reason for infrequent breedings despite the numbers shipped. They do best in good-sized aviaries; provide a wooden nest-box or small wooden barrel. Clutch 3–4; incubation 28 days. Some pairs (especially the males) are protective to the point of being savage when breeding. Provide bread and milk, soaked sunflower seed etc when chicks are being reared.

General management Although they are not as destructive as other species, house them in strongly-built metal, concrete and welded mesh accommodation with shelter and flight; the latter should ideally be 15 ft (4.5 m) long. A feeding hatch is valuable when breeding. They are hardy, intelligent and voluble birds. Hand-reared, they make excellent pets, but need time and attention if they are not to deteriorate.

Amazona ochrocephala (II)
Yellow-fronted Amazon Parrot, Yellow-crowned Amazon Parrot, Yellow-headed Amazon Parrot

14 in (36 cm). The plumage is mainly green, but paler and with a yellowish tinge on the underparts. The crown is yellow; the feathers of the neck and mantle are edged with black. The bend of the wing and the outer webs of the secondaries are red. The tail is green with a yellow terminal band. The bill is dark grey (orange on the sides of the upper mandible in some birds), the legs and feet grey. The sexes are alike.

Range Western Colombia eastwards through northern Brazil to Venezuela, the Guianas, Surinam; also Trinidad.

Note Nine subspecies are described; many differ from the nominate subspecies only slightly, but one of the most distinctive is *A. o. tresmariae*, the aviculturist's Double Yellow-headed Amazon Parrot. This is confined to the Tres Marias Islands off the western coast of Mexico. Also identified as Double Yellow-headed Amazons are *A. o. oratrix* (Mexico) and *A. o. belizensis* (Belize).

Status Variable; there are still good populations in some parts of the total range, but they are adversely affected by habitat loss and excessive trapping in other areas; *A. o. oratrix* is said to be in serious decline in Mexico.

Avicultural rating Popular and widely kept. This species has bred occasionally.

Feeding/Breeding/General management Similar to Blue-fronted Amazon Parrot. They are intelligent; and tame birds are often very talented mimics. They have more strident calls than the previous species.

Deroptyus accipitrinus (II)
Hawk-headed Parrot, Red-fan Parrot

---

14 in (36 cm). The forehead and crown are buff-white; the top and sides of the head brown with pale shaft streaks. The elongated feathers of the nape are dark crimson and edged with blue to form an erectile ruff; there is a similar colour pattern on the breast and abdomen. The upper surfaces are dark green; the tail is green, blue towards the tip, the underside is maroon and blue. The bill is black, the legs and feet dark grey. The sexes are alike but there is considerable variation of plumage between individual birds.

Range Venezuela, the Guianas, northern Brazil, north-eastern Peru.

Note A subspecies, *D. a. fuscifrons*, from central and north-eastern Brazil, is identified by its darker forehead and crown.

Status There is local distribution throughout the range; good populations in some areas but nowhere abundant.

Avicultural rating A highly-prized avicultural subject, it is occasionally imported but expensive.

Feeding Diet H with adequate grit. Also offer pellet foods (see p. 27), and a variety of fruit, vegetables, berries etc. Fruit is an important diet item with Hawk-heads.

Breeding Infrequently bred, this species needs a quiet, secluded aviary; it is possibly best when entirely enclosed on all sides. Otherwise, site the aviary against a tree or shrub background. The adults are often extremely savage when nesting. Provide a log or nest box permanently in the aviary for roosting as well as a potential nest site. Clutch 2–3; incubation 26–28 days.

General management This species needs careful, sympathetic handling. It seems susceptible to stress problems; aviaries should ideally be spacious so that the birds can retire from the proximity of human onlookers. A quiet location for shelter and flight accommodation is absolutely essential. They are invariably noisy and are not recommended for urban areas.

CUCULIFORMES

Musophagidae (Touracos)

The Musophagidae is a small family of arboreal, fruit-eating birds found only in Africa. There are several species in aviculture and breeding results are now very encouraging for the future of species in control. They do best in spacious aviaries which allow the opportunity for considerable agility, running through and along branches. They are tree-nesters, usually making a flimsy, fragile platform for 2 eggs. The chicks are able to climb before they can fly. There is a strong bond between most established pairs but the similarity of the sexes is usually the cause of many pseudo-pairs failing to reproduce. In the wild, Touracos live among the trees, from the plains up to evergreen forests 12,000 ft (3600 m) above sea level. The plumage colours are mainly blue, green, violet and purple; some are a duller grey-brown. Most are crested. In most respects they are excellent avicultural subjects; we now need to concentrate on improved breeding programmes.

Feeding Diet E, with the addition of one quarter (by volume) of Diet G, and a selection of berries when in season. Live food can be offered but many touracos are uninterested.

Breeding They were first bred at the Jersey Wildlife Preservation Trust in 1970. Provide a mesh foundation in likely nest sites (a tree or tall bush) and a few twigs to encourage interest. Clutch 2–3; incubation 19–20 days. The young are fed on regurgitated fruit etc.

General management They do best in spacious shelter and flight accommodation. They are hardy, but even acclimatised birds should be confined in a roomy frost-free shelter during periods of severe (cold) weather. They are generally active and intelligent birds. They have a loud call note from which their name is derived.

Corythaixoides concolor (III)
Grey Plantain-eater

20 in (51 cm). The plumage is entirely pale grey, including an upright pointing crest. The bill is black (green in the case of the female); the legs and feet are black. The sexes are alike except for the difference in the colour of the bill.

Range Southern Tanzania southwards to Zambia, Malawi, Zimbabwe to South Africa.

Status Abundant in many parts of the range.

Avicultural rating Not one of the most widely kept members of the genus, but it is occasionally available, and fairly hardy when acclimatised.

Tauraco hartlaubi [PLATE 75]
Hartlaub's Touraco

16 in (41 cm). The crown, crest and sides of the head are glossy blue-black; the throat, breast, ear-coverts, neck and mantle are green. The rump and tail are violet-blue; the wings are similar but with conspicious red primaries. The abdomen and under tail coverts are dusky; a spot in front of the eye and streak below are white. The bill is olive with a red tip, the legs and feet black. The sexes are alike.

Range Kenya, north-eastern Tanzania.

PLATE 75 *Tauraco hartlaubi*, Hartlaub's Touraco

Breeding Established pairs may nest in suitable aviaries. The flimsy platform nest may need a mesh base, but take care that disturbance does not inhibit laying. Wood or wire nest trays (as for doves and pigeons) can be fixed in position in likely nest sites to minimise later disturbance. Clutch 2; incubation 20–21 days. Offer bread and milk or sponge cake and honey water to supplement other foods for rearing.

General management This hardy species can remain outdoors throughout the year when acclimatised. A suitable roomy, well-lit shelter to enclose the birds in very severe (cold) weather is necessary. A lofty, well-branched aviary is ideal; the birds have the typical touraco bounding gait and are very arboreal. Their calls are loud but not discordant.

Tauraco leucotis
White-cheeked Touraco

16 in (41 cm). The head, neck, mantle, back and breast are green; the crest dark blue. The wing coverts and rump are blue-grey with a dark-blue gloss; the flights are red. The tail is blue, the abdomen dark grey. There is an area of white on the hind cheeks and in front of the eye. The bill is orange-red, the legs and feet black. The sexes are alike.

Range Ethiopia, south-eastern Sudan.

Note A subspecies, Donaldson's Touraco (*T. l. donaldsoni*), from south-eastern Ethiopia and western Somalia, is distinguished only by the dull red edge to the rear crest.

Status Fairly abundant in suitable habitat.

Avicultural rating Possibly the most widely-kept of touracos at present, and it is proving a fairly free breeder.

Feeding/Breeding/General management Similar to Hartlaub's Touraco but established pairs are more willing breeders.

Status Abundant where suitable habitat exists.

Avicultural rating These birds are increasingly widely kept and more breeding successes are being reported. They are easily managed and hardy, but need spacious accommodation.

Feeding Similar to Grey Plantain-eater. Fruit and berries are the mainstay of their diet but some insectivorous mixture (Diet G) should be mixed with diced fruit. High-protein trout pellets (soaked) can be added in moderate quantities (see p. 27). Live food should be offered although many birds show little interest except when breeding. Chopped green food (cress, lettuce, groundsel etc) may be taken by individual birds.

Cuculidae (Cuckoos, Malcohas, Anis, Coucals)

Few of these birds are likely to be encountered by aviculturists although Coucals are occasionally available. There are nearly 30 species; mainly in Africa, southern Asia and Australasia. They are stoutly-built and have a strong, hooked bill.

Centropus senegalensis
Senegal Coucal

---

15 in (38 cm). The top of the head, nape and tail are black, and there is a green gloss on the crown and nape; the tail has a purple gloss. The mantle and wings are chestnut, the under surfaces creamy-white. The bill is black, the legs and feet dark grey. The sexes are alike.

Range Senegal eastwards to southern Sudan, northern Uganda and Tanzania; also southwards to northern Angola.

Status Abundant in many parts of the range.

Avicultural rating Rarely available; established birds are not difficult but they are not winter-hardy and need slight heat during cold weather.

Feeding Diet G, with the addition of some minced raw beef, and adequate live food (crickets or locusts are recommended); tinned dog meat can be added to the diet from time to time. Any small dead mice, frogs, lizards etc will be eagerly consumed.

Breeding Only a few Coucals have been bred in captivity. They need a spacious, well-planted flight. The nest is untidy and bulky. Clutch 2–4; incubation 16 days. Abundant live food is needed for rearing. Cannibalism may be a problem if the diet is deficient.

General management Although not a strong flyer, this species needs plenty of space. A large outdoor aviary with access to a heated, well-lit shelter is ideal; alternatively the birds do best in indoor flights during the winter. Like other Coucals, it spends much time at ground level investigating vegetation etc. It is not a suitable companion for other species.

STRIGIFORMES

Tytonidae (Barn Owls)

Although there are only 10 species, Barn Owls are regarded as warranting separate classification from typical owls. No less than 38 subspecies are described for the Barn Owl (*Tyto alba*), encompassing a huge range: North and South America, Europe, much of Africa, south-east Asia, Australia and the Pacific Region. As the subject of superstition, they were formerly much persecuted; now the Barn Owl's extraordinary ability as a killer of rodent pests (up to 95 per cent of the total diet) has led to protection measures in various countries. It is an excellent avicultural subject; easy to house and manage and a free breeder.

Tyto alba (II)
Barn Owl

14 in (36 cm). The upper surfaces are rufous-buff with fine grey and white vermiculations and mottling; the facial disc and underparts are white with a few rufous-grey spots on the sides of the body. The bill is white, the legs white-feathered, the feet brown. The sexes are similar.

Range West and central Europe, Africa (except Sahara and area of Congo Basin), Madagascar and Comoro Islands, Arabia, Syria, Iraq, India, Sri Lanka. Burma, Indochina, Malaysia to Java and Sundas, New Guinea, Australia, central and north-eastern USA, Central America, the Guianas and West Indian islands, South America (except extreme south).

Status Variable, but there are good populations in many parts of the range and numbers are increasing in some areas.

Avicultural rating An increasingly popular avicultural subject, which breeds well in aviaries.

Feeding Dead mice, young rats and day-old chicks provide a suitable staple diet for these birds; the use of wild rodents is not recommended because of the strong possibility of feeding contaminated (poisoned) food. Food should be fed whole. Deep-freezing is a convenient way of storing carcase stocks; ensure that they are fully thawed before use. Provide multi-vitamin supplement regularly.

Breeding Provide a suitable-sized nest box (half-front) with 7.5–10 cm (3–4 in) depth of wood chips in the bottom; site it in the aviary shelter above ground level. Clutch 4–6; incubation 33–34 days. Established pairs are often prolific.

General management Although they are hardy, draughty or damp conditions are not conducive to good health. Secluded aviaries are best; many successful owl-breeders have only the front of the flight wired, the rest is boarded and the roof is clad mainly in one of the glass substitute materials now available. The area designated as a shelter should have a single perch and a good depth of peat (15 cm, 6 in) beneath; the latter should be removed or replaced as soon as soiling demands. Flights can have either concrete or gravel bases; it is essential to maintain scrupulous cleanliness. Aviaries are not the sole means of accommodating these birds; they thrive in disused and converted farm buildings, garages etc, but these must be dry and damp-free as well as admitting the necessary amount of light.

Strigidae (Owls)

There are more than 130 species of owl distributed throughout much of the world from the northern Arctic to the Falkland Islands in the south. Like Barn Owls, they are mainly nocturnal; all are predatory and their food preferences range from fish to rodents, insects to birds. Most owls fly silently by means of a special adaptation of the flight feathers which muffle the normal swish of their wings. Several species are kept and bred in aviculture and have proved adaptable, easy to manage and, often, good breeders.

Bubo africanus (II)
Spotted Eagle Owl

19 in (48 cm). The upper surfaces are mainly grey, heavily mottled with black and white; there are some white spots on the mantle. The underparts are grey-brown with darker barring; the wings and tail are barred. The breast has heavy dark-brown spotting. There are prominent ear tufts. The bill is black, the legs grey-feathered, the feet dark grey. The sexes are similar but the female is usually larger.

Range Senegal and Guinea eastwards across Africa to Chad, Central African Republic, southern Sudan and Ethiopia, Uganda, Kenya, southern Somaliland; south to Angola, South Africa. Also in southern parts of Saudi Arabia.

Status Reasonably abundant in some areas of the range.

Avicultural rating Fairly widely-kept; they are hardy and often good breeders.

Feeding Rats, young rabbits and day-old chicks.

Breeding They usually nest among rocks, and occasionally fallen timber; they will use a 'nest box' made of old bricks or a large open box. Clutch 2; incubation 27–29 days.

General management They are not difficult to manage; acclimatised birds are usually hardy but should not be exposed to excessive damp. An aviary of a type similar to that suggested for Barn Owls will suit, but make it larger if possible: 3.6 × 1.8 × 1.8 m (12 × 6 × 6 ft) is the suggested minimum for this species. Provide seclusion and furnish with only one or two perches; provide peat or gravel in the shelter and flight if the aviary is not on a concrete base.

Athene noctua (II)
Little Owl

8½ in (22 cm). The upper surfaces are dark grey-brown mottled and barred with white; the underparts are grey-white prominently streaked with dark brown. There is a white eyebrow streak. The bill is horn-coloured, the legs grey-white feathered, the feet grey. The sexes are alike.

Range Europe and North Africa eastwards to Manchuria.

Status Abundant in suitable areas of habitat throughout the range.

Avicultural rating A popular avicultural subject; easily managed and a good breeder.

Feeding Mice and day-old chicks; individual birds may accept large insects such as locusts although they are usually ignored.

Breeding Provide a hollow log or a suitable nest box; the site can vary from on the floor to several feet above the ground, but choose a secluded position. Clutch 3–5; incubation 27–28 days. The young are often very nervous after fledging.

General management Provide an enclosed aviary if possible with only one end open for observation; roosting and hiding boxes are necessary for this species.

Ciccaba woodfordii [PLATE 76] (II)
Woodford's Owl, African Wood Owl

PLATE 76 *Ciccaba woodfordii*, Woodford's Owl

13 in (33 cm). The upper surfaces are chocolate brown (the shade is often variable between birds from different areas) with white spots on the wing coverts and scapulars. The underparts are dark brown barred and mottled with white; there is usually a conspicuous white mark over the eyes. The bill, legs and feet are yellow. The sexes are alike.

Range Sierra Leone south to northern Angola; then Central African Republic eastwards to Sudan and Ethiopia and southwards to Kenya, Tanzania, southern Zaire to the Cape.

Status A familiar resident where suitable wooded habitat occurs.

Avicultural rating A good avicultural subject, although rarely available. It is hardy after acclimatisation and easily managed.

Feeding Large locusts, crickets and beetles are important items. Small dead rodents (mice, young rats) and dead day-old chicks can also be fed; dust the exposed flesh of food with a vitamin supplement.

Breeding They are infrequently bred. They will use a box or hollow log. Clutch 1; incubation about 28 days. Supply insects and pinky mice for rearing.

General management This species prefers secluded accommodation and does not appreciate prolonged exposure to damp or wet conditions; some natural cover in and around the aviary is appreciated. The bird likes to roost (in daylight hours) on a high perch out of direct light and screened by plant growth. Provide a comfortable shelter with block and perch. It can be noisy.

APODIFORMES

Trochilidae (Hummingbirds)

There are more than 300 species of hummingbird with specifically New World distribution. Most are tropical but others live in the temperate zone and one species breeds in south-eastern Alaska. They are so named because of the speed of their wingbeats which cause a high-pitched 'hum' in flight. They range in size from Princess Helena's Hummingbird from Cuba, which is 2½ in (6 cm) in length to the Giant Hummingbird of western South America which measures 9 in (23 cm) in length. Most have iridescent plumage; several have ornate feather decorations such as crests, frills, ruffs, gorgets, long tail feathers, racquet-tails etc. They are generally extremely pugnacious to each other, but with their human owners are confident to the point of being fearless. They were once regarded as difficult to maintain in good health for long periods; but modern, high-protein diets have eliminated many of the problems and with good husbandry and correct environment much greater longevity is now being achieved regularly. Most species are insectivorous and rely on *Drosophila* (fruit-flies) for additional protein intake; some species are especially insectivorous.

They can be housed in roomy cages, but do best in a spacious flight or tropical house environment. Many are surprisingly hardy, and access to outside flights is recommended when the weather is suitable; during the summer or warm spells many will spend most of the daylight hours outside if allowed. Long hours of darkness seem to have a debilitating effect on hummingbirds; so it is strongly recommended that a time switch and dimmer facility is installed in the accommodation to extend the hours of 'daylight' between late autumn and early spring where necessary. Mixed groups or groups of the same species must have adequate space, or constant fighting will result in losses. Hummingbirds are the only birds to undergo regular periods of torpidity when their body temperature is reduced and the birds become inactive; this occurs when roosting, resulting in energy-saving. When overcrowded or in a poor environment, torpidity may be an indication of stress-related problems caused through persecution by dominant rivals. Abundant feeding stations may help reduce this problem, but not if the population is too dense in the space available.

Colibri coruscans [PLATE 77]
Sparkling Violet-eared Hummingbird, Gould's Violet-eared Hummingbird

5 in (13 cm). The plumage is mainly shimmering, iridescent green; the ear-coverts, chin and a patch in the centre of the breast are metallic violet-blue. The tail is green with a subterminal dark blue band. The bill, legs and feet are black. The sexes are alike.

Range Colombia and Venezuela south to western Bolivia, north-west Argentina; it also extends into the Guianas.

Status Abundant in many parts of the range.

Avicultural rating One of the most widely-kept members of the family. It is easily managed, and has bred in aviaries on a number of occasions.

Feeding Diet L, with regular supplies of *Drosophila*; nectar mixtures must always be replaced if

PLATE 77 *Colibri coruscans*, Sparkling Violet-eared Hummingbird

cultures to prevent too close investigation by birds. Hummingbirds feed only on the wing when fit and in good feather condition; supply nectar in tubes with red-tipped spouts. Suspend these so that the birds can hover without being impeded by nearby obstacles. In addition to the suggested Diet L, there are now some excellent complete nectar foods on the market (see p. 28).

Breeding The Sparkling Violet-ear is one of a limited number of species being bred fairly regularly by experienced European and American aviculturists whose efforts are to be applauded. In warm-climate countries captive-breeding programmes with a wide range of species have achieved success. One of the world's leading hummingbird experts, Dr Augusto Ruschi, has bred more than 60 species in his aviaries at Espirito Santo, Brazil. They do best in dual indoor and outdoor accommodation (or a tropical house) and without competition from other hummingbirds. Sexing is a major problem with this and other species. The hen builds a small nest, usually choosing fine plant fibre, cobwebs, filaments of wool etc. In the wild, the nest is usually fixed securely in a fork; captive birds may construct (or attempt to) in a variety of locations, but usually the nest is made among vegetation. The male sometimes excites or distracts the female intent on building, laying and incubating, but only remove him as a last resort. Clutch 2; incubation 15–16 days. Provide a constant supply of *Drosophila* from the time the chicks hatch. Watch for aggression from the adult male as the chicks fledge; if he is still with the female he is likely to persecute the young very quickly, so remove the chicks as soon as they are independent, or remove the cock until the young are ready to be moved to their own accommodation.

there are signs of deterioration (through excessive heat etc) and the birds should never be without food during daylight hours for longer than it takes to replace the feeding tubes. A plain sugar-water mixture is adequate late in the day when the birds will soon settle to roost and expend little energy; it is important that a high-protein diet is available to them again at first light when the day's activity begins. There is some evidence to suggest that the use of honey in nectar mixtures gives rise to fungal conditions affecting the tongue etc. A supply of fruit flies must be maintained; it is best if cultures (easily established in suitable-sized plastic containers using over-ripe bananas etc) are rotated frequently. Keep a fine mesh over the

General management Hummingbirds are undoubtedly at their best in a tropical house setting against a background of tropical and sub-tropical plants, but they also thrive in shelter and flight accommodation; it is essential that the shelter has a heat source and that the birds can be totally enclosed in a well-lit environment during severe (cold) weather. They also live well in spacious indoor flights; but less well in even the most roomy cages. If using the latter, there should be only one bird per cage. Most hummingbirds enjoy rain, spray or leaf bathing. Although they are combative among themselves, hummers are good companions for zosterops, small tanagers etc. Remember the need to lengthen short winter days which is of paramount importance. Birds which have become torpid often revive in the warmth of a cupped hand; the process will be accelerated by allowing the bird to take food from a tube. Release them into normal quarters as soon as the condition has passed. Frequent or regular torpidity is a sign of health problems or inadequate food intake. Ensure that food is always available and that there is a maximum of eight hours of darkness in their quarters.

Popelairia popelairii
Wire-crested Thorntail Hummingbird

$4\frac{1}{2}$ in (11 cm). The crown, crest, chin and throat are glittering green, the upper surfaces bronze-green with a white band across the rump. The breast is black, the lower underparts grey-brown; the sides of the neck and the flanks are coppery-green. The tail, which is $2\frac{1}{2}$ in (6 cm) long, is dark blue with the undersides of the elongated shafts white. The bill, legs and feet are black. The female lacks the long wire-like crest and extended tail feathers; she has coppery-green upper surfaces, and is grey-black below. The band on the rump and patches on the flanks are white.

Range Eastern areas of Colombia, Ecuador and Peru.

Status There are good numbers in areas of suitable habitat, but the species is not abundant.

Avicultural rating Rarely available and regarded as one of the more 'difficult' species.

Feeding/Breeding/General management Basically similar to the Sparkling Violet-eared Hummingbird, but more insectivorous. They were formerly fairly short-lived in aviculture but modern diets have done much to simplify their management and improve prospects of longevity with this and other rare and delicate hummingbirds.

Thalurania furcata
Fork-tailed Woodnymph Hummingbird

$4\frac{1}{4}$ in (11 cm). The crown is an iridescent dark green; the rest of the upper surfaces are bronze-green with a band of purple-blue across the lower back. The secondaries and flights are brown. The chin and throat are glittering green, the lower underparts violet-blue. The thighs are white, the forked tail dark blue. The bill, legs and feet are black. The female is bronze-green above, her under surfaces grey-white.

Range The Guianas, north-eastern Brazil.

Note There are 23 subspecies described, exhibiting a number of plumage differences. Between them they range from Mexico, through Central America and over most of the sub-continent as far south as Bolivia, Paraguay and south-eastern Brazil.

Status Widespread and abundant in many parts of the range.

Avicultural rating Only occasionally available, but a popular avicultural species. They are hardy and easily managed.

Feeding/Breeding/General management Similar to Sparkling Violet-eared Hummingbird. These and allied species will bathe in a shallow dish of water, splashing along the surface like a ditching helicopter. They are pugnacious and not trustworthy with other small species in a restricted area.

Amazilia tzacatl
Rufous-tailed Emerald Hummingbird, Rufous-tailed Amazilia

$3\frac{3}{4}$ in (9.5 cm). The upper surfaces are bronze-green, the throat and breast a glittering emerald green. The upper tail coverts and tail are chestnut, the abdomen grey. The bill is red, black towards the tip; the legs and feet are black. The female is similar but duller.

Range Mexico south through Central America to western Colombia and Ecuador, western Venezuela.

Status Abundant in many areas.

Avicultural rating A popular and widely-kept species which is easily managed and fairly hardy.

Feeding/Breeding/General management Very similar to Sparkling Violet-eared Hummingbird. In outdoor aviaries, they are particularly adept at catching small flying insects. Like the Sparkling Violet-ear, they are usually long-lived when properly housed and fed.

COLIIFORMES

Coliidae (Mousebirds)

These are aptly-named birds that 'creep' along tree branches or through vegetation. All 6 species are found in Africa; they are well known in aviculture although not widely kept. Among their peculiarities is the fact that all four toes can be turned forward; their very sharp claws provide a good grip on tree-bark etc. Colies habitually sleep with head up and tail down, suspended like inverted bats. All are crested and have long tails. They are essentially frugivorous and destroyed as crop pests in fruit-growing areas of the range. Generally good avicultural subjects, they are sociable and good mixers with other species.

Colius indicus
Red-faced Mousebird, Red-faced Coly

13 in (33 cm). The plumage is entirely grey-green, more fawn on the under surfaces. An area of bare skin on the face is red. The base of the upper mandible is red, the rest of the bill black; the legs and feet are dark red-brown. The sexes are alike.

Range Southern Tanzania southwards through Malawi, Zambia to South Africa.

Status Abundant in some areas, but numbers have been reduced in areas of cultivation.

Avicultural rating Interesting and easily managed birds.

Feeding Diet E, with a small quantity of Diet G mixed with fruit. Some live food (mealworms, crickets etc) should be provided each day. Colies are known to steal and eat other birds' nestlings and it is probable that other items of animal matter as well as occasional small quantities of raw, minced beef would be welcomed.

Breeding They are fairly willing breeders in the correct environment. They build a loose cup-shaped nest of grass, plant stems, etc on a platform base of twigs; the nest is usually lined if materials are available. Clutch 2–4; incubation 13–14 days. The parents regurgitate food for the chicks; provide a very varied fruit diet plus berries and buds, if available. Live food is also important when chicks are present.

General management They fare best in small groups in a tropical house or outdoor aviaries with access to a comfortable shelter with a heat source in severe (cold) weather. They can be housed with other similar-sized birds but it is best if they are not put with frugivorous species because of the competition for food items. If in outdoor aviaries, watch their habit of sleeping attached to the wire-netting, and inviting attack by predators; old bark-covered logs, placed vertically, in the sleeping quarters may persuade the birds to roost under cover. They are not suitable birds for close confinement.

Colius macrourus
Blue-naped Mousebird, Blue-naped Coly

14 in (36 cm). This bird is very similar to the Red-faced Mousebird except that it is slightly larger and has a distinctive turquoise-blue patch on the nape. The sexes are alike.

Range Senegal eastwards to southern Sudan and Ethiopia, western Uganda, northern Tanzania; Somalia.

Status Abundant in many parts of the range; reduced in fruit-growing areas.

Avicultural rating See Red-faced Mousebird.

Feeding/Breeding/General management Similar to the Red-faced Mousebird. They are very gregarious, with a fairly loud but musical call.

TROGONIFORMES

Trogonidae (Trogons)

Despite only 30-plus species the distribution of this family is wide: the New World, Africa and Asia. The best-known member is the Quetzal of Central America; it has been described as the most beautiful bird in the world. Few members of the family are now available to aviculturists, but the Quetzal, despite being a protected species, continues to appear from time to time (presumably smuggled) on the market. The following notes do not indicate any form of support for this illicit trade in birds, but are rather a means of helping prevent the premature death of birds in wealthy but inexperienced hands. The Quetzal is Guatemala's national bird.

Pharomacrus mocinno [PLATE 78] (I)
Resplendent Quetzal

48 in (122 cm) including 36-in (91-cm) tail feathers. The upper surfaces (including a ridged crest), chin, throat and upper breast are iridescent green, the lower breast, abdomen and under tail coverts scarlet. The coverts are golden-green, the concealed flights black. The upper tail coverts are green-gold with a purple-blue gloss. The central tail feathers are elongated and bronze-green. The bill is yellow, the legs and feet dark grey. The female is generally duller and with a shorter tail; most of her under surfaces are grey-brown but there is a band of dull green across the breast and her abdomen is dull pink.

Range Southern Mexico south through Guatemala and Costa Rica to Western Panama.

PLATE 78 *Pharomacrus mocinno*, Resplendent Quetzal

Status Nowhere abundant within the limited range.

Avicultural rating One of the most coveted and sought-after of all exotic species. Unfortunately it has a poor record in zoos and private collections. There is little hope of establishing captive-breeding strains at present, but if a legitimate opportunity of acquiring potential breeding stock arises in future aviculturists must look again at the problems confronting them and try to find means of maintaining and breeding them in a suitable captive environment.

Feeding Diet E provides the basis, but ensure that whole grapes are hollowed and filled with a high-protein insectile diet. Quetzals have a capacious gape, and are capable of swallowing whole small avocados in their wild haunts; quite large stoned fruits are clearly within their capacity. Probably one of the biggest difficulties is ensuring substantial protein intake; 'spiked' grapes may be a partial answer despite the time-consuming preparation needed.

Breeding There are few prospects of success at present. They need a tropical house environment, but moderate, rather than high, temperatures are best. Decayed tree trunks provide a potential nest site; if the decay is far enough advanced the bird excavates a hollow or otherwise occupies an existing cavity. The nest hole is usually at a considerable height—15–60 ft (4.5–18 m). Clutch 2; incubation 17–18 days.

General management A cool environment with high humidity is best. Provide spacious aviaries, although the birds are fairly sedentary in habit. In the wild, the birds take fruit off trees and shrubs while in flight.

CORACIIFORMES

Alcedinidae (Kingfishers)

The kingfisher family contains more than 90 species with worldwide distribution. Many are brightly coloured; all share the characteristic short neck, compact body and long bill. There are effectively two groups: fish-eaters and insect-eaters. The best known and most widely kept, especially in zoological and bird gardens, is the Kookaburra or Laughing Kingfisher (*Dacelo novaeguineae*) from Australia. Others, however, although only occasionally available, are good avicultural subjects, including the smallest members such as Malachite and Pygmy. They do best in a tropical house type environment, although the Kookaburra is hardy and thrives in outdoor flights throughout the year.

Ispidina picta [PLATE 79]
African Pygmy Kingfisher

$4\frac{1}{2}$ in (11 cm). The crown and most of the upper surfaces are ultramarine-blue; there is some black barring on the crown. The sides of the head and neck are tawny-orange with a lilac suffusion; the chin and throat are white, with the rest of the underparts orange. The bill, legs and feet are orange-red. The sexes are alike.

Range Wide distribution south of the Sahara. The nominate subspecies is found from Senegal eastwards to Ethiopia, Kenya, and in the west through Cameroon, Congo, to Angola. Other subspecies are found from southern Zaire and Tanzania to Natal; Pemba and Zanzibar.

Status Abundant where suitable habitat exists in many parts of the range.

PLATE 79 *Ispidina picta*, African Pygmy Kingfisher

Avicultural rating Rarely available and for experienced aviculturists only; it is a delightful subject when properly housed and when a suitable diet is established.

Feeding It is unlikely that any of the recommended inanimate diets will be of interest to this species; feeding is likely to be a painstaking process in the initial stages at least. Crickets and small locusts are accepted; also mealworms. The birds are less interested in gentles or pinkies, which in any case have little food value. Small crickets die quickly when moistened; larger specimens can be 'painted' with a honey mixture so that a small amount of high-protein insectile food adheres to them. Although this is not a fish-eating species, various food items supplied by pet shops for marine or tropical fish can be offered: freeze-dried shrimps etc, which are high in protein and fibre and low in carbohydrates. It is essential that vitamin and mineral supplements are included in the regular diet; if the birds are taking any form of dry insectile food a powdered supplement can be utilised—alternatively liquid vitamins can be injected into live food (mealworms are probably best) with a fine-needle hypodermic syringe. Eventually use Diet F.

Breeding These birds are a real avicultural challenge. They are tunnel-nesters in the wild and it is possible that peat bales might be utilised; otherwise provide an artificial earth bank. Clutch 3–5; incubation 15–16 days.

General management They are best in a compartment of suitable size in a tropical house environment, so that some control over diet etc can be exercised. A good environment may be a stimulus to feeding; therefore provide perch vantage points amid growing vegetation with food vessels below. Items such as crickets and locusts will be active away from the containers so ensure that the compartment is escape-proof. Although they are mainly insect-hunters, they are frequently found in the wild along small water-courses diving for fish; if a small, deep pool can be provided and water temperature can be maintained (by means of a pyrex electric aquarium heater) at around 18–20°C (65–68°F) the fry of guppies and other quick-breeding tropical freshwater fish could provide valuable additional food. A shallow pool is not recommended because of possible injury to diving birds.

Halcyon senegalensis
Woodland Kingfisher

9 in (23 cm). The upper surfaces are light blue, greyer on the head and neck; the underparts are grey-white with a suffusion of blue-green. The coverts and tips of the primaries are black; the rump and tail are blue-green. The upper mandible is red, the lower mandible black; the legs and feet are black. The sexes are alike.

Range Senegal eastwards to southern Sudan, Ethiopia, Uganda, western and central areas of Kenya. A subspecies, *H. s. cyanoleuca*, which has a slightly more blue head, ranges from Sudan southwards to Transvaal and eastwards into southern Angola.

Status Abundant in some parts of the range where suitable conditions exist.

Avicultural rating Only occasionally imported and not widely kept.

Feeding Not a fishing species; in the wild it captures insects, amphibians etc. Newly-imported birds are often difficult to accustom to an inanimate diet, but they quickly learn to pick up strips of raw meat, heart, chicken flesh; a small quantity of insectivorous mixture (Diet F) can be 'dusted' lightly on raw meat or heart strips, but the birds will probably refuse it if overdone. Some live food is essential: large locusts etc. The birds will also take any available dead frogs, newts or lizards, and may occasionally capture and kill mice. Regular use of a vitamin supplement is essential.

Breeding Difficult. They are hole-nesters, usually in hollow trees, termite nests or a river bank; they have used a box. Clutch 3–5; incubation 13 days. A high-protein rearing diet is important.

General management Established birds thrive in a suitable aviary or tropical house environment. They are not completely hardy, and are noisy at times. They are best not associated with smaller or weaker species. They have a swift, direct flight; and self-inflicted injury is possible in limited-space accommodation through collisions with wire etc.

Dacelo novaeguineae [PLATE 80]
Kookaburra

PLATE 80 *Dacelo novaeguineae*, Kookaburra

18 in (45 cm). The forehead, crown and a prominent eye-streak are dark brown; the back and wings are brown with pale blue spots on the shoulder. The tail is rufous, with a barred black and white terminal band. The sides of the head, neck and under surfaces are creamy-white; some pale blue-green feathers are present in the rump. The bill is long and heavy, the upper mandible black, the lower mandible white; the legs and feet are grey. The sexes are similar.

Range Australia and Tasmania.

Status Variable; abundant in some areas, much reduced elsewhere. It has been introduced into many areas from the original limited range.

Avicultural rating Kept by few aviculturists and only occasionally available, it is hardy, long-lived and a good breeder.

Feeding The diet in aviaries consists mainly of cut-up dead rats, mice and some minced raw meat; they will also consume casualty amphibians and reptiles including small snakes. Kookaburras normally show little enthusiasm for insectivorous mixtures but a small quantity of Diet G can be mixed with raw meat. A vitamin supplement is an important regular additive.

Breeding One of the main obstacles to successful breeding is the acquisition of true pairs, as the sexes are alike. The birds will utilise a box with a tunnel entrance; alternatively a long (91–122 cm; 36–48 in) box placed on its side and tilted so that the entrance hole is higher than the potential nest chamber will often be used. Clutch 2–4; incubation 25–26 days. Parents sometimes kill and eat the nestlings; if in doubt take the youngsters for hand-rearing. If hand-rearing, it is best to separate the chicks even at a few days old to prevent them injuring each other with their sharp bills. Feed on finely minced mice, raw chicken flesh and small live food; skinned food should be given during the first week then, very gradually, add roughage in the form of fur etc to encourage the production of pellets.

General management They fare best in roomy shelter and flight accommodation. They are hardy but do best in a shelter during prolonged severe (cold) weather. They much enjoy bathing, so provide an ample-sized pool. They are not suitable companions for less robust species.

Momotidae (Motmots)

This is a family of nine species, all from Mexico, Central and South America. The most interesting feature in some are the subterminal bare shafts of the two central tail feathers, caused inadvertently by the bird preening; sometimes the feather breaks off at the tip, in other cases vanes are rubbed away to leave a bare quill, giving rise to the alternative name of 'barber-bird'. If excited or upset, the long tail is swung back and forth like a pendulum; at rest the bird often discloses its presence among branches by flicking its tail. They are mainly forest-dwellers and tunnel-nesters.

Momotus momota [PLATE 81]
Blue-crowned Motmot

17 in (43 cm). The forehead and sides of the head are iridescent turquoise-blue, with black in the centre of the crown; there is a conspicuous black stripe through the eyes and a black spot in the centre of the breast. The remainder of the plumage is mainly green, lighter on the underparts, except for chestnut on the neck, and some light blue on the throat. The long tail is green at the base, becoming blue; there is a dark tip after the bare shaft. The bill, legs and feet are dark grey. The sexes are alike.

Range The nominate subspecies has a range extending from eastern Venezuela, and the Guianas, to northern Brazil.

Note A total of 22 subspecies are described and many exhibit marked differences in plumage patterns. They share an extensive range from northern and central Mexico throughout much of the South American sub-continent as far south as north-west Argentina and southern Brazil.

Status Still abundant in areas of suitable habitat, but they are casualties of development and deforestation programmes in many parts of the range.

Feeding Diet F, with some finely-minced raw meat, pinky mice and abundant live food such as locusts, crickets etc. They also take various small invertebrates, including snails.

Breeding With proper facilities, the species would probably have proved a more frequent breeder than in the past. They are tunnel-nesters and will create a suitable chamber months before the breeding season, and use it as a dormitory. Provide an earth bank; alternatively use bales of peat, but the latter tend to dry very quickly unless moistened with a fine spray from time to time. Clutch 2–3; incubation 21 days. Provide a high-protein menu including plenty of small live food; gentles and mealworms will *not* provide the basis of a suitable rearing diet, so use natural items (smooth caterpillars, grasshoppers, beetles, small snails, woodlice etc) to supplement crickets, locusts, pinky mice etc.

General management They do best in a tropical house type environment; they are not hardy, and

PLATE 81 *Momotus momota*, Blue-crowned Motmot

although happy outdoors during late spring, summer and early autumn, they will show clear signs of distress in cold or prolonged cool, wet conditions. If a large tropical house is available provide a sectioned-off area for motmots; they are not safe for communal liberty groupings which include smaller species. Although not especially aggressive, their habits indicate a willingness to capture and consume small bird species as well as invertebrates. They have a loud call, and are often difficult and delicate when first received from quarantine due to the wrong (frugivorous) diet.

Coraciidae (Rollers)

The Coraciidae is a family of 11 species, which ranges from Europe eastwards to the Pacific and Australia; the greatest concentration of species is in Africa. All are stoutly built with bright colours, and some have long tail streamers. Mainly insect-eaters, they also pick up odd small birds and rodents. They are high flyers with the ability to perform spectacular aerobatics in both their courtship and threat displays.

Coracias benghalensis
Indian Roller

13 in (33 cm). The forehead and crown are pale blue; the mantle, back and wings are light brown, the shoulder of the wings and the outer webs of the flights are light or dark blue. The sides of the head and neck are vinous; the underparts vinous-mauve. The lower under surfaces are green-blue. The bill is black, the legs and feet brown. The sexes are alike.

Range Parts of Saudi Arabia eastwards to India, Bhutan and south-east through Thailand to southern Malaysia, parts of southern China; Sri Lanka.

Status Abundant in many parts of the range.

Avicultural rating They have a tendency towards somewhat lethargic behaviour; often their only movement is an occasional trip from perch to food-dish, and this plus their frequent aggression towards each other does not endear them to many aviculturists. Although difficult when first received, they are relatively straightforward when established. They are hardy and long-lived.

Feeding Eventually Diet G, with minced raw meat, supplemented by abundant live food. Newly imported birds often refuse most items other than live food and 'meating-off' may be a protracted affair. Once established they present few problems. A vitamin supplement should be added regularly to the basic diet. Some birds enjoy minced raw fish.

Breeding This is difficult in aviaries, but the species bred successfully at Blackpool Zoo in England in the 1980s. In the wild, the birds utilise tree holes and other lofty cavities, and might use an open-fronted box. Clutch 4–5; incubation 19–20 days. The task of providing sufficient live food is a daunting prospect.

General management Lofty aviaries are ideal for these birds; it is impossible to provide sufficient space for the typical nuptial flight but they must have roomy quarters. They are hardy, although not frost-proof, and they do not enjoy periods of cold or wet weather. They are not trustworthy with companions, and have loud calls.

Eurystomus orientalis
Dollarbird, Eastern Broad-billed Roller

12 in (30 cm). The plumage is mainly slate-brown with a suffusion of turquoise-blue; the underparts dull sea-green; the chin and throat are bright blue. There is some blue-green on the mantle and a pale-blue patch at the base of the primaries, which appears silvery in flight, giving rise to the popular name. The bill, legs and feet are red. The sexes are alike.

Range A total of 12 subspecies have a wide range in south-east Asia, the Pacific and Australasia including Nepal south-east to Thailand, Indochina,

Malaysia, Sumatra, Borneo, Java, New Guinea, Australia; also the Celebes, Andaman, Simalur, Lombok, Damar, Moluccas, Kei, Feni and Solomon Islands; Korea, north-east China, Japan.

Status Abundant in many areas.

Avicultural rating Interesting, but not a popular avicultural subject.

Feeding/Breeding/General management Similar to Indian Rollers. They are mainly crepuscular in habit. Breeding is unlikely but clutch 3–5; incubation 18–23 days (according to subspecies). They are mainly hole-nesters, and also use termite mounds. Wooden nest boxes are used by urban birds in some parts of the range in the Far East.

Upupidae (Hoopoes)

There is only one species of Hoopoe, but 10 subspecies are described, with an enormous range: Europe, the whole of Africa (except desert and rain forest), the Middle East to South-east Asia etc. They are very exotic-looking with an erectile fan-shaped crest and distinctive plumage. They are not difficult to maintain in aviaries, although they were formerly regarded as among the more delicate species. It has been said that the Hoopoe has the foulest nest in the bird world: a combination of rotting, uneaten food, an accumulation of droppings and a strange odour given off by the incubating female are the reason for this unusual accolade.

Upupa epops
Thailand Hoopoe

There is some difficulty in establishing which subspecies emanates from Thailand, as northern populations migrate south for the winter; but the birds from this area are likely to be either *U. e. orientalis*, *U. e. saturata* or *U. e. longirostris*. 11 in (28 cm). This bird has mainly salmon-pink plumage and a distinctive black-tipped crest. The wings and tail are banded with black and cream-white; the lower under surfaces are grey-white with a white rump. The bill, which is long and curved, is grey-black; the legs and feet are grey. The sexes are alike.

Range The three subspecies mentioned range over north-west India and Assam, eastern Siberia and north-east China southwards to Indochina, Malaysia and Sumatra.

Status Adaptable and therefore still fairly abundant in many parts of the species' wide range.

Avicultural rating Only rarely available. The Hoopoe has proved easily managed and long-lived once the initial diet problems have been overcome and if a good aviary or tropical house environment is provided. It has bred on several occasions in both public and private collections.

Feeding Initially live food will be the main item of diet; but the birds adapt quickly and soon pick up fortified insectivorous mixture (Diet G) and finely-minced raw meat. They enjoy using their long bill to probe into soft earth; it is well worth having a soil-box made for them which is about 9 × 9 × 4 in (23 × 23 × 10 cm). Fill it with soft earth (one of the peat-based composts from a garden centre is ideal), keep it moist and stocked up with cleaned earthworms and other damp-resistant live food (mealworms succumb quickly in a moist environment). A vitamin supplement should be given to the birds regularly. Food should be of a coarse consistency to assist their feeding technique.

Breeding Sexing difficulty often inhibits breeding, unless a small group is obtained allowing self-selection pairing. They are hole-nesters and will use a box. In China the birds are called 'Coffin Birds' because of their opportunist habit of entering holes in exposed coffins! Supply damp peat in the base of a box or a hollow log. Clutch 4–8; incubation 17–19 days. Very small live food is essential for rearing whilst the nestlings are very young; tiny crickets are ideal if they can be prevented from escaping until utilised by the parents. Eventually give larger items such as locusts and grasshoppers.

General management They do best in spacious accommodation such as an aviary or tropical house. They are fairly hardy (but not completely), so a slightly-heated shelter is needed if they are kept outdoors throughout the year. They do well in a temperate bird house and, as with other insectivorous species from the tropics, extra hours of artificial light in winter are recommended. They enjoy dust-bathing. They feed after the manner of toucans, picking up a morsel of food in the tips of their mandibles, and then throwing the head back to allow the item to fall into the throat.

Bucerotidae (Hornbills)

The Bucerotidae contains nearly 50 species which are distributed throughout the Old World tropics, but not Australia or Madagascar. They range from crow to turkey size; and have mainly sombre plumage with occasional colour in their casques or, in a few species, wattles. There is a very strong pair bond and in some species the female is walled up in a nest cavity from the time of laying to the young fledging—sometimes as long as three months. It is mainly the small species that are available to aviculturists. Most are easily managed and fairly hardy, although they are extremely susceptible to frostbite. There have been some breeding successes in aviaries, particularly with the small species, although the bigger representatives of the family also reproduce from time to time. They need spacious accommodation, especially the very large types which are probably beyond the scope of most private aviculturists. Their dietary requirements range from fruit to meat and insectivorous food.

Tockus erythrorhynchus [PLATE 82]
Red-billed Hornbill

17 in (43 cm). The upper surfaces are blackish with a long white stripe down the centre of the back and abundant white spots on the wing coverts; the under surfaces are grey-white. There is a white stripe above the eye. The central tail feathers are black, the outer feathers black and white. There are white bars on the wing. The bill is red; the legs and feet grey. The adult male has a black base to the lower mandible which is often absent in the female.

Range Senegal and northern Nigeria eastwards across Africa to southern Sudan, Ethiopia, Uganda, Kenya and Tanzania.

Status Abundant in many areas.

Avicultural rating Occasionally available. They fare best in experienced hands; they need a spacious aviary with frost-free sleeping quarters, and some heat is essential until they are properly acclimatised.

Feeding Basically Diet G with some minced raw meat and sectioned dead rats or mice; also provide a variety of diced fruit, small berries, and some live food. Bread in sweetened milk is a good occasional item, but the main diet must be based on insectile food, meat and live food.

Breeding True pairs will breed in a good aviary environment. Provide a box or hollow log fairly high up. If using a box, ensure that the aperture is small enough to be plastered over by the nesting pair; mud etc is used to enclose the female when the clutch is complete. In the wild, the female remains walled up for about 12 weeks; after she has broken out, the nestlings assist in re-building the barrier with mud brought by their parents, and remain in the nest chamber for a further 3–4 weeks. Clutch 3–6; incubation 29–30 days. Plenty of live food (crickets, locusts, some mealworms, pinky mice) is needed for rearing.

General management Not difficult when established. They can be housed in outdoor aviaries throughout the year, but they do not appreciate prolonged cold, wet conditions; they are also very susceptible to frostbite, and their feet are easily damaged if the birds are allowed out in frosty conditions. They can be housed with other medium-

PLATE 82 *Tockus erythrorhynchus*, Red-billed Hornbill

sized birds, but it is best not to put them with smaller species. Their calls are fairly loud and occasionally monotonous.

Tockus flavirostris
Yellow-billed Hornbill

18–20 in (46–51 cm). This species is similar in appearance to the Red-billed Hornbill, but has a slightly bigger, yellow bill and a broad subterminal black band on the outer tail feathers. The sexes are similar but the female is often smaller and with a less massive bill.

Range Ethiopia and western Somalia to southern Kenya.

Status Variable, but abundant in a few areas.

Avicultural rating See Red-billed Hornbill.

Feeding/Breeding/General management Similar to the Red-billed Hornbill, but it is less often available.

PICIFORMES

Capitonidae (Barbets)

This is a large family of some 80 species. The heaviest concentration is in Africa but distribution is worldwide except for the polar regions, Antilles, Australasia and Madagascar. They have heavy bodies and stout bills. They feed mainly on fruit and berries; but some are more insectivorous. The term 'barbet' is derived from the French *barbu*, meaning 'bearded', and refers to the characteristic bristles around the base of the bill. Many species are gregarious. They have distinctive calls.

Eubucco bourcierii [PLATE 83]
Red-headed Barbet

PLATE 83 *Eubucco bourcierii*, Red-headed Barbet

6½ in (17 cm). The whole of the head, throat and upper breast are scarlet; the rest of the upper surfaces are green, apart from a narrow band of pale blue on neck. The underparts are orange fading to yellow on the lower breast; the flanks are yellowish streaked with green and white. The bill is yellow-green, the legs and feet grey-green. The female has most of her upper surfaces green; the cheeks are blue, the throat grey. The forehead is black, the forecrown golden-yellow; the nape is olive. A band across the breast is yellow, the lower breast is olive with a tinge of yellow; the abdomen and flanks are streaked dull yellow-green.

Range Central and eastern Colombia. Apart from the nominate, a further five subspecies are described; these range from Costa Rica and Panama to west, central and eastern Colombia and western and eastern areas of Ecuador.

Status Local distribution throughout most of the range; there are good populations in some districts, but they are not abundant.

Avicultural rating A very popular little barbet which is occasionally offered and unusual in that, apart from the vivid scarlet head and upper breast, the female is perhaps even more handsome than the male. It is not difficult to keep but does not tolerate prolonged exposure to cold or damp conditions. Some heat is required in winter. It is an ideal tropical house resident.

Feeding Diet E plus Diet F and some live food. In the wild these attractive little barbets are fairly insectivorous and should have a regular supply of small insects in addition to their fruit and insectile mixture staple food. In season various berries are much enjoyed.

Breeding Like others of the family this species

nests in cavities, usually a tree-hole. Provide a box or hollow log. Clutch 2–4; incubation 14–15 days. Both male and female share the incubation and rearing; provide abundant small live food, such as small crickets or locusts, wax moth larvae, smooth caterpillars etc when nestlings are present.

General management Excellent in a tropical house environment, they are not hardy and do best in outdoor aviaries only during the summer months. Ideally they should be housed in a well-planted environment. Provide a nest-box or hollow log for roosting.

Semnornis ramphastinus
Toucan Barbet

8½ in (22 cm). The face, forehead, crown and nape are glossy-black, the cheeks, throat, breast and sides of the neck blue-grey. There is a white stripe on the sides of the head above and behind the eye; and a black vertical stripe on the sides of the neck. The mantle, back, rump and lower underparts are brownish-olive; there is a broad band of scarlet across the lower breast. The bill is yellow-green with a black tip, the legs and feet grey-green. The female is similar.

Range Western Colombia, western and central Ecuador.

Status Local distribution; nowhere are they abundant. Deforestation in some areas is reducing the population.

Avicultural rating Very rarely available, but a much-prized occupant of tropical house accommodation. They fare best in experienced hands.

Feeding/Breeding/General management Clutch 2–4; incubation probably 15–16 days. Abundant live food is needed when the nestlings are small but eventually their parents will start to feed them on fruit. Larger insects such as crickets and small locusts together with pinky mice should also be provided as the chicks grow. Not suitable companions for other smaller occupants of a tropical house, therefore they are best kept in a compartment within a suitable environment. There is usually a strong bond between established pairs.

Psilopogon pyrolophus [PLATE 84]
Fire-tufted Barbet

PLATE 84 *Psilopogon pyrolophus*, Fire-tufted Barbet

11 in (28 cm). The upper surfaces are green, the crown, eye-stripe and nape maroon-red. The cheeks are blue-grey, the chin and most of the underparts green. The throat is yellow with a narrow black band below; there are conspicuous black and red bristles at the base of the bill. The bill is yellow-green with a black band on the upper and lower mandibles; the legs and feet are grey-green. The sexes are similar except that the maroon-red markings on the head and nape are duller in the female.

Range Malaysia, Sumatra.

Status Abundant in many parts of the range but loss of habitat is now reducing numbers in some areas.

Avicultural rating These are large and easily-managed barbets which thrive in a tropical house environment, but eventually become hardy and can be housed outdoors. Slight heat is best in the winter; do not allow them out in severe (cold) weather because of the danger of frostbitten feet.

Feeding Use a mix of Diets E and G, with the addition of berries in season. They also enjoy live food, including an occasional small, dead mouse.

Breeding Provide a hollow log or box. Clutch 2–3; incubation 19–20 days. Abundant live food is needed when rearing, such as crickets, locusts, mealworms, pinky mice. The young leave the nest after 7–8 weeks. There is a strong pair bond.

General management These birds are easily-managed and ultimately hardy, but are not good companions for smaller or weaker birds, and non-pairs are often aggressive between themselves. If not kept in a tropical house, they do best in a shelter and flight which is spacious, and with the facility to enclose them within a well-lit shelter during bad weather. They are too big for over-wintering in cage accommodation. Provide a box or log for roosting.

Megalaima asiatica
Blue-throated Barbet, Blue-cheeked Barbet

9 in (23 cm). The forehead and crown are red with a black band across the forecrown; the cheeks, chin and throat are blue. The upper surfaces are grass-green, the underparts yellowish-green. There are crimson spots at the base of the bill and on the sides of the neck. The bill is grey-green, the legs and feet greenish. The sexes are alike.

Range Northern India, Assam, Burma eastwards to southern China, Thailand; northern Borneo.

Status Abundant in many areas of the range.

Avicultural rating A popular avicultural subject which is now available only infrequently.

Feeding Use a mix of Diets E and G, with the addition of berries in season. Also provide live food. These and similar-sized barbets often prefer certain items of fruit (apple, pear etc) supplied in large sections rather than diced. Individuals enjoy occasional raw minced beef.

Breeding These birds are hole-nesters. They may use a box, but a hollow log is more likely. Clutch 3; incubation 15–17 days. The nestling period is up to 8 weeks long. Provide abundant live food for breeding birds.

General management Newly imported birds are often in poor plumage and need care to acclimatise, but eventually become hardy. They are excellent in a compartment in a tropical house; established birds thrive in outdoor planted aviaries. Provide shelter and flight accommodation with a means of enclosing the birds within a well-lit shelter during severe (cold) weather. They can be over-wintered in a spacious cage indoors. Hand-reared examples are usually very tame; wild-caught adults are often nervous, and need patient, skilled handling. They are great bathers and generally noisy, cheerful birds, but their calls can be monotonous.

Megalaima haemacephala
Coppersmith Barbet, Crimson-breasted Barbet

6 in (15 cm). The forehead and a band across the breast are red; the area above and below the eye, the chin and throat are yellow. The upper surfaces are dark olive-green; the underparts are grey-white with conspicuous dark-green vertical streaking. There is an area of sooty-black from the base of the bill through the eyes, crown, hind-cheeks and sides of the neck. The bill is black, the legs and feet red. The sexes are alike.

Range Western Pakistan, north-west India eastwards to south-west China, Thailand, Malaysia, Sumatra, Java, Bali and Philippines. The nominate subspecies, *M. h. haemacephala*, is found only on the islands of Luzon, Mindoro, Mindanao, Samar and Leyte.

Status Abundant in many parts of the range.

Avicultural rating Now rarely available. This species is always regarded as one of the more delicate barbets until well established.

Feeding Use a mix of Diets E and G with berries in season and live food. Wild birds feed mainly on various wild figs.

Breeding This is a challenge. They would probably use a hollow log, or a section of rotten tree trunk which the birds can excavate. Clutch 3–4; incubation 13–14 days.

General management They need care when newly imported, and are best in experienced hands. They are ideal in a tropical house environment; keep in an outdoor aviary during the summer months only, as they are not winter-hardy. Provide a hollow log or a box for roosting. They have a monotonous, oft-repeated metallic call from which the popular name is derived. They often lose some colour after successive moults in cage or aviary.

Lybius torquatus [PLATE 85]
Black-collared Barbet

---

6 in (15 cm). The face, cheeks, chin and throat are red, the crown, nape and chest-band black. The upper surfaces are grey-brown, the secondaries edged with yellow. The lower underparts are yellow. The bill is black, the legs and feet dark-grey. The sexes are alike.

Note There is considerable variation in the plumage, particularly in the intensity and area of red on the face, chin and throat, among the 8 subspecies described. For example, *L. t. albigularis*, from south-west Tanzania, totally lacks the red; the head, throat and chest are black with white flecks.

PLATE 85 *Lybius torquatus*, Black-collared Barbet

Range Angola eastwards through southern areas of Zaire, Burundi, Kenya, Tanzania and south to Mozambique, north-east Zambia, Zimbabwe, eastern Cape Province and Natal.

Status Variable, but generally abundant in many parts of the range where suitable habitat exists.

Avicultural rating They are occasionally available and good avicultural subjects. They need care when first imported, but eventually are easily managed and fairly hardy.

Feeding Use a mix of Diets E and G with live food. They also enjoy tinned fruits (fruit cocktail, sliced pears or peaches). Provide an occasional pinky mouse.

Breeding Provide a substantial hollow log, or a rotten tree trunk or bough, for the birds to excavate their own nest-chamber. Clutch 3; incubation 18 days. Abundant live food is needed for rearing; wax-moth larvae are probably valuable in the early stages. In the wild, this species is parasitised by the Lesser Honeyguide (*Indicator minor*).

General management They probably are best not exposed to prolonged cold weather or frost. They are good in a tropical house; also in shelter and flight accommodation with a facility for enclosure in well-lit shelter with slight heat in severe (cold) weather. They have a loud whistling call. Like others of this family, they are aggressive and a risk in mixed company. They love bathing.

Trachyphonus vaillantii [PLATE 86]
Levaillant's Barbet, Crested Barbet, Levaillant's Ground Barbet

PLATE 86 *Trachyphonus vaillantii*, Levaillant's Barbet

8½ in (22 cm). The forehead, cheeks, chin, throat, lower breast and underparts are yellow, spotted and streaked with crimson; the mantle, wing coverts and band across the upper breast are black. The upper tail coverts are red, the tail black with white markings. There is an erectile black crest. The bill is yellow-green, the legs and feet dark-grey. The sexes are alike.

Range Angola eastwards to Tanzania; southwards to Zimbabwe, Botswana, eastern areas of South Africa.

Status Variable, but there are good populations in areas of suitable habitat.

Avicultural rating Occasionally available. They are easily managed once acclimatised, but are best not exposed to frost or prolonged cold or wet weather especially during the winter months.

Feeding Use a mix of Diets E and G with berries and regular supplies of live food. Individual birds often exhibit a preference for a mixture of live food and insects. Offer the occasional small amount of scraped or minced raw beef to birds which will take it.

Breeding Provide a substantial hollow log or a decayed portion of a tree trunk; fix it in place (upright) in the highest part of the aviary but sheltered from both the elements and predators. Clutch 3; incubation 17–19 days. Some plant cover around the nest-site is advantageous. Live food is needed for rearing.

General management Similar to the Black-collared Barbet. This species is by no means completely hardy and will not thrive in cold and wet conditions; even a spell of chilly, unseasonal summer weather seems to reduce its vitality. It is armed with a powerful bill and is not a safe companion for smaller birds (or even those of similar size), especially when approaching breeding condition. It has a fairly musical call and two birds will often perform 'duets'.

Trachyphonus darnaudii
D'Arnaud's Barbet

6 in (15 cm). The upper surfaces and tail are brown with prominent white spotting; the forehead and crown are black heavily spotted with

orange or yellow. The chin, throat and breast are yellow with black spots, the lower underparts yellow-white. The rump is orange-yellow. The bill is yellow-horn, the legs and feet grey. The sexes are alike.

Range Southern Sudan and Ethiopia, Uganda, Kenya, southern Somaliland, south-western Tanzania.

Status Abundant in areas of suitable habitat throughout most of the range.

Avicultural rating Occasionally available. They are not difficult when acclimatised but can never be regarded as totally hardy.

Feeding Use a mix of Diets E and G with additional berries as available; also some live food.

Breeding These birds are tunnel-nesters and may utilise an artificial earth-bank or peat bales. Birds in breeding condition in natural aviaries will sometimes tunnel downwards into the earth. Clutch 2–4; incubation 17–18 days. Provide extra live food when breeding, including various larvae.

General management They are best housed in a tropical house environment; they are fairly hardy but should not be over-wintered without heat nor should they be allowed out during severe (cold) weather or periods of prolonged cold and wet conditions. This barbet is generally a good mixer with other species and appears much less belligerent than others of the family.

Ramphastidae (Toucans)

There are nearly 40 species in the toucan family, which are confined to the New World: southern Mexico, Central and South America to northern Argentina, southern Brazil. The greatest concentrations are in the forested areas of Amazonia. These birds are characterised by the size and shape of the bill, which often has striking patterns and colours. In some of the larger species the bill length equals the body length, giving the bird a top-heavy appearance. The bill is very light, the outer shell being very thin and enclosing a spongy, honeycomb interior structure. Many are good avicultural subjects, which fare best in lofty aviaries providing them with ample space for flight. Several are reasonably hardy and can winter outdoors in warmer, southern areas, provided they have a comfortable box or hollow log as roosting quarters. In colder parts they are best housed indoors for the duration of the winter, but should not be kept in the restricted space of even a large cage. Bigger species are adept at using their huge bills as a precision feeding tool: they pick up a food item between the tips of the mandibles, toss it in the air and catch it in the open bill. Although fruit is a major item of diet, toucans are generally much more omnivorous; if the opportunity occurs they will catch and eat small reptiles, birds and mammals as well as a variety of insects.

Aulacorhynchus haematopygus [PLATE 87]
Crimson-rumped Toucanet

16 in (41 cm). The upper surfaces are green, more olive on the upper back, brighter on the lower back and wings and with a chestnut suffusion on the neck and mantle; the rump is crimson. The underparts are bright green with an indistinct blue band across the breast (individual birds show a greater or lesser area of blue). The tail is blue-green with a terminal chestnut band. The bill is dark red, with part of the ridge and an area in the centre black; the base of the upper and lower mandible is white. The legs and feet are grey-green. The sexes are alike.

Range Ecuador, northern, western and southern areas of Colombia, north-western Venezuela.

Status Reasonable numbers in some parts of the range, but they are losing ground as deforestation destroys their habitat.

Avicultural rating Infrequently available. They are reasonably hardy, and when established true pairs breed fairly freely.

PLATE 87 *Aulacorhynchus haematopygus*, Crimson-rumped Toucanet

Feeding Diet E, with the addition of some Diet G plus berries in season. Also offer some pellet food (see p. 27), live food and, if the birds are interested, sweetened bread and milk and rice pudding; it is essential that the latter items are not allowed to sour, so remove them if not eaten quickly, especially in hot weather. A small dead mouse will be appreciated; also larger items of live food such as locusts.

Breeding Provide a suitable hollow log; a box may be used but a log is better. Clutch 2–3; incubation 16–17 days. There is a long nestling period of up to 6 weeks. Provide a varied fruit and live food diet for rearing and ensure that the diet items are dusted with a multi-vitamin preparation.

General management They are not suited to restricted accommodation, but are fairly hardy and can live outside when acclimatised. Use tropical house or shelter and flight accommodation, the latter with a means of enclosing the birds within the shelter during severe (cold) weather. They are not suitable companions for other birds. They are usually fairly long-lived (up to 15 years) if housed and fed correctly.

Aulacorhynchus prasinus
Emerald Toucanet

12–14 in (30–35 cm). The plumage is mainly grass-green, paler on the underparts and with a yellow suffusion on the mantle and sides of the body; there is a tinge of bronze on the head and neck. The throat is white, grey, blue or black, depending on the subspecies. The tail coverts and terminal band on the tail are chestnut. The bill has a yellow-green ridge, with the rest of the upper mandible black; the lower mandible is similar but there is an area of chestnut towards the base. A white line outlines the base of the bill. The legs and feet are dark-grey. The sexes are alike.

Range Mexico and Central America to northern and western areas of South America.

Note A total of 15 subspecies are described, most of them exhibiting differences in the colour of the throat, or in the pattern or colour of the bill.

Status Small but stable populations in many parts of the range; loss of habitat is affecting this species adversely.

Avicultural rating Occasionally available; it is a good avicultural subject when established.

Feeding/Breeding/General management Similar to Crimson-rumped Toucanet. Clutch 3–4; incubation 15–16 days. Incubation usually commences after the second egg has been laid, and young of different ages in the same brood are frequent.

Pteroglossus inscriptus
Lettered Aracari

11 in (28 cm). The entire head is glossy black, the upper surfaces, wings and tail dark green. The underparts are yellowish, browner towards the abdomen. There is some chestnut on the thighs and the rump is red. The upper mandible of the bill is buff-white with a black ridge and tip; there

are several narrow black vertical lines along the sides. The lower mandible is black. The legs and feet are grey. The female is similar but has a chestnut head.

Range Central and southern areas of Brazil.

Status Habitat loss in some southern areas of the range is reducing the population.

Avicultural rating Formerly a popular avicultural subject, it is now rarely available.

Feeding Use a mix of Diets E and G with live food (locusts, crickets etc). Give a small dead mouse weekly. Berries are much enjoyed.

Breeding This bird nests in a box or hollow log. Clutch 2–3; incubation 20–21 days. The fledgling period is up to 6 weeks. Insects form the bulk of the rearing food for the first couple of weeks; thereafter more fruit is fed. There is evidence that in the wild the previous year's offspring of a pair roost in the nest cavity after the new chicks hatch and assist in feeding them.

General management They are probably best suited to tropical house or conservatory accommodation; they are not as hardy as some members of the family and should not be wintered in outdoor aviaries except in mild climates. They are best not housed with other species, especially smaller birds. Provide a box or hollow log for roosting.

Selenidera maculirostris
Spot-billed Toucanet

---

10 in (25 cm). The forehead, crown, nape and underparts are black. The rest of the upper surfaces are dark green. The ear-coverts and a collar are yellow. The bill colour may vary from (upper mandible) blackish with an olive-green tip to grey-white with a black ridge and black spots; the lower mandible varies between dark and light grey with some obscure darker markings. There is an area of bare skin extending from the base of the bill to the cheeks which is pale blue. The legs and feet are grey. The female has the top of the head, nape and underparts chestnut-brown; the back and wings are dark greenish-black and the bare skin on the cheeks is usually of a slightly more greenish hue than that of the cock.

Range Central, south and south-eastern Brazil.

Status Habitat loss is having an adverse effect.

Avicultural rating These are good avicultural subjects, but unobtainable at present.

Feeding/Breeding/General management Similar to the Crimson-rumped Toucanet. They are not especially hardy and best suited to tropical house or conservatory conditions; although frequently immobile, they need adequate space. Choose companions, if any, with care.

Ramphastos vitellinus
Channel-billed Toucan

---

18 in (45 cm). The plumage is mainly glossy-black, except for the lower cheeks, chin, throat and breast which are sulphur-yellow; the rump is red. An area of bare skin around the eye is blue. The bill is black with yellow at the base of the upper and lower mandibles. The legs and feet are grey. The sexes are alike.

Range Northern South America.

Note 6 subspecies are described. They are:
  R. v. culminatus (upper Amazonia);
  R. v. citreolaemus (central Colombia);
  R. v. vitellinus (Venezuela, Trinidad, the Guianas, northern Brazil);
  R. v. ariel (central and southern Brazil);
  R. v. pintoi (south-eastern Brazil);
  R. v. theresae (north-eastern Brazil).
Apart from the nominate subspecies, the best known in aviculture is probably the Ariel Toucan (*R. v. ariel*), but it is now among the least likely to be available because of export restrictions in Brazil.

Status Most toucan species are birds of the tree-canopy. They, more than most, are much affected by loss of habitat, due almost entirely to rapid deforestation of some areas.

Avicultural rating Habitat loss may be making many larger members of the Ramphastidae more vulnerable than is realised. Captive-breeding programmes could be vital to the survival of some species. They are generally good avicultural subjects and an increasing number of species are breeding in control.

Feeding Use a mix of Diets E and G with the addition of seasonal berries and regular live food. Some minced raw beef is desirable; it is often best to make up large pellets combining insectivorous mixture, diced fruit and minced beef. The birds show little interest in small live food, possibly because of the difficulty in picking it up; large mealworms, locusts and crickets are accepted. Pinky mice and small dead adult mice are enjoyed.

Breeding Provide a substantial log with the appropriate-sized cavity and aperture. Clutch 2–4; incubation 16–17 days. The nestling period is up to 7 weeks. Offer more animal matter if young are present, including suitable items of live food; pinky mice are likely to be important.

General management The need for spacious, lofty aviaries cannot be over-emphasised; they also thrive in a tropical house environment, but are uncertain companions for other species unless considerable space is available. They are not winter-hardy and should be housed indoors during prolonged periods of cold weather, although cage accommodation is not recommended.

Ramphastos toco [PLATE 88]
Toco Toucan

---

21 in (53 cm). The plumage is mainly glossy-black, except for the white throat, breast and rump; the under tail coverts are red. An area of bare skin around the eye is blue, with a yellow

PLATE 88 *Ramphastos toco*, Toco Toucan

outer ring. The bill is golden-orange with a large oval area of black at the end of the upper mandible; the base of both mandibles is black. The legs and feet are dark-grey. The sexes are alike.

Range The Guianas, northern, eastern and southern Brazil, Bolivia, Paraguay, northern Argentina.

Status See Channel-billed Toucan.

Avicultural rating Still available from time to time; otherwise see Channel-billed Toucan. This is one of the best-known and most popular species.

Feeding/Breeding/General management Similar to the Channel-billed Toucan. Because this species has been a focal point for former advertising campaigns for a certain alcoholic beverage it has captured the imagination of the public; many birds have been sold at a high price into a brief life in centres of entertainment, clubs etc. This process is to be deplored, especially in view of the increasing pressures on the species in the wild. There have been limited captive breeding successes, notably in the USA.

Picidae (Woodpeckers)

The Picidae comprises more than 200 species, with worldwide distribution. Most are birds of forest or woodland areas; a few, such as the piculets and the Ground Woodpecker from South Africa, are terrestrial. Most are very insectivorous; a few consume fruit as well as insects. Despite their specialist way of life and the initial difficulties of establishing them in a captive environment, several are good avicultural subjects, but they are presently mainly unobtainable.

Melanerpes candidus
White Woodpecker

---

9 in (23 cm). The head and underparts are white; there is some yellow on the nape and abdomen. The mantle, back and wings are black; the tail is black and white. The eye streak is black, an area of bare skin around the eye pale yellow. The bill, legs and feet are black. The sexes are similar but some females are said to have a paler yellow abdomen patch and to lack yellow on the nape.

Range Northern and central Brazil south through eastern Bolivia, Paraguay, to northern Argentina.

Status Uncertain but there are believed to be good populations where their habitat is intact.

Avicultural rating Occasionally available and a good avicultural subject which has bred quite frequently.

Feeding Use a mix of Diets E and F, with the addition of extra chopped (whole) hard-boiled egg, berries as available and a variety of live food items. One of the high protein pellet foods can be included in the diet but many of these birds show a preference for a largely (but not entirely) frugivorous diet.

Breeding Provide a substantial hollow section of tree trunk, or a decayed portion for excavation by the birds. Some pairs will use a substantial box hung in the highest part of the aviary; in locating this, observe the need for some protection from the elements, especially heavy rain and strong sun, and from predators. Put some wood chippings in the base of the nesting receptacle. Clutch 2–4; incubation 16–17 days. The nestling period is 5 weeks. Provide additional live food for rearing. The fledglings are similar to the parents, but the bare skin around the eye is blue in juveniles.

General management Like others of the family, these birds need spacious accommodation; a tropical house compartment is ideal, otherwise use a shelter and flight with a means of enclosing the birds in a warm shelter during colder periods. Furnish with logs or boughs, with natural bark on part of the shelter walls. Their accommodation must be of substantial construction as the birds will destroy unprotected timber; so housing should be of brick, concrete or metal framing. It is best if they are the sole occupants of the accommodation.

PASSERIFORMES

Eurylaimidae (Broadbills)

This is a family of 14 species, resident in the Old World tropics, which are essentially forest dwellers. Characterised by stout bodies and short wings, they also have short, flat bills. Green is the predominant colour, with areas of red or pink in some species, otherwise they have black markings. Only one species is fairly well known in aviculture but it is now rarely available. Broadbills are lethargic, but generally do well in a tropical house environment.

Calyptomena viridis [PLATE 89]
Lesser Green Broadbill

6 in (15 cm). The plumage is mainly iridescent emerald-green; there are small areas of black above the eye and on the hind cheek. Black on the secondaries and primaries forms distinct bars on the closed wings. A short crest extends forward almost to cover the deep, wide bill. The bill is grey-black, the legs and feet grey. The female is similar but lacks the black markings and has a green area around the eyes.

Range Southern Burma, Thailand to Malaysia; Sumatra, Borneo and some adjacent islands.

Status Although rarely seen in many areas of the range, there are probably still good populations; habitat loss has had an effect in some parts.

Avicultural rating A much-prized avicultural subject which does best in experienced hands, but is not difficult when established.

PLATE 89 *Calyptomena viridis*, Lesser Green Broadbill

Feeding Use a mix of Diets E and F with live food. At first it may be necessary to produce rudimentary pellets from the two main diet items to ensure good nutrition, as some birds will opt for a staple fruit and live food diet. In mixed groupings including nectar feeders broadbills also often select liquid food as their staple diet, which must be prevented.

Breeding Breeding may be possible in a suitable environment. It is unlikely that the birds will accept a nesting receptacle; the natural nest is a beautifully-constructed pendulous creation, suspended (often over water) by a narrow, woven neck. Clutch 3–5; incubation not recorded. A well-established tropical house with luxuriant plant growth is the most likely setting for attempts at breeding this species.

General management They undoubtedly fare best in a tropical house; this is not an outdoor aviary species, except during warm summer weather. It is generally a good mixer with other birds. Although relatively inactive for long periods, their main activity periods are usually at the beginning and end of the day. They have a fairly placid temperament and are invariably very tame. Warm, humid conditions are essential for a lengthy tenure of captive accommodation.

Cotingidae (Cotingas)

This family includes many spectacular avicultural gems, now, alas, rarely seen. They are essentially birds of the New World tropics; there are more than 80 species, including such as becards, fruit-crows, umbrella birds, bellbirds and cocks-of-the-rock. There is great variation in size (sparrow to crow size); some are undistinguished in appearance but many exhibit a great diversity of plumage with bizarre ornamentation. Several species continue to be available in Continental Europe etc, despite export restrictions in their countries of origin.

Rupicola rupicola (II)
Guianan Cock-of-the-Rock, Orange Cock-of-the-Rock, Golden Cock-of-the-Rock

13 in (33 cm). The plumage is mainly bright orange; the wings and tail are black. There is a white speculum on the primaries; the feathers of the back and rump are elongated and with forked or square tips, as are the secondary feathers. The disc-like crest is orange, edged with purple-brown; it extends forward almost to obscure the bill. The bill is orange-yellow, the legs and feet yellow. The female is dull orange with a shorter crest.

Range Eastern Colombia through southern Venezuela to the Guianas, northern Brazil.

Status Uncertain, but numbers will clearly be adversely affected by the loss of forest habitat in some areas.

Avicultural rating A prized avicultural subject; sadly, many that have been in private hands have been destined primarily for a 'career' as bird show exhibits. Some breeding activity is taking place in control, but it is still a rare event. They are generally fairly straightforward when acclimatised, but do best in experienced hands. They are now very expensive.

Feeding Use a mix of Diets E and G in the proportions of 3 parts fruit to 1 part insectile mixture. Also provide trout pellets (see p. 27), chopped (whole) hard-boiled egg and some live food. Some finely minced raw beef can be formed into pellets with the insectivorous mixture and offered alongside other items.

Breeding In the wild the cocks perform elaborate displays at specific locations in the forest, similar to a Blackgrouse lek; it may be that communal activity of this nature is an essential stimulus to more frequent captive breeding, although the high cost of birds almost certainly precludes this. The natural nests are mud and plant-fibre cups attached to a rock or a cliff face in forested areas. It is difficult to reproduce a good replica nesting environment, although fibreglass 'rocks' can be purchased. Provide an open box in a suitable location. Clutch 2; incubation believed to be 19–20 days. Little is recorded of rearing diets, but they probably should consist of a combination of fruit and live food.

General management They are best segregated in a separate compartment in tropical house type accommodation; they can also be housed in spacious outdoor shelter and flight accommodation.

Enclose them inside a warm shelter during severe (cold) weather. Outdoor accommodation should not be contemplated, even during the summer, until the birds are properly established, but there should be a good flow of air through the otherwise warm tropical house. This species is not a good mixer; the males often severely injure or kill the females in a restricted space. Most imports are hand-reared birds, and are usually fairly confident and steady which reduces the chance of stress-related problems with expensive stock.

Rupicola peruviana [PLATE 90] (II)
Andean Cock-of-the-Rock, Scarlet Cock-of-the-Rock

PLATE 90 *Rupicola peruviana*, Andean Cock-of-the-Rock

15 in (38 cm). Very similar to Guianan Cock-of-the-Rock, but the orange feathers are replaced by scarlet and there is no edge to the crest. The feathers of the lower back and rump are elongated, square-edged, pearl-grey; the wings and tail are black. The bill, legs and feet are orange-yellow. The female is generally duller and lacks the large, disc-shaped crest.

Range Colombia to western Venezuela: south to eastern and western Ecuador, Peru and northern Bolivia.

Status See Guianan Cock-of-the-Rock.

Avicultural rating Probably even more highly-prized than the previous species.

Feeding/Breeding/General management Similar to Guianan Cock-of-the-Rock. Little is known of their breeding activity. When acclimatised, they will almost certainly do better in a temperate rather than tropical house setting.

Pipridae (Manakins)

This New World family of more than 50 species is well known for its spectacular courtship displays. Manakins are mainly South American, but some range as far as Central America and Mexico. They are forest birds; essentially berry and fruit eaters, but some live food is taken. They are rarely available to aviculturists, but some White-bearded Manakins reach the market occasionally. The Blue-backed Manakin, once regularly imported, is even less frequently available. They do best in experienced hands, but are good avicultural subjects and are usually long-lived when properly established.

Manacus manacus [PLATE 91]
White-bearded Manakin, Black and White Manakin

$4\frac{1}{2}$ in (11.5 cm). The forehead, crown, nape, mantle, wings and tail are black; the sides of the head, chin, throat and under surfaces are white. The neck collar is also white, the rump grey. The bill is black, the legs and feet red. The feathers of the throat are elongated into a 'beard'. The female is mainly drab-olive with a grey-white throat,

PLATE 91 *Manacus manacus*, White-bearded Manakin

and yellow-green lower underparts. There are 11 subspecies.

Range Colombia, Ecuador, northern Peru eastwards to Venezuela, the Guianas, Trinidad, Brazil (except eastern areas), eastern Paraguay and north-eastern Argentina.

Status Abundant in many areas.

Avicultural rating Only occasionally available; they are generally not easy to establish, but eventually are not difficult to manage.

Feeding Use a mix of Diets E and F, plus berries (frozen cultivated berries help to maintain the supply through the year) and some small live food. Soaked raisins or currants should be given in an emergency only. Mix the fruit and insectile mixture together to ensure a balanced intake. Provide artificial nectar only if needed as an occasional pick-me-up or if the birds are ill.

Breeding Spectacular and complicated displays, involving several males, precede mating. The female alone constructs a nest of plant fibres etc, fairly low down in suitable vegetation and usually suspended below a branch or fork. Clutch 2; incubation 18–19 days.

General management They are ideal occupants of a tropical house, but are best in a separate compartment, as their small size renders them inconspicuous at liberty. They are never completely weather-resistant, and need sheltered accommodation throughout the year. Provide food stations at various elevations, as they do not like to come to the ground for feeding. They are gregarious, and associate well with other small birds. They are lively and active.

Chiroxiphia pareola
Blue-backed Manakin

$4\frac{3}{4}$ in (12 cm). The plumage is mainly black except for the crown (red) and the upper back (blue). The bill is black, the legs and feet red. The female is mainly olive-green, the under surfaces paler. The legs and feet are flesh-coloured.

Range Colombia, Venezuela, the Guianas, northern and eastern Brazil, eastern Ecuador and Peru; Tobago.

Status Still numerous in many parts of the range.

Avicultural rating See White-bearded Manakin.

Feeding/Breeding/General management Similar to White-bearded Manakin except, if anything, slightly more difficult to establish. Eventually they do well and can be long-lived.

Pittidae (Pittas)

The Pittas are colourful, long-legged, short-tailed denizens of the Old World tropics; the biggest concentration of species is in South-east Asia, plus Australia and Africa. There is a total of 26 species in all; these are mainly thrush-sized with the Giant Pitta the biggest at 11 in (28 cm). The plumage is mainly green with vivid splashes of crimson, purple, yellow etc. Alternative names are Jewel Thrush or Painted Thrush. They are popular avicultural subjects, although now rarely available. Sometimes difficult to establish, they always need careful, sympathetic handling; they are not suitable for cages, and fare best in a planted tropical house environment.

Pitta guajana [PLATE 92]
Banded Pitta, Blue-tailed Pitta

PLATE 92 *Pitta guajana*, Banded Pitta

8½ in (21.5 cm). The head is black with a golden-buff stripe from the base of the bill, above the eye, to the nape, with a brighter orange colour on the neck; the rest of the upper surfaces are dark brown. The chin and throat are buff-yellow, the underparts dark blue with orange-brown barring which is particularly pronounced on the breast and sides of the breast. The bill is black, the legs and feet grey-brown. The female is similar but with duller, narrower barring on the breast and sides.

Range Southern Thailand, Malaysia, Sumatra, Borneo, Java, Bali, Bangka Island.

Status Abundant in many parts of the range; often resident in suitable habitat near inhabited areas (plantations etc).

Avicultural rating Popular avicultural subjects, but now rarely available. They need care, and are often difficult to establish. Specialist accommodation is required.

Feeding Diet E, with some finely-scraped raw meat (mixed with insectile food) and a variety of live food. Mealworms and gentles are by no means suitable *regular* items of diet; but provide crickets, small locusts, any available beetles etc. Snails are especially appreciated. One of the high-protein foods (see p. 27), such as trout pellets, can be fed in moderation alongside other diet items.

Breeding Pittas have a reputation for aggression between the sexes and this has often inhibited their breeding potential, especially in limited-size accommodation. They need ample space such as a tropical house offers, but must also be regarded as unsafe companions for similar-sized or smaller birds. The large domed nest is constructed low down among vegetation with an entrance at the side. Clutch 4; incubation 18–19 days. Abundant live food is needed for rearing.

General management Pittas have very tender feet and will not survive contact with harsh, abrasive, unyielding floor conditions; they do best in a planted greenhouse, conservatory or tropical house, although many species enjoy the summer in suitable sheltered outdoor aviaries. They are not hardy, and are untrustworthy with other species (I have had a bird of this species kill several Diamond Doves). Care of their feet is all important; a poor diet can also affect them. Ensure that the birds are kept on a good depth of moist, resilient peat or leafmould, with decayed logs for use as vantage points. Pittas roost at varying heights; since the roosting perch will be used more than any other, ensure that it is of a fairly yielding quality and of the correct diameter (1¼ in/3 cm), preferably bark-covered and not allowed to dry out.

Pitta sordida
Hooded Pitta

---

8 in (20 cm). The whole of the head, chin, throat and nape are sooty-black except for the dark-brown crown. The upper surfaces are mainly grass-green; the lesser wing and upper tail coverts are shining-blue. The breast is green with some black on the lower underparts; the abdomen and under tail coverts are red. The tail is black with a blue tip; there is some white in the primaries which is mainly conspicuous when the wings are spread. The bill is black; the legs and feet are flesh-coloured. The sexes are similar.

Range India eastwards to Thailand, Indochina, Malaysia, Sumatra, Java, Borneo, New Guinea, Philippines, Nicobar and Sunda islands.

Note The nominate subspecies is from the Philippines; a further 12 subspecies are described.

Status Still abundant in many parts of the range, but habitat loss is affecting some populations.

Avicultural rating See Banded Pitta.

Feeding/Breeding/General management Similar to Banded Pitta. The first captive breeding took place during the 1930s.

Pitta brachyura (ssp. *nympha* II)
Blue-winged Pitta, Bengal Pitta, Indian Pitta

7½ in (19 cm). The head is black with a broad buff stripe above the eye; the mantle, back and wings are green. The rump, upper wing and tail coverts are shining-blue; the chin and throat are white, the rest of the underparts biscuit-brown except for a patch of red on the abdomen. There is some white on the primaries; the tail is black. The bill is dark horn, the legs and feet flesh-brown. The sexes are similar.

Range India north-eastwards to eastern areas of China, Korea and Japan; south-east to Burma, Thailand, Malaysia, Greater Sundas, Philippines; an occasional vagrant to Australia.

Status Abundant in many parts of the range.

Avicultural rating See Banded Pitta.

Feeding/Breeding/General management Similar to Banded Pitta. Individual birds are sometimes especially difficult to establish after importation, because of their total unwillingness to consume an inanimate diet; ensure that the insectile mixture is 'laced' with small, active gentles ('pinkies') that have been thoroughly cleaned to encourage the birds to eat.

Alaudidae (Larks)

There are more than 80 species of lark, mainly of Old World distribution. They are not widely known in aviculture, with a few exceptions. Finch-larks are frequently offered for sale. Smaller than a sparrow and with distinctive markings, their conical bill suggests a closer affinity with typical seedeaters, but they are much more lark-like in behaviour. In the wild they usually inhabit dry, arid regions and are extremely terrestrial in habit. They are by no means 'easy' avicultural subjects and it is important to create a suitable artificial environment for them or their feet will quickly suffer.

Eremopterix leucotis
Chestnut-backed Finch-lark, Chestnut-backed Sparrow-lark

4½ in (11.5 cm). The head, neck and under surfaces are black and there is a distinctive white cheek patch. The upper parts, wings and tail are mainly chestnut whilst the rump is grey-brown. The bill is grey; the legs and feet are brown. The female's plumage is duller and she lacks the distinctive markings on head and cheeks.

Range Senegal eastwards to Ethiopia, southwards through Kenya, Tanzania, Mozambique, to South Africa.

Status Abundant in many parts of the range.

Avicultural rating Available from time to time; these are interesting avicultural subjects, but need care.

Feeding Diet B with some small live food, including fly pupae. They normally show little interest in green food. Grit is essential.

Breeding These birds might be persuaded to breed in suitable quarters but it is unlikely that the sort of cool, damp conditions that often prevail outdoors in temperate countries would suit them. They scratch out a small, but often fairly deep, hollow in loose dry ground and line it with bits of dry grass and other plant materials. Clutch 1–2; incubation 12–13 days. Provide plenty of small insects for rearing. The chicks fledge after about two weeks.

General management Dry accommodation is important for these and related species; they often thrive best indoors but can be housed, during summer, in a sheltered outdoor aviary which has protection from driving rain. In indoor quarters they need an ample quantity of *dry* sand on the floor of their quarters; a dusting will not suffice—they should run about over a depth of not less than 1 in (2.5 cm). A few small stones or pieces of bark provide 'cover' and vantage points for the birds. It is important that their feet remain in good condition, and if housed for any length of time in small quarters (cages) scrupulously clean conditions must be maintained; it is also important to ensure that the sandy floor-covering is kept dry and does not become impacted.

Campephagidae (Cuckoo Shrikes and Minivets)

This family comprises more than 70 species with a wide distribution throughout the Old World from Africa through southern Asia to Australia and the Pacific regions. It also includes flycatcher shrikes and wood shrikes; but only minivets are of substantial interest to aviculturists. They are now rarely available. Elegant and colourful, they are among the most difficult of insectivorous birds to maintain in good health; many are in poor condition on arrival from Asia and further debilitation and losses often occur through incorrect feeding whilst the birds are in quarantine. Nevertheless, minivets are desirable subjects, particularly if well managed and in good health.

Pericrocotus brevirostris
Short-billed Minivet

$7\frac{1}{2}$ in (19 cm). The head, neck, upper back, chin, throat and wings (except for a band of red) are glossy-black; the central tail feathers are black. The rest of the plumage is scarlet. The female has a pale orange forehead; the rest of her upper parts and wings are grey-green except for the rump and upper tail coverts which are lemon-yellow. There is a broad band of yellow through the wing; the under surfaces are yellow. The tail is black and yellow. The bill, legs and feet are black.

Range Himalayas, Nepal, eastwards to northern areas of Burma, Thailand and Laos, north Vietnam, southern China.

Status Still good populations in some areas; deforestation has eliminated both habitat and birds, particularly in parts of Indochina.

Avicultural rating Unquestionably a 'difficult' species, which should not be tackled by other than widely-experienced aviculturists. These birds are very beautiful, but the cock's brilliant scarlet plumage may fade after moulting without an artificial colour-correcting agent.

Feeding Diet E, with varied live food. Crickets are valuable; gentles, allowed to complete their cycle, will produce flies, greenbottles or bluebottles fairly quickly if kept at about 75–78°F (24–25°C). Although a useful protein source, they also provide interest and encouragement to partake of other items of diet including inanimate mixtures. Small, well-cleaned housefly gentles ('feeders') can be mixed with Diet E to simulate movement. No matter how varied the selection of artificially-produced live food, deficiency problems frequently ensue in the medium to long term with these birds; therefore I strongly recommend the acquisition and use of as much natural live food as possible. The purchase of a moth trap will help; also collect smooth caterpillars, small grasshoppers, spiders etc. Minivets are genuinely among the most difficult of all avicultural subjects because they are virtually totally insectivorous; it is, however, worth the extra effort to maintain them in good health over an extended period.

Breeding This is likely to be a very difficult undertaking; even if an adult pair is persuaded to nest, the task of obtaining *natural* small live food in sufficient quantity for the nestlings would be an enormous task. They make a substantial cup-shaped nest of rootlets and plant fibres, often at a considerable height from the ground, attached to a tree branch. Clutch 2–4; incubation believed to be 16–17 days.

General management A tropical house or warm conservatory is the best type of accommodation. They are generally very arboreal in habit; they enjoy hawking for flying insects and are adept at capture on the wing. They mix well with other small softbills. Ensure that food vessels are well above ground level.

Pericrocotus flammeus
Scarlet Minivet

8½ in (21.5 cm). Very similar to the Short-billed Minivet, except for its larger size; there is more red in the wings of the male (more yellow in the case of the female).

Range India, Sri Lanka, eastwards through much of south-east Asia to southern China, Malaysia, Greater Sunda and Philippine islands.

Note A total of 20 subspecies occupy this extensive range. The nominate subspecies, *P. f. flammeus*, is from southern areas of India and Sri Lanka.

Status Abundant in parts of the range where suitable habitat exists.

Avicultural rating See Short-billed Minivet.

Feeding/Breeding/General management Similar to Short-billed Minivet.

Pycnonotidae (Bulbuls)

There are nearly 120 species of bulbuls, with a wide range throughout the Old World tropics, from Africa to the Philippines. Most are frugivorous to a degree, although insects and other items are consumed by many species. Many are long-standing avicultural favourites, generally cheerful, confident birds, although some are shy and secretive. Most become hardy and are long-lived. Some are free breeders in suitable aviaries.

Pycnonotus jocosus
Red-whiskered Bulbul, Red-eared Bulbul

---

8 in (20 cm). The upstanding crest, top of the head and a stripe from the lower mandible to the shoulder are dark brown; a patch on the cheek is white, with a conspicuous tuft of red extending backwards from below the eye. Most of the upper surfaces are brown; the underparts are white with a buff-brown suffusion on the flanks and towards the abdomen. The bill, legs and feet are black. The sexes are alike.

Range India, Nepal, Burma, south-west China, Thailand, northern areas of Vietnam, Indochina, northern Malaysia, Andaman Islands. Introduced into USA, Australia, Mauritius.

Note There is some variation of size among the 9 described races and I have, at various times, owned birds ranging between 6 in (16 cm) and approximately 8¼ in (21 cm) in length.

Status Abundant in many parts of the range; it is adaptable and occupies territory ranging from forest edges to cultivated areas.

Avicultural rating Very popular, hardy and easily managed.

Feeding Use a mix of Diets E and G with some live food. They enjoy berries. In mixed aviaries they will also pick up some seed, but this is not a regular item of diet for these birds.

Breeding Obtaining a true pair is the main obstacle to breeding success, but many have raised young in aviaries. Isolate the pair from other birds if nesting seems likely; they are very aggressive during the breeding season. They will nest in a small flight (9 × 6 × 3 ft, 2.7 × 1.8 × 0.9 m), preferably in a quiet, secluded situation; provide artificial cover (wired-in bunches of spruce, heather etc) if natural vegetation is not present. A cup-shaped nest of plant fibres etc is constructed; sphagnum moss seems to be a favourite item in construction, and the nest is lined with fine grasses. The birds may use an open wicker basket as a nest receptacle, occasionally with an old wild bird's nest as a base. The site is usually 3–6 ft (1–2 m) above the ground, dependent on the placing of suitable cover. Clutch 3–4; incubation 13–14 days. Provide ample live food for rearing, especially at first; smooth caterpillars, wax moth larvae etc are best. Watch for signs of aggression from the adult cock when the youngsters are independent; remove them from the aviary as soon as possible, usually after 5–7 weeks.

General management These are among the easiest of exotic softbills to establish; and are a good species for aviculturists with fairly limited experience. When properly acclimatised, they are usually completely weather-resistant, but an outdoor aviary should have a damp-free, draught-proof shelter attached. These birds can be housed with other species, but pairs are always best isolated; single specimens only in community groupings is a safe rule. They have cheerful, melodious calls.

Pycnonotus cafer [PLATE 93]
Red-vented Bulbul

PLATE 93 *Pycnonotus cafer*, Red-vented Bulbul

8½ in (21.5 cm). The whole of the head (including the crest), chin and throat are black; the rest of the upper surfaces are brown, the feathers of the wings and upper back have lighter margins, giving a scale-like appearance. The upper breast is dark brown, becoming lighter or whitish towards the abdomen; the lower tail coverts are red. The bill, legs and feet are black. The sexes are alike.

Range India, Sri Lanka, Pakistan, parts of western and eastern Himalayas, southern Assam, northern and central Burma, southern China.

Note The nominate subspecies described is from Sri Lanka only; a further 8 subspecies show differences mainly in depth and intensity of plumage colours.

Status Abundant in many parts of the range, including gardens in residential areas.

Avicultural rating See Red-whiskered Bulbul.

Feeding/Breeding/General management Similar to the Red-whiskered Bulbul. Clutch 2–4; incubation 14 days. Although individual birds get on with similar-sized companions, they are generally more robust in behaviour than the previous species; true pairs are not trustworthy in any company.

Pycnonotus nigricans [PLATE 94]
Red-eyed Bulbul

8 in (20 cm). The whole of the head, chin and throat are black, the upper surfaces dark brown. The underparts are brown, becoming grey-white towards the abdomen. The flanks are grey-brown, the under tail coverts yellow. A narrow ring around the eye is bright red. The bill, legs and feet are black. The sexes are alike.

Range Southern Angola, Namibia, Botswana, Zimbabwe, South Africa.

Status Abundant in suitable habitat in many parts of the range, although it is persecuted as a pest in some areas because of damage to commercial fruit crops.

Avicultural rating Occasionally available, hardy and easily managed.

Feeding Use a mix of Diets E and G. They should also have some live food; offer wild berries (bramble, elder, whinberries etc) in season.

PLATE 94 *Pycnonotus nigricans*, Red-eyed Bulbul

Breeding A typical bulbul nest is constructed, but it is usually slightly more bulky than those built by the Asiatic species; it consists of a deep cup lined with rootlets and fine plant fibres. Clutch usually 3; incubation 11–12 days. There is a short nestling period; the young usually fly within 2 weeks. Provide ample live food when the chicks are being reared.

General management Acclimatised birds thrive in outdoor aviaries, with shelter and flight; they are usually completely weather-resistant. Individual birds mix with others of similar size but can be spiteful; so segregate pairs.

Pycnonotus milanjensis
Stripe-cheeked Greenbul

7½ in (19 cm). The plumage is mainly dull olive-green in colour, the underparts with a yellowish suffusion; the chin and throat are darker green. The cheeks are whitish with indistinct grey barring. The bill, legs and feet are black. The sexes are alike.

Range South-east Kenya, Tanzania, Malawi, northern and western areas of Mozambique, Zimbabwe.

Status There is a wide distribution where suitable habitat (forest or scrub) occurs; they are abundant in some areas.

Avicultural rating This species is rarely offered by dealers, perhaps because the sombre colours diminish interest; they become hardy, but seem not to be as tough as the Asiatic or South African species. They are generally easy to manage.

Feeding Use a mix of Diets E and G, with some live food. Like others of the family, they enjoy berries when available.

Breeding Little is known of aviary breeding successes. The similarity of the sexes and infrequent importation reduce the opportunities for successful nesting. A typical bulbul cup-nest is built; and a secluded, planted aviary is best. Clutch 2; incubation 14–15 days. Provide similar rearing items to those suggested for the Red-whiskered Bulbul.

General management They are easy to acclimatise and establish, then do best in a suitable outdoor aviary; true pairs should be the sole occupants of the accommodation. A frost-free shelter is desirable for winter occupation. They exhibit generally quiet, secretive behaviour, especially if housed in a well-planted aviary.

Irenidae (Chloropsis, Iora, etc)

All 14 species of the Irenidae are distributed in the Orient—India to the Philippines; and this family includes some of aviculture's most popular softbills. The best known are several of the 8 species of Chloropsis or fruitsuckers and the handsome Fairy Bluebird. Although they are omnivorous, fruit predominates in their diet, except in the case of the rarely imported ioras which are more insectivorous. Although easy to manage, many chloropsis have enjoyed only a short life-span in controlled conditions; a faulty diet may be one of the reasons—they do not thrive if offered a surfeit of such items as artificial nectar, oranges and honey water and sponge cake mixtures. Properly housed and fed these birds live for several years. Although eventually fairly hardy, they are not recommended for year-round occupation of outdoor aviaries. They thrive in a tropical house environment, but are very pugnacious with others of their own kind and smaller or weaker companions; heated quarters are best for occupation during the colder months of the year.

They usually have a good temperament, and are inquisitive and friendly; so large cages or indoor flights are acceptable housing in winter, but scrupulous attention to cleanliness is important. Many of the Chloropsis are adept mimics of familiar sounds, mainly the song of other birds, but also such things as creaking hinges and other frequently-heard high-key sounds.

Chloropsis cochinchinensis
Blue-winged Chloropsis, Blue-winged Leafbird, Blue-winged Fruitsucker

7 in (18 cm). The forehead is yellowish. The rest of the upper surfaces are bright grass green. The face, chin and throat are black with a yellow border of varying intensity of colour; the moustachial streak is cobalt-blue. The underparts are lighter green suffused with yellow. There is a large area of bright blue on the flight feathers. The tail is blue or green. The bill, legs and feet are black. The female is similar but lacks the black and yellow markings on the face, chin and throat; there is also less blue on the wings.

Range India, Sri Lanka, Burma, eastern and southern Thailand, Cambodia, Indochina, Malaysia, Sumatra, Borneo, Java, Billiton and Natuna Islands.

Note A total of 10 subspecies are described. The nominate subspecies, *C. c. jerdoni*, from southern India and Sri Lanka, is Jerdon's Chloropsis; once imported in reasonable numbers, it is now rarely seen.

Status Abundant in many parts of the range.

Avicultural rating A popular and widely-kept species which is reasonably easy to manage, but is not recommended for inexperienced aviculturists.

Feeding Diet E, but many chloropsis prefer large sections of fruit (half pear etc) which they enjoy probing into. Also offer some Diet G, mixed with a little diced (or even pulped) soft fruit. Orange is not favoured for these birds, although they will consume it avidly if permitted. Many aviculturists supply an artificial nectar or sugar water mixture; the danger is that individual birds may reach the point where they consume it to the exclusion of other foods, with eventual fatal consequences. Liquid foods for these birds are a very desirable addition to the main diet, but watch for signs of 'addiction' and take steps to reduce the amount of nectar available—possibly supplying it only for a couple of hours each day. Soft berries are also much enjoyed by these birds, elder and blackberries being two favourites; they also appreciate cultivated kinds such as blackcurrant and raspberry. A selection of small, soft-bodied live food should be provided each day; many birds quickly learn to take a mealworm from the fingers, and there sometimes is a danger of the owner over-feeding in order to show off the bird's tameness. Mealworms are best limited to no more than half a dozen per day, if the birds are closely confined in a cage—a few more if in spacious aviary accommodation. In the aviary, feeding stations should be well above ground-level; these birds do not like to descend below perch level for food or water.

Breeding Only a few breeding successes have been recorded with members of the genus *Chloropsis*; the main problem appears to be incompatibility between male and female. Introductions are fraught with danger; it is best if the birds can be housed in adjoining aviaries with a removable (and easily replaceable in case things go wrong) wire-screen partition. Watch the cock for signs of aggression at all times, including when the female is incubating as well as during the nestling and fledgling period. Clutch 2–3; incubation 13–14 days. The nest is cup-shaped, usually in a fork and hidden by vegetation. An open wicker basket placed in cover may encourage or stimulate building. Provide a variety of foods if the young hatch, but with a predominance of small live food; it is essential to gather as much natural live food as possible during the nestling period as even the use of a multi-vitamin preparation with a mainly mealworm or gentle rearing diet could lead to deficiency problems.

General management Imported birds often arrive in very poor feather but, if no disease is present, they can be brought along without problems, and their plumage will obviously improve with the first moult. Recent imports need careful treatment until fully acclimatised, and if obtained in the summer months are best not allowed into outdoor accommodation until the following year. They are good occupants of a tropical house, but aggressive towards other species, therefore they need to be confined within their own compartment. They also thrive in planted aviaries during the late spring, summer and early autumn; they can be housed outdoors throughout the year provided that slight heat is available in the shelter with a facility to confine the birds during severe (cold) weather, frosts etc. Single birds make excellent pets; they become confident and tame very quickly. Nevertheless, keeping such birds as single pets is not recommended; efforts to establish pairs of this and other exotic species with a view—eventually—to overcoming the difficulties leading to indifferent breeding results must be a priority. These birds can occupy roomy cages (minimum 48 × 18 × 15 in, 122 × 45 × 38 cm) with natural branch perches of appropriate diameter. Provide free-running metal trays; use newspaper (replace once or twice daily) in the trays and ensure the cleanliness of wire fronts, perches, dishes etc, which quickly become soiled and sticky.

Chloropsis aurifrons
Golden-fronted Chloropsis, Golden-fronted Leafbird, Golden-fronted Fruitsucker

7½ in (19 cm). The forehead and top of the head are bright yellow; the upper surfaces are bright green. The underparts are slightly paler. The face, chin and throat are black with a yellow border; the moustachial streak is purple-blue. There is some bright blue at the bend of the wing. The bill is black, the legs and feet dark grey. The female is similar but lacks the black and yellow markings on the face, chin and throat; the moustachial streak and blue on the wings are paler.

Range India, Sri Lanka, then eastwards to Burma, northern areas of Laos, Thailand, Cambodia, Vietnam, southern China, Hong Kong; Sumatra.

Status Abundant in many parts of the range.

Avicultural rating This is probably the best-known member of the family; it has been widely kept for many years and is a good avicultural subject.

Feeding/Breeding/General management Similar to Blue-winged Chloropsis. There are a number of records of this species having been kept at liberty during the summer months and they are reputed to be good 'stayers'.

Chloropsis hardwickei
Orange-bellied Chloropsis, Orange-bellied Leafbird, Orange-bellied Fruitsucker, Hardwick's Chloropsis/Leafbird/Fruitsucker

8½ in (21.5 cm). The entire upper surfaces, except for the black tail, are a bright olive-green; there are extensive areas of black on the face, chin, throat and breast. The moustachial streak is purple-blue. The lower breast and underparts are orange; there is some bright blue on the bend of the wing, with the primaries cobalt-blue. The bill is black, the legs and feet dark grey. The female is similar but lacks the black on the face, chin throat and breast, and the moustachial streak is paler. Her lower underparts are pale orange; there is little or no blue in the wings.

Range Eastern Himalayas, Burma, southern China, northern Thailand and Vietnam, Malaysia; Hainan.

Status Good populations in areas of suitable habitat (open forest) throughout the range.

Avicultural rating A much sought-after species which is occasionally offered for sale.

Feeding/Breeding/General management Similar to Blue-winged Chloropsis. Some breeding success has been recorded with this species. It is arguably more aggressive than its relatives, but is a beautiful, fluent songster.

Irena puella [PLATE 95]
Fairy Bluebird, Blue-backed Fairy Bluebird

10 in (25 cm). The top of the head, nape, shoulders, mantle, back, rump, upper and under tail coverts are a shining light blue; the remainder of the plumage is velvety-black. The eyes are ruby-red. The bill, legs and feet are black. The female is mainly slate-blue suffused with green; she has dark flight feathers.

Range India, Burma, Thailand, Malaysia, Sumatra, Borneo, Java, island of Palawan.

Status Abundant in many parts of the range.

Avicultural rating A favourite avicultural species which is fairly easy to manage and does best in spacious accommodation.

Feeding Use a combination of Diets E and G, with fruit offered diced. Regular live food is also important. Fairy Bluebirds have extremely small feet by comparison with their body-size; they are somewhat ungainly on the ground and therefore food and water stations should be at a higher level and close to the perches.

PLATE 95 *Irena puella*, Fairy Bluebird

Breeding Regular availability of imported stock may be the major reason that only a few breeding successes have been recorded, for individual pairs, in suitable accommodation, have proved fairly willing to lay eggs and rear young. The nest is somewhat lightweight and flimsy; usually sited among vegetation. An open box or wicker basket may be used in a suitable situation. Clutch 2; incubation 13–15 days. Provide a varied diet including plenty of live food (soft-bodied larvae such as smooth caterpillars, wax moth larvae, are excellent). The young fledge after 2 weeks, but the parents continue to feed for a further 3–5 weeks. Although generally less pugnacious than chloropsis, a breeding pair is clearly best housed in their own section of a tropical house, or as the sole occupants of an aviary.

General management Although gregarious in the wild, Fairy Bluebirds are best kept as pairs in controlled conditions; the female will often prove to be the dominant member of a pair, driving the cock from the food dish etc. Normally this behaviour does not lead to more strenuous exchanges and compatibility between established pairs is usually good. Newly-imported birds often arrive in poor plumage as a consequence of their close confinement in transit and mainly frugivorous diet while in quarantine; they need care at this stage but are not usually 'difficult' to bring on. Provide bathing facilities, but spray if the birds are unwilling to use a dish; individual birds are sometimes reluctant to enter water except in a more natural environment. They fare best in tropical house conditions, although they are relatively hardy; they also thrive in a planted outdoor aviary, but are not weather-resistant and need either a heated shelter or alternative indoor flight accommodation in the winter. They have loud, mellow, ringing calls.

Bombycillidae (Waxwings)

There are only 3 species of waxwings, but a wide range between them; they are northern hemisphere birds, extending across North America, northern Europe and much of north-east Asia. They are sleek and handsome with fine, soft plumage. Very gregarious in the wild, their main haunts are pine forests. Although the main diet is fruit and berries, they switch to become almost completely insectivorous when feeding their young; mosquitoes form an important part of the rearing diet of the Bohemian Waxwing in some parts of the range. They are good avicultural subjects but need careful treatment; attention to diet is important in view of their invariably lethargic habits.

Bombycilla cedrorum [PLATE 96]
Cedar Waxwing

6 in (15 cm). Most of the plumage is a warm olive-brown, paler on the underparts. There is a black mask from the base of the bill through the eye.

PLATE 96 *Bombycilla cedrorum*, Cedar Waxwing

The crest is usually flat and backward-curving but can be raised. There is a yellow tip to the dark tail. The flight feathers have up to half a dozen bright red, wax-like tips, the purpose of which is unknown. The bill, legs and feet are black. The sexes are alike.

Range Canada, United States, Mexico, Central America to Colombia and Venezuela.

Status Variable, but there are generally good populations in most parts of the range.

Avicultural rating These are good avicultural subjects, hardy and easily managed, but they need a carefully regulated diet. They are difficult to breed.

Feeding A small quantity of Diet G, but mainly diced sweet apple, a few currants (soaked for 24 hours before use) and whatever berries are available; I store various wild berries in the deep freeze for use throughout the year as required for these and other berry-feeders. Rowans are a favourite berry in the wild; others which are enjoyed include elder, blackberry and various cultivated types. Live food is a debatable addition to the diet; the larger items such as mealworms and gentles seem to pass straight through the gut. I prefer to allow the gentles to pupate and feed a few to the birds in this inanimate form if they will take them.

Breeding This is far from easy and it may be best if a number of birds can be housed together in a suitable-sized aviary; they are gregarious birds in the wild and self-selection of mates may be a stimulus to breeding activity. Provide plenty of natural or artificial cover in the aviary and place shallow boxes or wicker baskets in suitable sites. The nest is constructed of plant fibres and moss, with a soft lining of down or wool. Clutch 4–6; incubation 14–15 days. They are very insectivorous when feeding nestlings, so provide wax moth larvae and live ant pupae; small flying insects are taken on the wing and a moth trap may help. The key to successful rearing is undoubtedly the provision of suitable small live food in sufficient quantity.

General management They fare best in an outdoor aviary (shelter and flight) accommodation. They are hardy and able to winter outside, but should have access to a dry shelter. This species normally mixes well with other birds but breeding birds are best housed with their own kind only. They tend to be somewhat lethargic in habit, and if closely confined for any length of time problems of obesity can arise; in such circumstances it is advisable to cut down the quantity of soaked currants used in the daily diet, and provide more apple.

Prunellidae (Accentors or Hedge Sparrows)

Most of the 12 species of Accentors live in the mountainous regions of Europe, North Africa and

Asia. But in Britain the Dunnock is at home in gardens and along hedgerows, well away from mountains. In Nepal, one race of the Alpine Accentor (*Prunella collaris nipalensis*) is known to breed at altitudes up to 18,000 ft (5500 m). Most members of the family are insect-eaters, although they switch to a mainly seed diet during winter.

Prunella modularis
Dunnock, Hedge Accentor, Hedge Sparrow

5¾ in (14.5 cm). The sides of the head (including the eyebrow stripe), chin, throat and underparts are slate-grey; the upper surfaces are rufous-brown with darker streakings. The ear-coverts are brown; there is an indistinct buff-brown wingbar. The bill is dark brown, the legs and feet flesh-brown. The sexes are alike but the female is more brown.

Range Europe eastwards to Lebanon, eastern Caucasus, Iran.

Status Abundant in many parts of the range.

Avicultural rating A popular subject; a drably-coloured bird but an attractive and pleasant songster.

Feeding Diet G, with some live food. These birds also enjoy picking through a wild seed mixture.

Breeding Usually a free breeder, the Dunnock will build a cup-shaped nest among vegetation. Clutch 3–5; incubation 13 days. Provide abundant small live food for rearing. Watch for aggression from the adult cock when the young fledge; remove the latter as soon as they are seen to be feeding themselves.

General management They do best in an outdoor shelter and flight; good natural vegetation (shrubs, bushes) is desirable in part of the flight, as these birds are secretive in habit and spend much time among cover. They are hardy and can be wintered in outdoor accommodation, but need sheltered dry roosting places.

Note In the UK only aviary-bred and closed-ringed birds of this species can be offered for sale under the provisions of the Wildlife and Countryside Act, 1981 (Schedule 3, Part 1).

Muscicapidae—Turdinae (Thrushes, Chats etc)

The subfamily Turdinae comprises more than 300 species, distributed throughout the world except the Arctic; it includes some of bird world's finest vocalists—Nightingale, Shama etc. Most are insect-eaters, but others supplement their diet with fruit and berries, as well as seeds from time to time. Many are excellent avicultural subjects, which are hardy, easily managed and fairly willing breeders.

Erithacus calliope
Siberian Rubythroat

5½ in (14 cm). The upper surfaces are olive-brown; there is a prominent white stripe above the eye and a white moustachial streak. The chin and throat are bright red edged with dark grey; the remainder of the underparts are grey-brown to grey-white on the abdomen. The bill is black, the legs and feet dark flesh-coloured. The female is similar but lacks the red throat, and the other markings are less clearly defined.

Range Siberia; this species winters in India and southern China.

Status There are believed to be good populations in many parts of the range.

Avicultural rating Only very occasionally available, and much prized by aviculturists specialising in insectivorous species.

Feeding Diet F, and some small live food. I prefer to feed only limited amounts of living food to birds of this type as a stimulus to investigate an inanimate diet; blowfly pupae are fed in generous quantities to individuals who will accept them—

which most do after a short time. Wax moth larvae and crickets are acceptable items of live food for these and similar small thrushes.

Breeding Compatibility between male and female is a frequent problem; it is best to house them in adjoining aviaries and to introduce them carefully. Provide abundant natural cover in the flight; the nest is usually fairly low down among vegetation and made of rootlets, plant fibres etc. Clutch 4–6; incubation 13–14 days. Abundant small live food is essential for rearing; small crickets, locusts, wax moth larvae and whatever can be collected from the wild will help supplement cleaned 'feeder' gentles and small mealworms.

General management They thrive in shelter and flight accommodation; they are hardy and weather-resistant, often sleeping outside even during severe weather. It is important to have suitable evergreen shrubs in the flight to provide cover. They are not suitable companions for other birds and even pairs do not get along well together outside the short breeding season.

Cossypha niveicapilla
Snowy-headed Robin Chat, Snowy-crowned Robin Chat

---

8 in (20 cm). The forehead and crown are white; the cheeks, mantle, back and wings dark slate. The sides of the neck and whole of the under surfaces are rufous-orange. The tail is black and orange. There is a variable amount of rufous-orange on the lower back, rump and upper tail coverts. The bill, legs and feet are black. The sexes are alike.

Range Widespread distribution from Senegal, Guinea, Liberia, Ivory Coast etc, eastwards through Ghana, Nigeria to Central African Republic, Zaire, southern Sudan, Uganda, western Kenya, south-west Ethiopia and north-west Tanzania.

Status Generally abundant where suitable habitat exists.

Avicultural rating Only occasionally available; they are excellent inhabitants of a tropical house, and there have been some breeding successes.

Feeding Diet F, with the addition of live food (crickets, small locusts, occasional mealworms, wax moth larvae etc). Individual birds may take a little fruit, especially if in a community with frugivorous species, but an insectile diet is staple.

Breeding The cup-shaped nest is usually well hidden in vegetation, and some mud is often combined with plant fibres in construction; there is a soft lining of fine rootlets, grasses etc. Clutch 2–3; incubation 13–14 days. Like relatives they are inclined to be aggressive, especially when breeding, and do best in separate accommodation. Nevertheless, a successful breeding occurred in a tropical house at Harewood Bird Gardens, Leeds, England, in 1984 when a breeding pair tolerated companions ranging from zosterops and small waxbills to orioles. Provide abundant live food for rearing; various larvae including those of the wax moth are valuable.

General management These are lively active birds that thrive best in spacious accommodation; a tropical house is ideal as they are not weather-resistant, and fare best indoors from late autumn to late spring. They also flourish in planted outdoor aviaries during the summer months. They are usually good mixers in roomy accommodation, but not trustworthy with smaller or weaker companions in a limited space. They need care when first imported, but if properly acclimatised and established are usually trouble-free in experienced hands.

Copsychus saularis
Magpie Robin, Dhyal Thrush, Dhyal Bird

8½ in (21.5 cm). The whole of the head, upper surfaces and breast are glossy blue-black; the lower breast and underparts, wing bar and outer feathers of the tail are white. The bill is black, the legs and feet dark flesh-coloured. The female is mainly dark grey and grey-white.

Range Western Pakistan, India, Sri Lanka eastwards to Burma, Thailand, Indochina, southern China, Malaysia, Hainan, Philippines, Andaman and Greater Sunda islands.

Note 18 subspecies are described.

Status Abundant in many parts of the range.

Avicultural rating A popular avicultural favourite, although it is now much less freely available. Some aviary-bred birds are usually offered for sale each season, however.

Feeding Diet F, with a variety of live food. Individual birds may show an interest in fruit and wild berries.

Breeding The Magpie Robin has bred fairly frequently, and the establishment of aviary-bred strains should be possible. Breeding pairs are very aggressive and need separate accommodation, but have shown a willingness to breed in small shelter and flight accommodation (flight 8 × 6 × 6 ft, 2.4 × 1.8 × 1.8 m). Some breeders allow the parents the freedom to forage in gardens and the countryside when young are in the nest; a small hatch allowing exit and entry to and from the aviary. The birds are enclosed permanently again just before the chicks fledge. Clutch 3–5; incubation 13–15 days. Abundant small live food is essential for rearing, especially during the first critical days.

General management Single birds usually mix fairly well with companions. Introduction of the male to the female for breeding is often difficult; house them alongside each other in adjoining aviaries if possible before allowing together. Acclimatised birds are usually hardy, but must have a comfortable, frost-free shelter in winter. They are fluent songsters, and usually become tame and confiding.

Copsychus malabaricus [PLATE 97]
White-rumped Shama, Shama

PLATE 97 *Copsychus malabaricus*, White-rumped Shama

11 in (28 cm). The upper surfaces and wings are a glossy blue-black; the chin, throat and breast are black, the rest of the underparts are chestnut except for a conspicuous white rump. The tail is long and graduated; it is black, with the outer feathers white. The bill is black, the legs and feet dark flesh-coloured. The female is usually of slighter build and has a much shorter tail; her colour varies from shades of grey to grey-brown with rufous-buff lower underparts.

Range India, Sri Lanka, Nepal, Burma, Thailand, Indochina, south-west China, Malaysia, Hainan Island, Andaman Islands, Greater Sundas. Introduced into Hawaii.

Note As in the previous species, 18 subspecies are described.

Status Abundant in many areas of suitable habitat, including gardens.

Avicultural rating One of the most popular and widely-kept of all exotic softbills; it is a splendid songster and mimic. It is not a beginner's species, but is fairly easily managed; breeding successes are becoming more frequent.

Feeding Diet F, with a variety of live food. Individual birds may enjoy fruit and berries; I have one cock Shama at the time of writing that eagerly seeks out elder and blackberries in season. Shamas will 'sell their souls' for mealworms; but beware not to overfeed in a mistaken attempt to win a bird's confidence—Shamas are very confident and need little in the way of inducements to establish relationships with human owners.

Breeding Achieving compatibility between male and female is usually the biggest obstacle to progress in the breeding season, but more Shamas are now being bred in aviaries. They do best in spacious planted aviaries with abundant cover for the female to escape the attentions of the dominant cock; adjoining aviaries help to overcome introduction problems. Provide open wicker baskets or open boxes fixed into cover as potential nest sites. Depending on predator levels in the area, releasing the parents immediately the chicks hatch (confining them again just prior to fledging) is probably the best way of achieving successful rearing. One major advantage of this method is that the adults feed natural live food, which eliminates the deficiency problems that often arise when artificially-produced insects or larvae are used for rearing, even though vitamin additives are also utilised. The disadvantages are the risk of accident to free-flying adults through pests, predators or 'sportsmen'. Clutch 4–5; incubation 13–14 days. Abundant live food is vital if the young are to be reared successfully. When fledged, check carefully to establish when the juveniles are self-supporting, and remove them at once before the adult male injures or kills them—a characteristic Shamas share with Magpie Robins and other close relatives.

General management They fare best in a shelter and flight, although many birds live in smaller accommodation as pets. Properly acclimatised birds are hardy and will winter outside, with access to a comfortable frost-free shelter. Single cock Shamas can be housed with other birds, but choose companions with care, as their confident, aggressive demeanour may lead to domination of less robust species with consequent stress problems. Cock Shamas are among the bird world's finest songsters; the quality of vocalisation varies between individuals—many are addicted to mimicry and add notes of other less musical species (sparrows or starlings) to their natural song. Although I prefer to see the full utilisation of individual birds to make up potential breeding pairs whenever possible, I also freely admit that keeping an individual cock Shama as an enchanting tame, confident pet is most tempting. Few birds are more delightful companions.

Phoenicurus leucocephalus [PLATE 98]
White-capped Redstart

$6\frac{1}{2}$ in (16.5 cm). The crown and nape are white; the remainder of the upper surfaces, chin, throat and breast are black. The lower breast, underparts and rump are chestnut; the tail is chestnut with a black terminal band. The bill, legs and feet are black. The sexes are alike.

Range Mainly Himalayas, northern Burma, Indochina to eastern China.

Status Local, but abundant in parts of the range which provide suitable habitat; streams and rocky outcrops of the Himalayan foothills are well populated.

Avicultural rating Now very rarely available, but this was formerly a popular avicultural subject; it does best in experienced hands.

Feeding Diet F, with some live food. It is worth

PLATE 98 *Phoenicurus leucocephalus*, White-capped Redstart

exploring the contents of an aquarists' shop for food for this species; live food such as mosquito larvae is likely to be a valuable addition to the diet of a bird which spends much time in the wild by the waterside.

Breeding This is difficult and likely to be successful only in suitable accommodation with running water. The nest is usually a bulky structure in a cleft, fissure or hole in a bank, invariably alongside water. Clutch 3–5; incubation 12–13 days. Abundant live food is needed for rearing.

General management Correctly acclimatised birds eventually become hardy and can winter outside with access to a frost-free shelter. Although a tropical house environment may suit, they do not thrive in warm, humid conditions; a densely-planted outdoor aviary with a stream is probably better. They can be pugnacious in mixed company. They are very attractive in a suitable aviary, with their typical flickering redstart tail; in the wild they are almost inseparable from running streams.

Zoothera citrina
Orange-headed Ground Thrush

8½ (21.5 cm). The head, neck and underparts are rufous-orange, lighter towards the abdomen; the mantle, back, wings and tail are blue-grey. There is an indistinct blue-grey streak behind the eye, and a small white wing patch. The bill is black, the legs and feet flesh-coloured. The female is smaller and more slightly built; her rufous-orange feathers are duller and the slate-blue is replaced by olive-brown.

Range Western areas of Pakistan, India, Sri Lanka, northern Burma, south and south-eastern China, Malaysia, Hainan, Nicobars, Andamans and Greater Sundas.

Note 12 subspecies are described.

Status Abundant in many parts of the extensive range.

Avicultural rating Easily managed and hardy, but aggressive.

Feeding Diet G, with a small amount of minced raw meat each day. Some fruit may be taken and live food of all kinds should be included in the day's menu; these birds are well able to tackle medium to large sized locusts and they can have up to half a dozen mealworms each per day.

Breeding Breeding successes have been recorded over a fairly long period. A pair should have separate accommodation, unless in a very spacious aviary; they are extremely aggressive, especially to species with terrestrial habits sharing the same quarters. The cup-shaped nest is usually made in vegetation, often in a fork; they may use an open box or a wicker basket as a foundation. Clutch 3–5; incubation 14–15 days. Abundant live food is needed for rearing. Remove fledged chicks from the parents' aviary as soon as they are self-supporting.

General management Acclimatised birds are

winter-resistant and can be housed outdoors in a shelter and flight throughout the year; provide dry, draught-proof sleeping quarters. They are good songsters and occasionally mimic other bird-song. They spend much time at ground level, and conflict over territory requires careful choice of companions; it is best not to put them with birds that occupy lower levels.

Turdus olivaceus
Cape Thrush, Olive Thrush

9 in (23 cm). The cheeks and the whole of the upper surfaces are slaty-olive; the chin and throat are grey-white with darker streaking. The breast is olive, shading to rufous-orange on the abdomen and flanks. The bill is orange-yellow, the legs and feet orange-brown. The sexes are alike.

Range Senegal eastwards to Sudan, Ethiopia, Uganda, then south to South Africa; Central African Republic, Zaire, Angola.

Status Abundant in many parts of the range; this species is regarded as a crop pest because of damage caused to commercial fruit plantations.

Avicultural rating Hardy and easily managed, but not a popular avicultural subject.

Feeding Use a mix of Diets E and G, in the proportions 1–3; offer live food of all kinds, including snails.

Breeding It has proved a willing breeder in some collections. The cup-shaped nest is made of plant materials, and is usually hidden in vegetation; they may use a wicker basket or an open box. Clutch 2–5; incubation 13–14 days. Provide ample live food for rearing.

General management They are extremely hardy when acclimatised and can remain outdoors throughout the year if a suitable shelter is available; some birds may choose to roost in an evergreen thicket despite wintry weather. Choose their companions with care; individual birds will mix with others of similar size and temperament. Pairs should be segregated except in very spacious aviaries or a tropical house. They have typical thrush-like habits, and are generally aggressive.

Turdus merula
Blackbird

10 in (25 cm). The entire plumage is glossy-black. The bill is orange-yellow, the legs and feet brown. The female is dark brown, mottled; her underparts are slightly paler. The bill is brown.

Range Europe, North Africa, Middle East eastwards to Central Asia, Pakistan, India, Sri Lanka, southern China; introduced to New Zealand.

Note 16 subspecies are described.

Status Abundant in many parts of the range.

Avicultural rating A popular avicultural subject, which is easily managed and is a willing breeder.

Feeding Diet G, with some fruit and regular live food: mealworms, small locusts, beetles etc. These birds also enjoy a few soaked raisins and currants; and they will eat a variety of other household items ranging from grated leftover cheese to shredded green vegetables.

Breeding A pair should be the sole occupants of an outdoor shelter and flight with ample natural or artificial cover; introduce the birds to each other with great care as serious fighting between male and female sometimes occurs. Provide an open-fronted nest-box or a shelf; the nest is built of coarse plant fibres, leaves etc. Provide a small area of mud for the birds to use as a binding agent (not a lining) in nest-building. Clutch 3–5; incubation 13 days. Provide abundant live food for rearing.

General management These are handsome birds and excellent songsters, but not good mixers. They are best kept permanently in an outdoor shelter and flight; they are very hardy but need a sheltered area for roosting.

Note In the UK only aviary-bred and closed-ringed birds of this species can be offered for sale under the provisions of the Wildlife and Countryside Act, 1981 (Schedule 3, Part 1).

Turdus philomelos [PLATE 99]
Song Thrush

PLATE 99 *Turdus philomelos*, Song Thrush

9 in (23 cm). The upper surfaces are olive-brown; the underparts buff-white, more rufous on flanks, with dark brown spots on the breast, paler towards the abdomen. The bill is dark brown, the legs and feet flesh-brown. The sexes are alike.

Range Europe eastwards to Central Asia; some birds from northern areas of the range winter in North Africa.

Status Abundant in many parts of the range.

Avicultural rating A good avicultural subject which is hardy, easily managed and a free breeder.

Feeding Diet G, with items of live food and some fruit. As well as mealworms, locusts etc, these birds enjoy garden snails.

Breeding Similar to the Blackbird. They do best in a secluded situation; although these birds breed readily, in my experience they will desert the nest if subjected to disturbance. Clutch 3–6; incubation 13–14 days. Mud is necessary for nest lining.

General management Similar to Blackbird.

Note In the UK only aviary-bred and closed-ringed birds of this species can be offered for sale under the provisions of the Wildlife and Countryside Act, 1981 (Schedule 3, Part 1).

Pomatorhinus erythrogenys [PLATE 100]
Rusty-cheeked Scimitar Babbler

PLATE 100 *Pomatorhinus erythrogenys*, Rusty-cheeked Scimitar Babbler

9½ in (24 cm). The upper surfaces are earth-brown, with the forehead and sides of the head, mantle, wings and abdomen chestnut; the underparts are white with streaking and spotting on the throat and breast. The bill, legs and feet are yellow-brown. The sexes are similar.

Range Himalayas, Nepal, Burma, north-west Thailand and eastwards to south and south-east China; Taiwan.

Note 14 subspecies are described.

Status Good populations in some parts of the range; the status of the eastern subspecies is uncertain.

Avicultural rating Easily managed after acclimatisation, but not a popular avicultural species.

Feeding Diet G, plus live food. Some individuals appreciate a few berries from time to time but normally little fruit is consumed.

Breeding This babbler builds a large domed nest, usually concealed among low-growing vegetation. Clutch 3–4; incubation 14 days. Potential breeding pairs should be housed in secluded, planted aviaries without companions.

General management Acclimatised birds are hardy and can be accommodated in a shelter and planted flight outdoors throughout the year. They are generally secretive, but active. They are not trustworthy with smaller or weaker companions. They have a distinctive, loud call and song.

Garrulax albogularis
White-throated Laughing Thrush, White-throated Jay-Thrush

12 in (30 cm). The upper surfaces and a band across the breast are olive-brown; the chin and throat are white. The lower underparts are rufous-brown. The long tail has a white or grey terminal band. The bill is brown, the legs and feet grey. The sexes are similar.

Range Pakistan, northern India and Himalayas eastwards to north-west Vietnam, south-west China; Taiwan.

Status Abundant in many parts of the range.

Avicultural rating A popular and easily managed large softbill which becomes hardy.

Feeding Diet G, with chopped or minced raw meat (dust with multivitamin powder) daily. Most kinds of live food are accepted and the birds also appreciate the occasional small dead mouse. Also offer Diet E.

Breeding They build a bulky cup-shaped nest. Clutch 3–4; incubation 13–14 days. Many nestling jay-thrushes in aviaries are killed and eaten by their parents; so it is essential that a) abundant live food for rearing is available at all times and b) adult birds are provided with diversionary interests in the aviary, such as a deep 'litter' of leaf-mould or dead leaves, pieces of rotting natural timber, soil boxes etc.

General management Acclimatised birds are totally weather-resistant. They are generally untrustworthy with other birds; and are intelligent, with jay-like habits. Their size and active disposition demand spacious accommodation; shelter and flight accommodation is best. They are very confident and have noisy calls.

Garrulax leucolophus
White-crested Laughing Thrush, White-crested Jay-Thrush

12 in (30 cm). The whole of the head (including the crest), chin, throat and upper breast are white; there is some grey-white in the feathers at the back of the crest and head. A distinctive streak from the base of the bill through the eyes is black. The mantle, flanks and under tail coverts are rufous-brown; the remainder of the plumage is olive-brown. The bill is black, the legs and feet dark grey. The sexes are similar.

Range Himalayas south-eastwards to Burma,

parts of Thailand, Indochina, south-west China; Hainan.

Status Abundant in many areas.

Avicultural rating See White-throated Laughing Thrush.

Feeding/Breeding/General management Similar to White-throated Laughing Thrush. These highly intelligent birds have a restless, inquiring nature and do best in a large planted aviary. They are not safe with small or medium-sized companions. Pairs, or small groups, live well together. They are noisy.

Garrulax chinensis
Black-throated Laughing Thrush, Black-throated Jay-Thrush, Chinese Jay-Thrush

---

11 in (28 cm). The forehead (including a small, bristle-like crest), face (including the area around the eyes), chin and throat are black; there is a conspicuous white cheek patch. The upper surfaces are olive-brown, greyer on the back of the head and nape; the underparts are grey, with the flanks and under tail coverts olive-brown. The tail is olive with a black terminal band. The bill is black, the legs and feet dark brown. The sexes are similar.

Range Burma, parts of Thailand eastwards through Indochina to southern China; Hainan.

Status Abundant in many parts of the range.

Avicultural rating Although first imported more than 100 years ago, only within recent years has this bird become a fairly popular avicultural subject; it is hardy and easily managed.

Feeding/Breeding/General management Similar to White-throated Laughing Thrush, but this species is of a quieter disposition and mixes with other medium-sized species more easily. It is not as active and inquiring in habit as the previous species. It has bred in aviaries.

Garrulax erythrocephalus
Red-headed Laughing Thrush, Red-headed Jay-Thrush, Chestnut-crowned Laughing Thrush

---

10½ in (26.5 cm). The forehead and crown are chestnut, the upper surfaces and tail olive-brown, except for the golden-olive flight feathers, a suffusion of golden-olive on the tail feathers, and chestnut and black near the shoulder of the wing. The chin is dark brown, the throat and breast rufous, the feathers with lighter edges; the lower underparts are buff-grey. The bill is black, the legs and feet brown. The sexes are similar.

Range Western Himalayas, Nepal, Sikkim, Bhutan eastwards to south-west China; peninsula Thailand, Malaysia.

Status Reasonably abundant in many areas of the range.

Avicultural rating Becoming increasingly popular; there are few importations, but the species breeds well in suitable accommodation.

Feeding/Breeding/General management Similar to White-throated Laughing Thrush. This species has a generally quieter disposition than others of the family and is a fairly good mixer. It is inquisitive and spends much time at ground level. It is seen at its best in a well planted outdoor aviary which offers the possibility of much regular investigation by the birds.

Leiothrix argentauris
Silver-eared Mesia

---

7 in (17.5 cm). The forehead, chin, throat, upper breast and neck are orange-yellow; the crown and sides of the face are black with large silver-white ear patches behind the eyes. The upper surfaces, including the tail, are brown; the rump and tail coverts are red. There are conspicuous areas of red and golden-orange in the wings. The lower underparts are buff-brown. The bill is yellow, the legs and feet flesh-coloured. The female is less

bright and has an orange-buff rump and tail coverts.

Range Himalayas, Burma, Thailand, Indochina, Malaysia, southern China, Sumatra.

Note There is some variation of colour among the 8 described subspecies.

Status Good populations in many parts of the range.

Avicultural rating A colourful 'up-market' relative of the Pekin Robin; it is now only occasionally available, but is hardy and easily managed.

Feeding Use a mix of Diets E and G, with regular items of live food. Various berries are enjoyed as available. Some birds like picking through a dish of mixed seeds.

Breeding This is a generally secretive species, despite its striking plumage. A secluded, well-planted aviary, or a tropical house environment, is best for breeding. If in a small aviary there should be no other occupants; like the Pekin Robin, it frequently disturbs other nesting birds sharing the same accommodation. It builds a cup-shaped nest, usually low down and hidden among vegetation. Clutch 3–4; incubation 13–14 days. Abundant live food is essential for rearing; a shortage of suitable items often leads to rejection of the nestlings by the parents.

General management Acclimatised birds usually prove extremely hardy and can over-winter in shelter and flight accommodation. This species prefers a quiet, well-planted environment. It usually mixes well with similar-sized birds except during the breeding season. It has a pleasant song.

Leiothrix lutea
Pekin Robin, Red-billed Leiothrix, Pekin Nightingale

---

6 in (15 cm). The top of the head and the nape are bright olive; the mantle, back and rest of the upper surfaces are grey-olive. The chin and throat are yellow with an orange-red border. The rest of the underparts are yellow, with more grey on the flanks. There is a prominent grey moustachial streak, with some yellow on secondaries and primaries. The bill is red, the legs and feet light brown. The females are similar, but many are much paler. Visual sexing is difficult as some variation in plumage among the six subspecies is confusing. The male's song is the best guide.

Range Western Himalayas eastwards to Burma, Assam, northern areas of Vietnam, southern China. Introduced to Hawaii.

Status Generally abundant in many parts of the range.

Avicultural rating Arguably the most popular of all softbills, it is easy to manage and an excellent first choice for the aviculturist wishing to gain experience with this group of birds.

Feeding Use a mix of Diets E and G. Regular supplies of live food are also important. Many Pekin Robins consume quantities of seed and the species has been maintained on a staple seed diet alone; such spartan fare is not advised and the birds will be happier and healthier on the mixed diet recommended.

Breeding These are fairly willing breeders, but there are many instances of nestlings being abandoned or ejected from the nest within a few days of hatching; a shortage of suitable live food is the usual cause. Provide an abundant supply of items including some mealworms (cut up at first), wax moth larvae, crickets, small locusts, smooth caterpillars etc. A shovelful of manure placed inside the breeding aviary usually attracts a large numbers of flies in the summer which the adult birds will catch. Clutch 3–4; incubation 14 days. Adults fare best as the sole occupants of a breeding aviary, if possible.

General management Acclimatised birds are extremely hardy and can be housed permanently in outdoor aviaries; provide a dry, draught-proof shelter, although many birds prefer to roost outside (usually in an evergreen bush) throughout the year. They mix well with other birds but are egg-thieves and will also take nestlings of small species. Very active, they are almost constantly on the move and if they escape are very difficult to catch, even in a confined space. The cocks are in a minority among dealer stocks and probably 80 to 90 per cent of 'pairs' sold are both female; many people mistakenly believe that the hen's attractive liquid call is 'song'. True song cannot be mistaken and is a good guide to positive sexing.

Minla cyanouroptera
Blue-winged Siva, Blue-winged Minla

6½ in (16.5 cm). The forehead and crown are grey; the rest of the upper surfaces are olive-brown with areas of suffused blue. A streak above the eye is grey-white and there is an area of white around the eyes; the underparts are grey-white. There is some blue in the flight and tail feathers. The bill is brown; the legs and feet are flesh-coloured. The sexes are similar.

Range Central and eastern Himalayas, Assam, Burma, Thailand, south-west China, Malaysia.

Status Abundant in some parts of the range.

Avicultural rating This attractive species is only occasionally available. Acclimatised birds are hardy, but do best in slight heat during the colder part of the year; they are excellent tropical house birds.

Feeding Use a mix of Diets E and G, with some live food. In a mixed community sivas will take artificial nectar if available but this is not recommended except if the birds are debilitated after importation and quarantine.

Breeding Several breedings have been recorded. They do best in secluded accommodation. A cup-shaped nest of rootlets and plant fibres is made; fine grass and hairs are used as lining. The nest is usually hidden among vegetation. Clutch 4–5; incubation 13–14 days. Provide abundant live food for rearing as a shortage may lead to desertion or ejection of the nestlings.

General management They thrive in either a tropical house or shelter and flight accommodation, although if the latter has an unheated shelter move them to a suitable indoor flight in late autumn. They are usually good mixers but will take eggs from other nesting pairs. There is a close pair-bond between well-established male and female. They have a pleasant, quiet song.

Heterophasia capistrata [PLATE 101]
Black-headed Sibia, Black-capped Sibia

9 in (23 cm). The whole of the head (including a short, bushy crest) is black; the nape and upper tail coverts are rufous. The back is grey-brown, the wings blue-grey. There is some black and rufous, plus a white bar, in the wings. The whole of the under surfaces are rufous-brown. The tail, which is long and graduated, is black and rufous with a broad dark grey terminal band. The bill is black, the legs and feet brown. The sexes are alike.

Range Himalayas, east to west.

Status Abundant in many parts of a somewhat limited range.

Avicultural rating A popular avicultural subject which is easily managed and hardy.

Feeding Use a mix of Diets E and G, with some finely scraped raw meat from time to time. They should have some live food each day. Some birds enjoy berries. Nectar is taken if available but is not recommended as a regular item of diet.

Breeding Identification of true pairs is a problem. House in a tropical house compartment or a secluded well-planted outdoor aviary. They construct a fairly small cup-shaped nest of grass, rootlets etc; and will use sphagnum moss as a building material. Clutch 2–3; incubation 13–15 days.

PLATE 101 *Heterophasia capistrata*, Black-headed Sibia

Provide abundant live food for rearing.

General management Acclimatised birds are very hardy and can remain in an outdoor aviary throughout the year if a suitable dry, draught-proof shelter is available. They are active and usually very visible in a large planted flight; this is not a secretive, skulking species. They are good mixers with birds of comparable size and temperament, but probably an unsafe companion for smaller or weaker birds. They have pleasant calls but no real song.

Yuhina nigrimenta [PLATE 102]
Black-chinned Yuhina

---

4 in (10 cm). The forehead, crest and chin are black; the sides of the head and nape are grey. The rest of the upper surfaces and tail are olive-brown, the underparts grey-white or buff. The upper mandible is black, the lower red; the legs and feet are orange-yellow. The sexes are alike.

Range Himalayas, Assam, northern Burma and Indochina, eastwards to south-eastern areas of China.

Status Variable, but there are good populations in some areas.

Avicultural rating These interesting and attractive small birds are not advisable for inexperienced hands, but acclimatised examples are not difficult to manage.

Feeding Mainly Diet E, with some Diet F, and small live food in a separate container. They also need nectar, and Diet L will suffice. If the birds enjoy relative freedom (in a planted aviary or tropical house) a small amount of good quality sponge cake can be soaked in nectar for them each morning. Various items of small live food are required; fruit flies are useful for these birds.

PLATE 102 *Yuhina nigrimenta*, Black-chinned Yuhina

Breeding The small cup-shaped nest is usually sited amid dense vegetation. Clutch 2–3; incubation 12–13 days. Abundant small live food is essential for rearing, which is likely to be easier in a tropical house than a wire-mesh aviary. Fruit flies plus house flies (hatched from 'feeder' maggots) are important in the early stages of rearing; also collect as much untainted live food as possible from gardens or the countryside—blackfly, aphids etc are suitable.

General management They need care when first imported. They are eventually easily managed, although exposure to prolonged cold and wet conditions is not acceptable. Slight heat should be provided during the winter months, although they can have access to outside flights if a comfortable shelter is available. If in outdoor aviaries during the winter it is essential that artificial light is available to extend short days for extra feeding. They mix well except when in breeding condition. They have a squeaky, amusing song.

Muscicapidae—Panurinae (Parrotbills)

All but one of the 19 species of parrotbills are found in the Orient; the exception is the Bearded Reedling (*Panurus biarmicus*), which has a range extending from Great Britain across Europe to Russia, Iran and Manchuria. Although their bills bear a slight resemblance to those of parrotlike species, there is no relationship. They are mainly gregarious; large flocks forage (and sleep) in reed beds and bamboo thickets. Some species construct nests of plant material bound together with cobwebs. Shipments of parrotbills reach the market at random and usually lengthy intervals. One species, the Grey-headed, is the main importation although others have been seen from time to time. They are generally unsatisfactory subjects. Although omnivorous and apparently easy to manage, they are usually not long-lived.

Paradoxornis gularis
Grey-headed Parrotbill

$7\frac{1}{2}$ in (19 cm). The top of the head, the nape and hind-cheeks are blue-grey; the forehead, eyebrow stripe and bib are black. The lores, eye-ring and underparts are white, grey-white on the flanks and abdomen; the mantle, back, wings and tail are rufous, with some black or dark brown in the flight feathers. The bill is orange, the legs and feet flesh-coloured. The sexes are similar.

Range Eastern Himalayas, Assam, northern areas of Burma, Thailand and Indochina, South Vietnam and southern China.

Status Variable, but large numbers are recorded in some parts of the range; eastern populations are indeterminate.

Avicultural rating An unusual and interesting species, but it is often difficult to maintain in good health over a long period.

Feeding They are omnivorous, and arriving at an acceptable diet is a process of trial and error. Among items consumed by birds formerly in my possession were Diets E and G together with plain canary, white millet and sunflower seeds, pine nuts, various berries (blackberries, elder, rowans etc) and live food. There is often some reluctance to take the latter items to begin with.

Shelled peanuts are also consumed, but should be offered only in a limited quantity.

Breeding A cup-shaped nest is made among reeds, tall grass etc. Clutch 3–4; incubation 13–15 days. Results might be achieved if a small colony of birds was established in a suitably planted aviary.

General management They are frequently in poor condition after importation and quarantine, but usually recover well on a varied diet. Long-term good results with these birds are still rare. They are hardy and thrive in shelter and flight accommodation when acclimatised; it is best not to put them with smaller or weaker companions. These are active, agile and interesting birds.

PLATE 103 *Hypergerus atriceps*, Oriole Babbler

Muscicapidae—Sylvinae (Old World Warblers)

The Sylvinae is a very large subfamily with nearly 400 species ranging in size from about 3½ to 10 in (9–25 cm). They are mainly of muted shades, but a few tropical species have brighter colours. They are mainly solitary or found in pairs, and are insect-eaters. Species that breed in northern areas migrate to Africa for the winter where food (insects) is plentiful; a few include fruit in their diet. A very limited number are of interest to aviculturists, although several African and European migrant species are kept, mainly in Continental Europe. The example described here is not a typical representative of the subfamily and is available to aviculturists only spasmodically, but is a beautiful and interesting subject for experienced bird-keepers. The Oriole-Babbler's closest relatives include Camaropteras and Eremomelas.

Hypergerus atriceps [PLATE 103]
Oriole Babbler, Oriole Warbler, Moho, Warbling Moho

8 in (20 cm). The head, nape, chin, throat and upper breast are black, the feathers margined with silvery-white to give a scale-like effect; the upper surfaces are a bright olive-green, with some black in the flight feathers. The underparts are yellow, greyer towards the abdomen; the under tail coverts are dull yellow. The tail is green, yellowish below. The bill, long and curved, is black, the legs and feet light brown. The sexes are similar.

Range Senegal eastwards to Central African Republic.

Status Rare throughout the limited range. Over 50 years ago, C.S. Webb, the eminent collector, noted their rarity in West Africa, but managed to take a few to England in 1936.

Avicultural rating These are very rarely available and extremely difficult to establish. I obtained a pair in 1981 which, initially, resolutely refused *all* inanimate food.

Feeding Eventually the birds must be persuaded to accept Diet F (or similar) if they are to survive and eventually thrive. They are clearly highly insectivorous; the importer reported that my pair had come through quarantine on nothing other than mealworms and water—a situation in urgent need of change when the birds were received. Provide a high-protein insectivorous mixture and add a *very small* quantity (half a teaspoon to a cup of mixture) of melted honey to produce a slightly sticky consistency. Put some live food in the dish containing the mixture (I used very clean feeder maggots) to promote the bird's interest; some mixture will adhere to the live food when the latter is consumed. The birds eventually preferred pupae (chrysalids) to live food! Offer other types of small live food including crickets, spiders etc.

Breeding Nothing is known about captive breeding. The wild birds are said to build a nest suspended from a branch or twig; it is large, domed and with a side entrance. Clutch 2–3; incubation not recorded.

General management These are ideal occupants of a tropical house; acclimatised birds also flourish in an outdoor aviary during the summer months when their ability as insect-catchers can be seen. My birds displayed a bold, confident demeanour; they are very active and reminiscent of Honeyeaters in some behavioural characteristics, and have lovely liquid calls. It is best to put them in indoor quarters in the winter. Choose their companions with care, except in very spacious accommodation.

Muscicapidae—Malurinae (Australian Wrens)

Only 9 members of the genus *Malurus* are likely to be of interest to aviculturists. All are confined to Australia, Tasmania and some offshore islands except for the Black and White Wren (*Malurus alboscapulatus*) from New Guinea. The main characteristics are their small size, long tails and a combination of bright colours. They are extremely rare in aviculture and difficult to maintain for long periods except in the most experienced hands.

Malurus cyaneus
Blue Wren, Blue Fairy Wren, Superb Wren-Warbler

6½ in (16.5 cm). The crown, cheeks and mantle are light blue; the nape, back, throat and tail black. A band across the breast is glossy-black, the lower under surfaces grey-white. The bill is black, the legs and feet light brown. The female's upper surfaces are brown, her underparts grey-white. Her eyebrow stripe is rufous, the bill, legs and feet light brown. Many males have a period of eclipse when they resemble the females, but they can be distinguished by their black bill; mature, dominant cocks retain their colour throughout the year.

Range South-eastern areas of Australia, Tasmania, Kangaroo and King islands.

Status A familiar and fairly abundant species in many parts of the range.

Avicultural rating These birds are rarely available and a real challenge even to experienced aviculturists. A good tropical house species but they do best in a separate compartment, so that diet control is possible.

Feeding They are insectivorous and need some live food, but use of any type of gentles or maggots is not recommended, although pupae (chrysalids) are a valuable item of diet. Diet F with a few (4–6 according to size) *cut up* mealworms. Offer fly pupae more or less ad lib and ensure the continuity of supply. Collect small natural live food (aphids etc). A small amount of trout pellets (see p.27)—up to 5 g per day—can be crumbled and mixed with insectile food. Fruit flies are invaluable.

Breeding This bird builds a domed nest of plant fibres, lined with feathers and wool. Clutch 3–4; incubation 13–14 days. Provide abundant small live food for rearing; several cultures of drosophila can be rotated and flies can be hatched from surplus housefly pupae. Use a multi-vitamin supplement to dust live food where practicable. Utilise as much small natural live food as possible.

General management They are difficult to establish and need constant care until established. They fare best in a planted indoor compartment (connecting with a small outdoor flight, if possible). Provide live food in the indoor section to help conserve the quantity, which disappears quickly in the wire-netting area. Provide artificial thickets of cover wired into the high corners of the indoor section for roosting.

Muscicapidae—Muscicapinae (Old World Flycatchers)

There are more than 300 species in the Muscicapinae, which includes several of the most desirable avicultural gems—niltavas and other Asiatic flycatchers, for example. They range from Europe, Africa, Asia to Australia and the Philippines. They are insect-eaters, and are adept at capturing prey on the wing, a process known as 'hawking'. Many are hole-nesters; others produce cup-shaped nests of plant material. They are generally among the more 'difficult' and delicate of avicultural subjects but in the hands of experienced bird-keepers many thrive over long periods and breeding successes have been recorded with a number of species. Some are good tropical house birds; others do better in shelter and flight units; all except a small number of Asiatic species need some heat during the colder months of the year. Many species have only short legs and weak feet, which is an indication that they do not easily adopt temporary terrestrial behaviour in order to feed; so ensure that food and water stations are by a perch.

Ficedula narcissina
Narcissus Flycatcher

$5\frac{1}{2}$ in (14 cm). The whole of the upper surfaces (except for the back and rump), cheeks, wings and tail are black; the back, rump and all underparts are yellow. There is a yellow eyebrow streak, and a white wing patch. The bill, legs and feet are black. The female has olive-brown upper surfaces; her underparts and rump are buff-yellow.

Range Sakhalin (Japan), Yakushima and Riukiu islands; the species migrates to south-east China, Hainan, Taiwan, north Borneo, Philippines.

Status Variable, but there are still good populations in many parts of the range.

Avicultural rating They are occasionally available and much sought-after by softbill enthusiasts. There are some establishment problems, but in general this species is a fairly good avicultural subject.

Feeding Diet F, with some small live food. It is best if the latter is used only to stimulate feeding from a high-protein inanimate diet, although some living food is desirable at all times. Large gentles (maggots) are not recommended; pupae are better when the birds learn to accept them. Use 'feeder' maggots to encourage interest in the insectile mixture. Natural live food is a valuable addition to the diet.

Breeding This is difficult. Clutch 3–8; incubation 12–13 days. They make a cup-shaped nest, but the birds will also nest in boxes; sites vary from tree-holes to dense vegetation. Insect larvae are the main rearing food. The nestling period is 15 days.

General management They are not recommended for tropical house accommodation, but thrive in a shelter and flight with some slight heat during the winter. They are fond of bathing, but ensure that only shallow water containers with small pebbles inside are used, to prevent accidental drowning; the short legs and weak feet can lead to problems of this kind unless precautions are taken. They are good mixers and generally peaceful.

Ficedula mugimaki
Mugimaki Flycatcher

$5\frac{1}{2}$ in (14 cm). The head, cheeks and upper surfaces are black; there is a white wing bar and white rump. The chin, throat and breast are rufous, the lower underparts white. There is a white eyebrow stripe. The bill, legs and feet are black. The female has brown upper surfaces, a white throat, and a pale rufous-buff breast and flanks; her lower underparts are white.

Range North-east Asia; this flycatcher migrates to China, Hainan, Taiwan, the Greater Sundas and Philippines.

Status Good populations in many areas.

Avicultural rating See Narcissus Flycatcher; this species is occasionally available.

Feeding/Breeding/General management Similar to Narcissus Flycatcher.

Ficedula sapphira
Sapphire-headed Flycatcher, Sapphire Flycatcher

$4\frac{1}{2}$ in (11.5 cm). The forehead, crown and nape are shining blue; the cheeks, sides of breast, upper surfaces and tail are a darker blue. The underparts are white with the chin, throat and centre of the breast pale orange. The bill, legs and feet are black. The female's upper parts are rufous-brown; her under surfaces are similar but brighter and paler, and whitish towards the abdomen.

Range Eastern Himalayas to southern China.

Status Reasonably abundant in some parts of the range.

Avicultural rating Occasionally available. This species needs skill to establish and acclimatise and is best in experienced hands.

Feeding/Breeding/General management Similar to Narcissus Flycatcher. Like others of the family it is an inveterate bather; use only a shallow dish, in view of their small, weak legs and feet, to prevent drowning tragedies. Also ensure that the perches are of the right diameter if the bird is confined to a cage; it is best if slender, natural perches are used at all times.

Cyanoptila cyanomelaena
Blue and White Flycatcher, Japanese Blue Flycatcher

$6\frac{1}{2}$ in (16.5 cm). The crown, mantle, rump and tail are shining blue, the rest of upper surfaces cobalt-blue; the forehead, cheeks, chin, throat and upper breast are black. The rest of the underparts are white. The bill is black; the legs and feet dark grey. The female is mainly olive-brown, brighter on the rump and tail; her underparts are pale olive-grey becoming whiter towards the abdomen.

Range Manchuria, Korea, Japan; it migrates to Hainan, southern Burma, Thailand, Malaysia, Java, Borneo, the Philippines.

Status There are good populations in many areas, including some urban districts in the breeding range.

Avicultural rating These popular avicultural subjects are now only rarely available in Europe. They need care during acclimatisation but eventually are fairly robust.

Feeding Diet F, with regular items of live food (see Narcissus Flycatcher).

Breeding This species builds a nest of moss, plant fibres etc, often in a hole or cavity. Clutch 3–5; incubation 13–14 days. Abundant small live food is needed for rearing.

General management Although these birds (and other flycatchers from the Himalayas, eastern Asia etc) are often housed in tropical houses, when acclimatised they are better in outdoor shelter and flight accommodation if slight heat can be provided in winter; they should also be enclosed within a shelter during periods of severe (cold) weather and frost. They need care until established, then are reasonably straightforward but do best in experienced hands.

Niltava sundara [PLATE 104]
Rufous-bellied Niltava

PLATE 104 *Niltava sundara*, Rufous-bellied Niltava

6 in (15 cm). The crown, nape, a small area on the lower cheeks, shoulder of wings and rump are iridescent cobalt-blue; the forehead, chin, throat, sides of head and neck are black. The mantle, back and tail are a dark violet-blue; the breast and under parts rufous-orange, paler towards the abdomen. The wings are black and blue. The bill is black, the legs and feet dark brown. The female's upper surfaces are brown, paler below with a grey-white band across the throat and iridescent blue spots on the sides of the neck; her tail is chestnut.

Range Himalayas eastwards to south-west and southern China; northern Laos, Thailand, Malaysia.

Status Good populations in many areas, but they are difficult to observe in the forest undergrowth which is their main habitat.

Avicultural rating A connoisseur's bird, now only occasionally available, it needs care until acclimatised, then is extremely hardy but still requires careful management.

Feeding Diet F, with a selection of small live food; also utilise fly pupae for these birds. Limit mealworms to about 6 per day. Wax moth larvae are valuable. Individual birds will occasionally take soft fruit.

Breeding A shallow, loosely-built nest of plant fibres etc is made, usually low down and hidden among the vegetation. Clutch 4; incubation 12–14 days. Abundant live food is the secret of successful rearing, which is unlikely to be accomplished unless an adequate and varied supply is maintained. Some birds will utilise open boxes or wicker baskets for nesting purposes. A breeding pair should be the sole occupants of an aviary.

General management This is another species which does better in a shelter and flight than in a tropical house-type environment; warm humid conditions do not suit these birds. They are hardy after acclimatisation; individual birds I have owned have regularly scorned the shelter to roost outside in a bush, including throughout the winter. They usually occupy the lowest part of the aviary; they are quiet, somewhat secretive, but the male's bright colours ensure visibility. They are often active late into the evening; they hawk for flies well into the dusk period, and a moth trap (black beam) may be a good acquisition to attract insects to the aviary.

Niltava tickelliae
Tickell's Flycatcher, Tickell's Niltava, Tickell's Blue Flycatcher

6 in (15 cm). The forehead and eyebrow stripe are a shining mid-blue; the rest of the upper surfaces are cobalt-blue, the lores, wings and tail blue-

black. The chin, throat and breast are orange-rufous, then there is a marked division between breast and the white lower underparts. The bill is black, the legs and feet dark brown. The female's upper surfaces are earth-brown, her chin, throat and breast rufous-buff with white lower underparts.

Range India, Sri Lanka, Burma, Thailand, Cambodia, Indochina, Malaysia, Sumatra.

Status There are good populations in a few parts of the range, but it is becoming rare in several areas.

Avicultural rating Occasionally available and much in demand. They need care to acclimatise and establish, but eventually are not difficult to manage and can be long-lived.

Feeding Diet F, with a selection of small live food. If housed in an outdoor aviary during the summer months, the birds will capture a good deal of natural live food; in a tropical house or the closer confinement of an indoor flight more artificially-produced live food is desirable, but limit the use of mealworms, and gentles are not recommended. Use wax moth larvae, crickets, fly pupae (when the birds are accustomed to them) etc.

Breeding Occasional breedings are recorded. They usually build a nest of plant fibres etc amid the vegetation, but will also use a box or wicker basket as a base. Clutch 3–4; incubation 12–13 days. Abundant live food is necessary for successful rearing; live ant pupae and wasp grubs have both proved to be valuable rearing items if available. The hen is less passive during courtship than related species; she often sings and sometimes performs a simple display to the cock as a prelude to mating.

General management This species is not as hardy as some previously mentioned flycatchers and niltavas, but is a good occupant of a tropical house. If kept in outdoor shelter and flight accommodation it is best if the birds are brought indoors by mid-autumn, unless the shelter is heated and can be enclosed. It is advisable to split pairs at the end of the summer or in autumn; if they are closely confined together, there is a strong possibility of injury through fighting.

Eumyias thalassina [PLATE 105]
Verditer Flycatcher

PLATE 105 *Eumyias thalassina*, Verditer Flycatcher

6½ in (16.5 cm). The entire plumage is greenish-blue (verditer-blue) except for a small black patch from the base of the bill to under the eyes. The bill, legs and feet are black. The female is similar but duller (more grey) and with some white mottling on the chin and throat.

Range Himalayas eastwards to northern Laos, southern China; also Malaysia, Sumatra, Borneo.

Status Abundant in many parts of the range.

Avicultural rating A popular avicultural subject but now only rarely available.

Feeding Diet F, with varied small live food. In the aviary, it enjoys 'hawking' to take flying insects on the wing.

Breeding This species builds a bulky cup-shaped nest of plant fibres, moss etc. It is usually well hidden in the vegetation, occasionally in a crevice in a building or under eaves. Clutch 3–5; incubation 13–14 days. Abundant live food is vital if the young are to be reared successfully. Female Verditer Flycatchers are much less often seen in aviculture than the males.

General management They need care during the acclimatisation and establishment process; they may be difficult to persuade onto an inanimate diet, but this important transition, in part at least, should be achieved, for the birds will not thrive on a staple diet of artificially-reared live food, e.g. mealworms, gentles, wax moth larvae etc. They are better in a shelter and flight than a tropical house environment and eventually are hardy enough to winter outside; provide a comfortable shelter, but the birds may prefer to roost outside at all times. Watch for problems with perches and bathing caused by their very small, weak feet.

Dicaeidae (Flowerpeckers)

There are more than 50 species of flowerpeckers, which range from 3 to nearly 8 in (7.5–20 cm) long. Berries form an important part of the diet of some species. They have an important role as pollinators of flowers, but also create problems by spreading mistletoe; the berries are a favourite item in their diet and the seeds are spread through their droppings. There is an interesting adaptation of the intestines and gizzard: the nutritional element in the berries is extracted quickly and the seeds bypass the gizzard to be excreted within minutes of being swallowed. Insects etc reach the gizzard in normal fashion. These birds are generally 'difficult' avicultural subjects, but many of the problems encountered may be due to incorrect diet; the importance of berries as a regular diet item is indicated by the intestinal arrangements mentioned above. Nectar foods are usually offered as a near-staple food, and the consequences of this are rapid falling-away in condition followed by death. There is little excuse nowadays for not improving on such time-honoured but unsatisfactory diets for these birds. Mistletoe berries are not vital, but many suitable types ranging from elders to blackcurrants can be gathered in season and stored for months in a deep-freeze; this may not solve all the difficulties encountered with these interesting little birds but should prove a step forward in their husbandry.

Dicaeum trigonostigma
Orange-bellied Flowerpecker

$3\frac{1}{2}$ in (9 cm). The entire upper surfaces are slate-blue, the chin and throat grey. The under surfaces are orange, and there is an area of the same colour in the centre of the back; the rump is orange or yellow in some subspecies. The bill is black, the legs and feet dark brown. The female's upper surfaces are olive-green, her underparts buff-yellow.

Range Eastern India, Burma, Thailand, Malaysia, Greater Sundas, the Philippines.

Note A total of 16 subspecies are described, and considerable variation in plumage patterns is exhibited. The nominate subspecies, *D. t. trigonostigma*, described, is from southern Thailand.

Status Still fairly abundant, but loss of habitat is affecting populations in some parts of the range.

Avicultural rating These are interesting avicultural subjects, but rarely successful. Great skill is needed to manage them over an extended period.

Feeding Diets E, F and L should all be included in regular menus for these birds, together with the all-important regular supply of berries. Left to their own devices, flowerpeckers of most species will feed on a liquid nectar mixture to the exclusion of other foods. Various theories have been advanced as to how to overcome this kind of problem, including supplying nectar only in the morning, thus forcing the birds to consume other items later in the day. It is probably the most sensible way round the problem, but I personally prefer to offer nectar during the morning and again, for a short period, before nightfall. Fruit should be supplied finely diced; small berries are offered whole. The birds often tend to gorge on available berries and this may be another problem as there is a need to maintain continuity of supply

Breeding A difficult proposition in aviculture. The birds build a suspended nest, hanging from a slender branch by means of 'handles' fashioned from down and plant fibres. Clutch 2–3; incubation 11–12 days.

General management They are extremely difficult to establish and constant care is needed to achieve reasonable longevity. They are ideal occupants of a tropical house; as they are susceptible to a drop in temperature, which causes clear signs of discomfort, they are not recommended for outdoor aviaries at any time. They are frequently aggressive with their own kind, but good companions for other small softbills such as zosterops.

Dicaeum cruentatum
Scarlet-backed Flowerpecker, Red-backed Flowerpecker

3½ in (9 cm). The upper surfaces, from the forehead to the contrasting black tail, are bright red; the underparts have a central grey-white band, with the wings and sides of the body black. The bill, legs and feet are black. The female is olive-grey above, paler below, and with a red rump.

Range Northern India eastwards to Indochina and southern China, Hainan; eastern Thailand, Malaysia, Sumatra, Borneo and nearby islands.

Status Widespread and reasonably abundant in many areas.

Avicultural rating See Orange-bellied Flowerpecker.

Feeding/Breeding/General management Similar to Orange-bellied Flowerpecker.

Nectariniidae (Sunbirds)

Sunbirds are effectively the Old World equivalent of the New World hummingbirds; there are more than 100 species in Africa, South-east Asia and Australia. Most have long, down-curved bills, and are mainly nectar-feeders with some insects. Many species exhibit spectacular sexual dimorphism; in others, the male and female are similar in appearance. Many are excellent avicultural subjects; the African species are most frequently seen, the Asiatic types less frequently. The former group seem better and more reliable in aviculture; many Asiatic species travel badly, are difficult to quarantine and, as a consequence, are often in a very debilitated condition when available for resale. Asiatic sunbirds are often much more insectivorous than their African counterparts. Most sunbirds are excellent tropical house subjects but are often aggressive and intolerant of either their own kind or related species; male-female relationships are also somewhat fragile except during the short courtship and breeding period. Some species are hardy and excellent in outdoor planted shelter and flight accommodation, but few aviculturists would risk all-year occupation with the prospect of dubious (prolonged cold and wet) winter weather. Several species can be bred in aviaries, but the event is rare; compatibility between the male and female is one problem, also the fact that in dimorphic species the more spectacularly-plumaged males usually greatly outnumber females in shipments.

Anthreptes longuemarei
Violet-backed Sunbird

5 in (13 cm). The upper surfaces are an iridescent violet; the underparts are a contrasting white. The pectoral tufts are yellow. The wings are darker blue. The bill is black, the legs and feet dark grey. The female's upper surfaces are grey-brown, her upper tail-coverts violet and underparts grey-white; there is a patch of yellow on her abdomen.

Range Senegal, Gambia, Guinea eastwards to southern Chad, Central African Republic, Zaire,

Sudan, Uganda and southwards through Tanzania, Malawi, Zambia to northern Mozambique and eastern areas of Zimbabwe.

Status Variable; a wide distribution, but nowhere are these birds abundant.

Avicultural rating Occasionally available, but one of the more 'difficult' of African sunbirds.

Feeding Diet L forms only part of the diet for these birds; the shorter, straighter bill indicates different feeding habits to others of the family. Soft fruit is an important diet item. Provide pears, tomatoes, grapes, etc in half-sections, not diced. They also need abundant small live food; *Drosophila* and other small flies are best. In the wild, only a moderate amount of nectar is consumed in comparison to other foods.

Breeding This species builds a domed nest of fine grass and plant fibres, lined with softer materials and usually placed low down among vegetation. Clutch 2; incubation 14–16 days.

General management A tropical house environment is ideal for this species; they are gregarious in the wild and, if space is available, a small group may be housed together. They are very active, agile and tit-like in behaviour, which is a further indication of different feeding habits. Up to the present time this is not a very successful species in aviculture.

Nectarinia senegalensis
Scarlet-chested Sunbird, Scarlet-breasted Sunbird

$5\frac{1}{2}$ in (14 cm). The forehead, crown and moustachial stripe are an iridescent green; the throat darker green. The upper surfaces, lower breast and underparts are dark brown, the throat and breast red. The bill, legs and feet are black. The female's upper surfaces are grey-brown, her underparts olive-yellow; there is some grey mottling on her throat.

Range Senegal to Nigeria and eastwards through Zaire, Central African Republic to southern Sudan, Ethiopia and southwards to Namibia and South Africa.

Status Widespread and abundant in areas of suitable habitat.

Avicultural rating One of the most popular sunbirds in aviculture, it is hardy and easily managed after acclimatisation; but has a tendency to lose the brilliant red colour if closely confined (caged) without the use of an artificial colouring agent. This problem rarely arises, however, in a good aviary or tropical house environment and with a suitable, varied diet.

Feeding Diet L, with a generous supply of small live food such as *Drosophila* and other small flies. Many aviculturists offer plain sugar-water late in the day; I prefer the recommended mixture for all feeds.

Breeding Successful captive-breedings have been recorded with this species. The first essential (as with many other sunbird species) is to achieve compatibility between the pair; they do best in well-planted accommodation, so that the female can escape the too-pressing or aggressive attentions of the male. The nest in the wild is domed and suspended; boxes or baskets are used in aviaries. Clutch 2; incubation 13–14 days. Small insects are the main rearing food and an abundant supply is essential to ensure success; drosophila cultures are important—also allow housefly pupae to complete their cycle and feed the resulting flies. These should be placed in glass jars, with transparent plastic kitchen film stretched over the neck and pierced with a protruding straw or twig; the emerged flies crawl up the twig and escape through the perforation in small numbers, rather than a wasteful mass exodus.

General management They are successful in a tropical house or outdoor shelter and flight accommodation, but not hardy enough to overwinter without some protection and warmth. They are generally reasonable mixers, especially in spacious quarters, but watch their behaviour with smaller or weaker companions.

Nectarinia sperata
Van Hasselt's Sunbird, Purple-throated Sunbird

3½ in (9 cm). The forehead, crown and nape are an iridescent gold-green; the mantle and shoulders are black, with the rest of the upper surfaces a very dark blue-green. The chin and throat are iridescent amethyst; the breast and underparts are maroon-red becoming blackish towards the abdomen. The bill, legs and feet are black. The female's upper surfaces are grey-brown, with a faint green suffusion on the mantle; her light yellow-olive underparts become paler towards the abdomen.

Range Eastern Pakistan, Assam, Burma, Bangladesh, Thailand, Laos, Cambodia, Vietnam, Malaysia, Borneo, Java, Sumatra and other islands of Greater Sundas, Palawan and the Philippines.

Note 12 subspecies are described.

Status Variable, but abundant in areas of suitable habitat, including cultivated areas and gardens.

Avicultural rating One of the smallest and most attractive of the Asiatic sunbirds, it is occasionally imported and a good avicultural subject, although not for the inexperienced.

Feeding Diet L, with ample small live food. *Drosophila* and other small flies are best.

Breeding This is a challenge but the species has been bred in aviculture. A small nest of moss, lichens and plant fibres is constructed. Clutch 2; incubation 13–14 days. Abundant live food is essential for successful rearing.

General management They fare best in a tropical house or sheltered outdoor accommodation in the summer only. They are fairly hardy when acclimatised but not weather-resistant. They are good companions for other small softbills such as zosterops, except when in breeding condition.

Nectarinia chalybea [PLATE 106]
Lesser Double-collared Sunbird, Southern Double-collared Sunbird

PLATE 106 *Nectarinia chalybea*, Lesser Double-collared Sunbird

4½ in (11.5 cm). The upper surfaces are an iridescent bronze-green; the chin and throat are similar, but blue-green or blue with a broad band of red below. The lower underparts are grey-white or grey-brown. The wings and tail are darker with an olive-green suffusion; the upper tail coverts are olive. The bill, legs and feet are black. The female's upper surfaces are grey-brown, her underparts pale grey.

Range Angola, Zambia, Tanzania, Malawi and southwards to Zimbabwe, South Africa.

Note 6 subspecies are described.

Status Good populations in most parts of the range where suitable habitat exists; they are also found in gardens, and cultivated areas.

Avicultural rating A popular avicultural favourite, although rarely available at present. They are easily managed, but not for the inexperienced.

Feeding/Breeding/General management Similar to Scarlet-chested Sunbird.

Nectarinia mariquensis
Mariqua Sunbird

5 in (13 cm). The upper surfaces are an iridescent bronze-green; the throat is similar but with a coppery tinge. The breast is metallic blue; there is a maroon-red band across the lower breast, and the lower underparts are black. The bill, legs and feet are black. The female's upper surfaces are grey-brown; she has a pale eye-stripe, and is buff-white below with some streaking. The juveniles are similar but with a black throat.

Range Southern Sudan, Ethiopia, Somalia southwards through Kenya, Uganda, eastern Zaire to Angola, Zambia, Zimbabwe, eastern areas of South Africa.

Status Reasonably abundant in many areas.

Avicultural rating A popular and easily managed sunbird.

Feeding/Breeding/General management Similar to Scarlet-chested Sunbird.

Nectarinia famosa [PLATE 107]
Malachite Sunbird, Yellow-tufted Malachite Sunbird

9 in (23 cm) including a 4-in (10-cm) tail. The

PLATE 107 *Nectarinia famosa*, Malachite Sunbird

plumage is mainly an iridescent emerald-green, the wings and tail darker. The pectoral tufts are yellow. The bill, legs and feet are black. The non-breeding plumage is grey-brown, paler below, but with a long tail. The female's upper surfaces are grey-brown; she is buff-yellow below, with a short tail.

Range South-eastern Sudan, Ethiopia, southwards through Kenya, Uganda, eastern Zaire to Zimbabwe, South Africa.

Status Scarce in many northern parts of the range, but more abundant in the south.

Avicultural rating A beautiful and highly-prized species; acclimatised birds are hardy. They are aggressive.

Feeding Diet L, with a plentiful supply of *Drosophila* and small flies.

Breeding Successful breedings in aviaries have been recorded with this species. It constructs a spherical nest, in the wild suspended from a bush or low tree. Clutch 1–2; incubation 13–14 days. An abundant supply of *Drosophila*, flies, aphids etc is vital for successful rearing.

General management This is an extremely belligerent species, so care is needed if housing with other species; it fares best with similar-sized (or slightly bigger) species which do not compete for food and territory. Pairs need especially careful scrutiny until some level of compatibility is achieved. They are not good occupants of a tropical house-type environment; a warm and humid atmosphere does not suit. Acclimatised birds can remain in outdoor shelter and flight accommodation throughout the year and with free access to an open aviary, provided that a comfortable and slightly warm shelter is available.

Aethopygia siparaja
Yellow-backed Sunbird, Crimson Sunbird

5½ in (14 cm) including a long, pointed tail. The forehead and crown are an iridescent green; the nape, mantle, back, cheeks, chin, throat and breast are crimson. The rump is yellow, the lower underparts olive. The moustachial streak is a metallic violet-blue. The wings are olive, the tail metallic green. The bill, legs and feet are black. The female is a dull olive-green above, paler grey-green below; she has a short, rounded tail.

Range Himalayas and northern India eastwards to Bangladesh, Burma, Thailand, Cambodia, Vietnam, Hainan; Malaysia, Sumatra, Borneo, Java, Nicobar, Celebes and Philippine islands.

Note Considerable variation of plumage occurs among the 16 described subspecies; the nominate subspecies, *A. s. siparaja* from Malaysia, Sumatra and Borneo, is probably the most likely to be encountered in aviculture.

Status Abundant in many parts of the range, although migratory movement causes some fluctuations.

Avicultural rating A brilliantly-coloured but delicate and 'difficult' avicultural subject, which is rarely available at present.

Feeding Diet L, with abundant small live food; in general, Asiatic sunbirds appear much more insectivorous than their African relatives and a constant supply of *Drosophila* and other small flies is essential.

Breeding This bird builds a pendent nest of fine plant fibres etc with an entrance at the side. Clutch 2–3; incubation 13–14 days.

General management It is likely that properly-acclimatised and fit Yellow-backed Sunbirds would do best housed in sheltered outdoor shelter and flight accommodation; but the tendency is (mainly because of their justified reputation as difficult to maintain) to accommodate them in a tropical house or conservatory environment, which is probably not to their advantage. A good compromise is the combination of a glazed conservatory-type building with an outdoor flight. They are often in poor condition after importation and quarantine, and then find it extremely difficult to recover full fitness.

Nectariniidae—*Arachnothera* (Spiderhunters)

This is a small genus of 10 species which are closely related to the sunbirds; they are distributed in many parts of Malaysia and other Oriental regions. They have dull plumage and long, curved bills. Almost exclusively insectivorous, they build suspended cup-shaped nests and attach them to the underside of large leaves with plant fibre 'stitches'.

Arachnothera affinis
Grey-breasted Spiderhunter

7 in (18 cm). The upper surfaces are olive-green; the underparts are grey-green with pronounced streaking on the throat and breast. The bill, legs

and feet are black. The sexes are alike.

Range Southern Burma, Thailand, Malaysia, Sumatra, Java, Borneo and Bali.

Status Abundant in many parts of the range, but habitat loss has reduced numbers in some areas.

Avicultural rating Occasionally available, they are difficult to establish. Many shippers and importers are unwilling to handle them because of their specialised diet and the difficulty of switching birds to a substitute diet.

Feeding Abundant small live food is essential to begin with; the birds can eventually be 'weaned' to soft fruits and small amounts of insectile mixtures (Diets E and F). Nectar (Diet L) should also be provided, but the emphasis on live insects as the major part of the diet is likely to continue. Feed soft fruits in large (half) sections and sprinkle them lightly with insectivorous mixture. Supply flies and other small insects.

Breeding In the wild the nest is built of moss, plant fibres, cobwebs etc, then stitched to the underside of a large leaf; banana leaves are a favourite. In some instances silk produced by spiders is used as a 'thread' to attach the nest to the leaf. Clutch 2; incubation 12–13 days. Abundant *Drosophila* and other small flies would be essential for successful rearing.

General management Despite the difficulty of persuading spiderhunters to sample inanimate diets, they are extremely rewarding birds for experienced aviculturists; they often become tame and confiding. A tropical house or conservatory environment is ideal for them. They usually mix well with other species but initial careful observation is recommended when introducing them into an established mixed community. Acclimatised birds thrive in a planted outdoor aviary (shelter and flight) during the late spring and early autumn periods, but are not weather-resistant.

Zosteropidae (Zosterops)

This family contains more than 80 species with a wide distribution throughout Africa, south-east Asia and Australasia. All are small-sized (average 4 in, 10 cm) and have characteristic green or yellow plumage with conspicuous white eye-rings. The sexes are alike. They are mainly arboreal in habit and feed on nectar, fruit and insects; they cause considerable damage in some fruit-growing areas. Escaped birds are now at pest proportions in parts of California; they are regarded as a pest, also, in fruit-growing districts in Australia. Many are popular avicultural subjects, easily managed and willing to breed, although the number successfully reared in aviaries is unsatisfactory so far. A mixed diet of nectar, fruit, insectile food and live food is necessary; they are eventually fairly hardy and will winter in an outdoor aviary provided that a comfortable shelter is available. They are excellent in a tropical house, but, if it is small, zosterops may damage some plants.

Zosterops erythropleura
Chestnut-flanked Zosterop, Chestnut-flanked White-eye

$4\frac{3}{4}$ in (12 cm). The upper surfaces are grey-green; the tail is darker green. The chin, throat and breast are yellow; the lower underparts are grey-white. There is a variable area of chestnut-brown on the flanks, and a white eye-ring. The bill is black, the legs and feet dark grey. The sexes are alike.

Range Manchuria, south-eastern areas of China, Korea; migrates to parts of Burma, Thailand and Cambodia.

Status Abundant in many areas.

Avicultural rating Less frequently available than Oriental Zosterop (*Z. palpebrosa*). An attractive and easily-managed species.

Feeding Use a mix of Diets E, F and L, with small live food. Fruit should not be diced but offered in half sections; pear is a great favourite and soft berries (elder, blackberries etc) are also greatly enjoyed. Unlike, for example, flowerpeckers (*Dicaeum*), they rarely exhibit a clearly defined preference for one type of food, and diets are therefore naturally varied; the birds will even pick through a dish of small seeds if housed with finches. Live food should consist of flies and small soft-bodied insects and larvae; many enjoy fly pupae (chrysalids).

Breeding A cup-shaped nest of grass, plant fibres etc is constructed, usually in vegetation, but in my aviaries this species and *Z. palpebrosa* have attached the nest to the wire-netting sides and divisions of accommodation. Clutch 2–4; incubation 11–13 days. They are very aggressive when nesting and do best in separate accommodation unless housed in a spacious aviary or tropical house. Many zosterops build and incubate successfully, only to abandon or eject the youngsters within hours or days of hatching. Live food (or rather not enough of it) is probably the reason for this behaviour; even for tiny birds such as these the amount required to feed a brood of 2–5 nestlings is substantial, and insects appear to be a near-staple food for the babies at first. Plan ahead and have several *Drosophila* and fly cultures ready for the approximate hatching date; if the birds are in an outdoor wire netting aviary, establish a system of feeding flies, so all do not disappear with the removal of container lid (see 'Breeding', Scarlet-chested Sunbird). Aphids, blackfly etc are also valuable rearing foods if not contaminated by toxic sprays etc.

General management Newly imported zosterops are often in poor feather condition, due to a combination of factors including overcrowding during shipping and quarantine, the sticky nature of their diet, the lack of bathing facilities etc. Such birds need care but can usually be acclimatised and established without problems. The species thrives in a tropical house or conservatory environment or an outdoor shelter and flight. Despite their small size and fragile build, acclimatised birds are hardy; they do best with a frost-free shelter. They are good mixers except when breeding.

Zosterops japonica
Japanese Zosterop

$4\frac{1}{2}$ in (11.5 cm). The upper surfaces are mid-green with an area of yellow on the forehead; the chin, throat and upper breast are yellow. The lower breast and belly are grey. There is a white eye-ring. The bill is black, the legs and feet grey. The sexes are alike.

Range Japan, eastern and southern China, Hainan, Burma, northern areas of Thailand, Laos, north Vietnam; Hong Kong, Taiwan.

Status Generally abundant in many parts of the range.

Avicultural rating Occasionally available and easily managed.

Feeding/Breeding/General management Similar to Chestnut-flanked Zosterop.

Zosterops palpebrosa [PLATE 108]
Oriental Zosterop, Oriental White-eye, Indian Zosterop, Indian White-eye

$4\frac{1}{4}$ in (11.5 cm). The upper surfaces are olive-green suffused with yellow; the chin, throat and breast are yellow, sometimes the area of yellow extends down the centre of the belly to the abdomen or, in some subspecies, the entire underparts are yellow. The forehead is yellow in some subspecies, olive-green in others. There is a white eye-ring. The bill is black; the legs and feet are dark grey. The sexes are alike.

Range Eastern Afghanistan, India, Sri Lanka, Bangladesh, Burma, northern Thailand to southern China; Malaysia and Greater Sundas; also Himalayas, Nepal, Bhutan.

Status Abundant in many parts of the extensive range.

Avicultural rating This is the most frequently

PLATE 108 *Zosterops palpebrosa*, Oriental Zosterop

imported of the Asiatic zosterops, it is easily managed and pairs are willing nesters in a suitable environment.

Feeding/Breeding/General management Similar to Chestnut-flanked Zosterop.

Zosterops atricapilla
Black-capped Zosterop, Black-capped White-eye

4¼ in (11.5 cm). The upper surfaces are grey-green; the forehead, face and crown are sooty-black. The underparts are silvery-grey. There is a white eye-ring. The bill is black, the legs and feet dark grey. The sexes are alike.

Range Northern, central and southern areas of Sumatra; north Borneo.

Status Good populations in some parts of the limited range.

Avicultural rating Importations are spasmodic, but they are valuable additions to the available range of zosterop species and subspecies in aviculture.

Feeding/Breeding/General management Similar to Chestnut-flanked Zosterop. It has a slightly finer bill than its relatives and this species also seems slightly more insectivorous; in general terms, it thrives on the same treatment as for others of the family. Modest numbers became available during the early part of the 1980s and have subsequently shown themselves to be willing breeders.

Zosterops senegalensis [PLATE 109]
African Yellow Zosterop, African Yellow White-eye

PLATE 109 *Zosterops senegalensis*, African Yellow Zosterop

4 in (10 cm). The upper surfaces are green (or yellow-green), the underparts bright yellow. The flights and tail feathers are darker. There is a white eye-ring. The bill is black, the legs grey. The sexes are alike.

Range The wide range extends from Senegal eastwards across Africa to Ethiopia; south to Angola in the west, Mozambique, Natal in the east; Great Comoro Islands.

Note 21 subspecies are described and there is some variation in the plumage; it is generally much more yellow than that of the Asiatic species available to aviculturists.

Status Abundant in many parts of the range and regarded as a crop pest in fruit-growing areas, but distribution is patchy; it is absent from the forests of central Africa (Congo).

Feeding/Breeding/General management This bird is slightly more insectivorous in habit than some Asiatic species; it is agile and tit-like, exploring leaves and twigs for insects. The birds are gregarious, as are others of the family, but prove to be aggressive defenders of breeding territory. They need secluded, planted accommodation to achieve breeding success; abundant live food is required for rearing. Sexing is a problem, but like all others of the family only the cock has the soft, attractive song which is an excellent sexing guide.

Emberizidae—Emberizinae (Buntings)

This subfamily comprises nearly 280 species of Old and New World distribution; it includes buntings, longspurs, juncos, song sparrows, warbling finches, grassquits etc. Many are familiar avicultural subjects. They are mainly seedeating, but some also include considerable amounts of live food in their diet, especially when rearing young. Most are terrestrial in habit; and are often aggressive in defence of their territory. They are generally good avicultural subjects as long as the need for a varied diet is observed. Breeding pairs do best in separate accommodation; individuals often mix well with other species.

Emberiza citrinella
Yellow Bunting, Yellowhammer

$6\frac{1}{2}$ in (16.5 cm). The head, chin and throat are canary-yellow; the nape, mantle, back and rump are rufous-brown or chestnut. The underparts are yellow to yellow-buff. The head, chin and throat have brown streaks and spots; the back and sides have rich-brown markings. The upper mandible is brown, the lower mandible horn-coloured. The plumage is less bright in winter. The female is duller and with heavier spotting and streaking.

Range UK eastwards across Europe to Russia, Siberia. Introduced to New Zealand.

Status Generally good populations, although there are some fluctuations due to habitat changes and climatic extremes.

Avicultural rating A popular avicultural subject.

Feeding Diet C, with regular items of live food. Natural and cultivated green food is important; cress, groundsel, chickweed, shepherd's purse etc are valuable. Wild seed mixtures are enjoyed. Adequate grit is essential.

Breeding These are nervous birds that need secluded, planted aviaries to breed successfully. It is best if the breeding pair are the sole occupants of small shelter and flight accommodation; if other aviaries adjoin, ensure that related species are not alongside or nearby, or screen with suitable dividers. Clutch 3–5; incubation 13–14 days. They are highly insectivorous when rearing and abundant supplies of natural and artificial live food are needed. Remove the young from the breeding aviary as soon as they are independent.

General management In my experience, they have slightly uncertain temperaments for mixed groupings, although single individual birds often integrate successfully. They are hardy and weather-resistant, but do best with a comfortable shelter and evergreen roosting sites in the flight. They will mix with other finches and buntings during the winter, but watch for signs of aggression as the birds come into breeding condition.

Note In the UK only aviary-bred and closed-ringed birds of this species can be offered for sale under the provisions of the Wildlife and Countryside Act, 1981 (Schedule 3, Part 1).

Emberiza tahapisi
Seven-striped Bunting, Cinnamon-breasted Rock Bunting

6 in (15 cm). The head, chin and throat are black

with white streaks down the centre of the crown, and above and below the eye; the rest of the upper surfaces are rufous. The underparts are cinnamon-brown, the wings and tail brown. The bill is horn-coloured, the legs and feet brown. The female's head is grey-brown with duller markings; the chin and throat are grey.

Range Sierra Leone eastwards across Africa to Saudi Arabia; south to north-eastern Uganda, north-west Kenya; Socotra.

Status Abundant in suitable habitat in many parts of the range.

Avicultural rating A handsome and easily-managed species which is usually a fairly willing breeder.

Feeding Diet A, with spray millet, seeding grasses, items of green food and some live food. Adequate grit is essential.

Breeding It builds an open cup of grass, rootlets and other plant fibres, usually low down in vegetation or at ground level. Clutch 2–4; incubation 13–14 days. Abundant live food is needed for successful rearing.

General management They are usually good mixers but breeding pairs are unlikely to tolerate companions in small aviaries. They are very active and spend considerable time at ground level. They become hardy but do not enjoy prolonged cold and wet conditions; if in shelter and flight accommodation, provide some heat in the shelter during very cold and wet weather.

Emberiza flaviventris [PLATE 110]
Golden-breasted Bunting

6½ (16.5 cm). The upper surfaces are rufous-chestnut (darker in some birds); the feathers of the wing coverts have white margins and tips. The chin is white, the throat and breast yellow. The crown and sides of the face are black with a white stripe along the centre of the crown and an area of white on the cheeks. The bill is horn-coloured, the legs and feet brown. The female is similar but duller; the breast is greyish.

Range Mali and Nigeria eastwards through central Africa to Ethiopia, Kenya, and south to Angola and northern Namibia.

PLATE 110 *Emberiza flaviventris*, Golden-breasted Bunting

Status Abundant, especially in central and eastern parts of the range.

Avicultural rating A popular and widely kept species, which is easily managed and hardy.

Feeding Diet A, with spray millet, seeding grasses in season, green food and regular live food. Grit is essential.

Breeding They do best as the sole occupants of a secluded planted shelter and flight. They build an open cup-shaped nest in vegetation, often supported in a fork. They may also use a wicker basket or an open box. Clutch 2–3; incubation 13 days. Abundant small live food is needed for successful rearing.

General management Acclimatised birds can remain outside in suitable accommodation, but do not usually cope with cold and wet conditions. They can be mixed with similar-sized birds but breeding pairs need segregation.

Lophospingus pusillus
Black-crested Finch, Pygmy Cardinal

5 in (13 cm). The top of the head (including an upstanding crest), chin and a broad band from the base of the bill, through the eyes to the sides of the neck, are black; the eye-stripe and sides of the throat and lower cheeks are white. The upper surfaces are grey-brown, the wings darker; the underparts are grey-white, paler on the abdomen. The bill is grey, the legs and feet light-brown. The female is paler and has a grey-white throat.

Range Southern Bolivia, Paraguay, Argentina.

Status Numbers appear stable in many areas.

Avicultural rating A popular avicultural subject, which is easily managed but not completely hardy. They do not thrive if exposed to prolonged cold and wet conditions.

Feeding Diet A, with millet sprays, seeding grasses, green food and some live food. Grit is essential.

Breeding This species is a very willing nester, but the young are difficult to rear unless abundant supplies of suitable live food are available throughout most of the nestling period. Wax moth larvae and fly pupae (if the adults are accustomed to the latter before nesting commences so much the better) are two of the easiest items to obtain; supplement these with as wide a variety as possible of other items. Clutch 2–3; incubation 12–13 days. The birds build a shallow cup of rootlets, grass etc, usually in a bush or bunch of artificial cover. Remove the young as soon as they are independent.

General management When acclimatised and firing on all cylinders, this bird is inclined to be a bully in mixed company; it is best not mixed with small or quiet birds, and not housed with any companions at all if the quarters are small. It is by no means a hardy species and probably should be kept in indoor quarters during the winter.

Poospiza ornata
Pretty Warbling Finch, Cinnamon Warbling Finch

5 in (13 cm). The crown and rump are dark grey, the remainder of the upper surfaces grey-brown. The eyebrow stripe is cinnamon-brown; there is a white spot below the eye. The underparts are biscuit-brown, darker and more rufous in the centre of the throat and breast, paler towards the abdomen. The bill is horn-coloured, the legs and feet pale brown. The female is similar but paler.

Range North-west Argentina.

Status Not known.

Avicultural rating These birds appear on the market only occasionally. They are a quietly-coloured but interesting species which has bred occasionally.

Feeding Diet A, with the addition of millet sprays and seeding grasses. Green food and some live food is also needed. Grit is essential.

Breeding It has bred only on a few occasions. It builds a cup-shaped nest of plant fibres among vegetation. Clutch 3; incubation 12–13 days. Abundant live food is needed for successful rearing. The breeding pair must have an aviary to themselves due to the aggressive behaviour of the male.

General management They do best in a small shelter and flight unit, but are not hardy enough to winter outside without some heat in the shelter. Individual birds, especially cocks, have proved very aggressive; they are not recommended for mixing with smaller or weaker species.

Poospiza torquata
Ringed Warbling Finch

5 in (13 cm). The upper surfaces are grey-brown; there is a broad band of black from the base of the bill to the cheeks. There is a white eyebrow stripe. The chin and throat are white; there is a broad black band across the upper breast. The lower underparts are grey-white; there is an area of chestnut on the vent. The bill is horn-coloured, the legs flesh-coloured. The female is similar but paler; she lacks the black markings on the cheeks and breast. There is some streaking on her underparts.

Range Bolivia, western Paraguay, northern and central Argentina.

Status See Pretty Warbling Finch.

Feeding/Breeding/General management Similar to Pretty Warbling Finch. Although, like the previous species, they spend some time at ground level, they are also active among vegetation at a higher level in the aviary; they are agile and tit-like in habit. Their temperament is slightly suspect; small doves or quail are recommended companions. In my experience, they are hardier than the Pretty Warbling Finch but do not enjoy exposure to cold or damp conditions.

Sicalis flaveola
Saffron Finch, Yellow Finch

5½ in (14 cm). The forehead and crown are orange, the upper surfaces yellow-green with some streaking on the back. The underparts are yellow. The wings and tail are darker and with olive-green edges to the feathers. The bill is horn-coloured, the legs and feet light brown. The female's upper surfaces are more olive and the streaking in the plumage is more pronounced.

Range North-west Peru, Colombia, Venezuela, the Guianas, Trinidad; northern Argentina, eastern Bolivia, Paraguay, south and south-eastern Brazil, Uruguay. The species has been introduced to parts of Central America and Jamaica.

Status Abundant in many areas.

Avicultural rating This is one of the most familiar of small seedeating species from South America. It is easily managed and hardy but does not enjoy cold and wet conditions. It is generally belligerent with other species.

Feeding Diet A, with millet sprays, seeding grasses, green food and some live food. Grit is essential.

Breeding Established birds are usually willing breeders. They make a fairly large and untidy cup-shaped nest, usually in a box or other receptacle. Clutch 3–5; incubation 13–14 days. Provide abundant live food and germinated seed for rearing; wild seeds are also appreciated at this time.

General management This is an attractive and easily managed species, but it is not a good mixer; temperament varies with individual birds, but they are generally best regarded as potential troublemakers in mixed company. Shelter and flight accommodation suits. It should be a good subject for captive-breeding programmes. The cocks have a pleasant song.

Sporophila albogularis
White-throated Finch, White-throated Seedeater

4¾ in (12 cm). The forehead, crown, nape and sides of the head are grey-black; the rest of the upper surfaces are grey. The chin, throat, sides of the neck and underparts are white except for a black band across the upper breast. The wings and tail are black or dark grey; there is a white wing bar. There is a small white spot between the base of the bill and the eye. The bill is yellow, the legs and feet grey. The female's upper surfaces are grey-brown with an indistinct white patch on the wings; the underparts are grey-white with a greyish band across the upper breast.

Range North-eastern Brazil.

Status Not known.

Avicultural rating This species is only occasionally seen in aviculture at the present time. It is easy to manage and becomes hardy after acclimatisation; but is generally extremely aggressive.

Feeding Diet A, with millet sprays, seeding grasses, green food and some live food. They enjoy soaked and germinated seeds. Grit is essential.

Breeding They build a cup-shaped nest. Clutch 3; incubation 13–14 days. It is essential a breeding pair are the sole occupants of a small shelter and flight unit. Abundant live food, seeding grasses and germinated seeds are needed for successful rearing. Remove the young birds from the breeding aviary as quickly as possible after they are seen to be self-supporting, because of the likelihood of attack by the adult cock.

General management The species is known to be long-lived. The males are usually spiteful and difficult in mixed company. The females are rarely imported. The cocks have a pleasant song. They are hardy but best not exposed to prolonged cold or damp conditions.

Sporophila bouvreuil
Reddish Finch, Capped Seedeater

4 in (10 cm). The forehead, crown and back of the head are black; the rest of the upper surfaces are rufous-cinnamon, darker towards the rump. The underparts are similar but slightly paler; there is some grey-white on the cheeks and a fine black stripe above the eye. The wings and tail are black or dark grey-brown with a white wing patch; the tail is black or dark grey-brown. The bill is black, the legs and feet dark brown. The female's upper surfaces are grey-brown, the face, sides of the neck, chin and throat warm-buff. Her underparts are buff-white.

Range North-east Argentina, south-east Paraguay, south and south-eastern areas of Brazil.

Status Not known.

Avicultural rating See White-throated Finch. Rarely available.

Feeding/Breeding/General management Similar to White-throated Finch. These are very small and attractive birds, which mix reasonably well with other small finches, but are aggressive towards other *Sporophila*. Those previously in my possession all proved to be cocks.

Sporophila castaneiventris
Chestnut-bellied Seedeater, Lavender-backed Finch

3¾ in (9.5 cm). The head and upper surfaces are blue-grey; there is a small area of white at the base of the lower mandible. The chin, throat and underparts are bright chestnut. The wings are black or dark grey; the tail is black or grey with an area of chestnut on the under tail coverts. The bill, legs and feet are black. The female has plumage of varying shades of olive-brown, paler on the underparts; her lower under surfaces and thighs are warm buff.

Range Eastern Ecuador, south-east Colombia, southern areas of Venezuela, the Guianas, western Brazil, northern Bolivia.

Status There are believed to be good populations in the northern parts of the range; elsewhere their status is not known.

Avicultural rating See White-throated Finch. Extremely rare in aviculture.

Feeding/Breeding/General management Similar to White-throated Finch. I have owned only one (true) pair which failed to breed. They appear more peaceful than other *Sporophila*. The pair in question settled down quickly in a small outdoor shelter and flight of which they were the sole occupants; whilst in my possession they were brought indoors each winter although the species is probably hardy when acclimatised.

Tiaris canora
Cuban Finch, Cuban Grassquit

4½ in (11.5 cm). The forehead, face, cheeks, chin and throat are black, the upper surfaces olive-green. The sides of the head and neck and a band across the lower throat are yellow; the rest of the underparts are grey or black, lighter towards the abdomen. The wings and tail are grey-brown. The bill is black, the legs and feet brown. The female is much duller; her face is dark chestnut-brown and the yellow areas are much paler.

Range Cuba and Isle of Pines; introduced to New Providence.

Status Fairly abundant in some parts of Cuba.

Avicultural rating Occasionally available but fairly expensive. These are very attractive birds, but aggressive. They are fairly willing breeders.

Feeding Diet A, with millet sprays, seeding grasses and some green food. Some birds acquire a taste for live food but this does not appear to be an important part of the diet. Grit is essential.

Breeding They are disruptive at the best of times, certainly so when nesting; pairs should be housed in individual units for successful breeding. Provide some cover and open-fronted nest-boxes or wicker baskets; the birds may use these receptacles or build a domed nest among vegetation using grasses, rootlets and other plant fibres with a soft lining of wool, feathers and fine grass. Clutch 3–4; incubation 12–13 days. Provide extra sprays of millet, seeding grasses and germinated seeds for rearing. Some pairs take small live food at this time. Remove the young birds as soon as they are independent.

General management These very self-confident little birds are prepared to bully much larger species sharing the aviary; therefore either segregate them or choose companions with care. They have a cheerful song; acclimatised birds become hardy but should not be exposed to cold and wet conditions during the winter. Some stocks are still maintained in Europe and an effort is now being made to consolidate them. It is best not to house breeding pairs in sight of each other or the cocks will spend most of the time threatening rival males.

Tiaris olivacea
Olive Finch, Yellow-faced Grassquit

4¾ in (12 cm). The upper surfaces are grey-green; the eyebrow stripe, chin and throat are yellow, edged with black and with a large area of black on the breast. The rest of the underparts are grey-brown to grey-white. The bill is black, the legs and feet brown. The female is paler and lacks the black markings on her underparts.

Range Eastern Mexico, Central America, Panama, Colombia, Venezuela, Greater Antilles.

Status Abundant in many areas.

Avicultural rating See Cuban Finch.

Feeding/Breeding/General management Similar to Cuban Finch, although this species is even less freely available at present. Many pairs have proved willing, even prolific, breeders, which makes the present shortage and high price of stock even more mystifying than can be explained by the export ban in many parts of their natural range.

Coryphospingus pileatus
Pileated Finch

5½ in (14 cm). The centre of the crown (or crest) is red, edged with black; the rest of the upper surfaces are grey. The cheeks, chin, throat, abdomen and under tail coverts are grey-white; the breast and flanks are grey. The bill is horn-coloured, the legs and feet grey-brown. The female is grey-brown above, her underparts whitish with some streaking; she has a black crown or crest.

Range Colombia, northern and western Venezuela, eastern and central areas of Brazil, Margarita Island.

Status Variable, but there are still good numbers in some parts of the range.

Avicultural rating Only occasionally available. These are attractive, easily-managed birds; established examples are usually willing nesters.

Feeding Diet A, with millet sprays, seeding grasses and various items of green food; provide some live food. Grit is essential.

Breeding They build a cup-shaped nest in natural or artificial cover; they will also use a wire-mesh 'cup' as a foundation, or a wicker basket. Clutch 3–4; incubation 12–13 days. Provide extra live food, germinated seed etc for rearing. Insects probably form the bulk of the rearing food at first.

General management Established pairs become hardy, but are best not exposed to prolonged damp or cold in winter; my birds usually roosted among cover in an unheated shelter throughout the winter. They are generally good mixers but some aggression may be shown, mainly when the cock is driving the female prior to mating. It is probably best if they are housed in separate accommodation for breeding. They spend much time at ground level in the aviary.

Coryphospingus cucullatus [PLATE 111]
Red-crested Finch

PLATE 111 *Coryphospingus cucullatus*, Red-crested Finch

5½ in (14 cm). The centre of the crown (or crest) is red with a narrow black edge; the head, nape, mantle, rump and underparts are red-brown. The chin is white, the wings and tail grey-black. The bill is lead-grey, the legs and feet grey. The female is mainly light brown, paler below; she has a smaller crest.

Range The Guianas, north-eastern Brazil; eastern Peru to Paraguay, southern Brazil, north-eastern areas of Argentina.

Status Variable, but there are still good numbers where the population is well established, including in cultivated areas.

Avicultural rating A good avicultural subject, although it is now less freely available than in the past.

Feeding Diet A, with millet sprays, seeding grasses, some green food and live food. Grit is essential.

Breeding Several breedings have been recorded. They build a cup-shaped nest among natural or artificial cover; and will use wicker baskets or open boxes. Clutch 2–3; incubation 11–12 days. Abundant live food is essential if rearing is to be accomplished successfully.

General management Acclimatised stock becomes hardy, but does best in frost-free accommodation during the winter, and does not enjoy exposure to cold or wet conditions. They are usually lively and assertive but can be mixed with other birds except when breeding, when separate accommodation is recommended. Like the Pileated Finch, they spend a good deal of time exploring the lower regions of their aviary.

Rhodospingus cruentus
Rhodospingus Finch, Crimson Finch

4½ in (11.5 cm). The head, nape, mantle, back, wings and tail are black; the under surfaces are crimson, more orange-red towards the abdomen. The centre of the crown is red. The bill is lead-grey, the legs and feet black. The female is mainly brown above and on the flanks; she has buff-yellow underparts and the feathers of the centre crown are buff-yellow.

Range Northern and western Ecuador south to northern Peru.

Status There are good populations, despite the limited range.

Avicultural rating These attractive and interesting birds are available only at infrequent intervals. They are not among the easiest species to establish.

Feeding Use basically Diet A, but also offer millet sprays; persevere with some Diet F and items of soft fruit such as pear, grape etc. Regular live food is an important item of the diet. A glance at the narrow, pointed bill gives some indication that this is not a wholly granivorous species, and they will not survive for long if fed only on mixed seeds etc. Some birds enjoy sprouted and germinated seeds. Grit is essential.

Breeding They build a cup-shaped nest of plant fibres; the birds may use a wicker basket or open box. It is best if nesting receptacles are concealed among cover. Clutch 2–4; incubation 11–12 days. Abundant live food is needed for successful rearing; insects may be the sole rearing diet until the young fledge, but this is accomplished in the wild in only 8–10 days despite the chicks being blind and naked on hatching. Rapid growth of the nestlings in the wild is a pointer to the quality of the rearing diet needed to achieve success in captivity.

General management Acclimatised birds can be housed in a shelter and flight during the summer months, but they are definitely not hardy and need some heat during severe (cold) weather, they should not be exposed to prolonged damp, wet or cold. They are good mixers, except when approaching breeding condition when segregation is necessary. Potential breeding stock does best outside in summer; a period of settled, hot weather may stimulate breeding activity.

Gubernatrix cristata [PLATE 112]
Green Cardinal, Yellow Cardinal

7½ in (19 cm). The forehead and crest are black, the upper surfaces olive-green spotted and streaked with black. The chin and throat are black, the underparts yellow-green, more yellowish on the sides of the throat and the lower under surfaces. The wings and tail are black and green. The bill is lead-grey, the legs and feet black. The female is generally duller in colour and lacks the areas of yellow around her head and sides of the neck.

Range Eastern Argentina, Uruguay, south and south-eastern Brazil.

PLATE 112 *Gubernatrix cristata*, Green Cardinal

Status Not known.

Avicultural rating Once freely available and widely kept, this species is now only occasionally offered for sale. It is hardy, easily managed and a willing breeder.

Feeding Diet A, with regular green food and some live food. They also enjoy picking through mixed seeds (Diet C) as well as seeding grasses. Groundsel is a particularly favoured item of green food. Grit is essential.

Breeding Established pairs are usually very willing to nest; they are best not kept with other birds when breeding. Clutch 3–5; incubation 13–14 days. Live food is vital for successful rearing; offer wax moth larvae, small crickets, smooth caterpillars, fly pupae etc. Also provide soaked and germinated seeds and extra bundles of seeding grasses if available, otherwise some soaked millet sprays.

General management Acclimatised birds are usually hardy and easily-managed; they thrive in outdoor shelter and flight accommodation. Outside the breeding season they usually associate well with birds of similar size and temperament, although not other related or near-related species.

Paroaria coronata
Red-crested Cardinal

$7\frac{1}{2}$ in (19 cm). The whole of the head (including an upstanding crest), chin, throat and stripe down the centre of the breast are red; the nape and sides of the neck and the whole of the underparts are white, grey-white on the flanks. The mantle, back, wings and tail are grey, the feathers of the wings with black margins. The bill is horn-coloured, the legs and feet lead-grey. The sexes are alike, but some hens can be distinguished by a slightly paler red on head and throat.

Range Extreme south-east of Bolivia, Paraguay, southern Brazil, Uruguay, northern and eastern areas of Argentina.

Status Good populations in some areas.

Avicultural rating These lively and spectacular birds are popular avicultural subjects. Eventually hardy, they are not safe with smaller or weaker birds and can be aggressive even in the company of larger birds.

Feeding Diet A, with occasional millet sprays and seeding grasses. They also enjoy green food and live food. Grit is essential.

Breeding Obtaining true pairs is one of the biggest obstacles to more frequent breedings with this species. It is willing to nest in modest-sized aviaries; making an untidy cup-shaped nest of plant fibres with softer and finer material as a lining, usually in vegetation or artificial cover. They will use baskets or boxes etc. Clutch 3–4; incubation 13–14 days. Plenty of live food is essential for successful rearing. Remove the young from the breeding aviary as soon as they are independent to prevent attack by the adult male.

General management It is best if they are the sole occupants of a shelter and flight. They can remain outdoors throughout the year when acclimatised. Plenty of cover is desirable as the birds are usually nervous and quick to take fright, although they are invariably conspicuous and active.

Paroaria capitata
Yellow-billed Cardinal

6½ in (16.5 cm). The whole of the head and chin are bright red; the throat and a narrow area down the centre of the breast are black. The mantle, back, wings and tail are dark-slate; the nape and sides of the neck and under surfaces are white, grey-white on the flanks. The bill is pale orange-brown, the legs and feet flesh-coloured. The sexes are similar.

Range South-west Brazil, Paraguay, south-east Bolivia, northern and central areas of Argentina.

Status Good populations, but declining in some areas following loss of habitat.

Avicultural rating Now rarely available. They are eventually hardy; good aviary birds, they mix better than their relatives.

Feeding/Breeding/General management See Red-crested Cardinal. Breeding pairs are best housed in separate accommodation, but they usually get on with similar sized companions in non-breeding periods.

Spiza americana
Dickcissel

6 in (15 cm). The forehead and crown are grey-brown with darker flecking; the rest of the upper surfaces are chestnut-brown, darker on the tail and wings and with pronounced streaking. The eyebrow-stripe is warm buff; there is a narrow buff-white eye-ring. The cheeks are grey and the chin and throat are white, bordered with black; the latter shade extends down to the centre of the breast. The underparts are yellow, with grey on the flanks and towards the abdomen. The bill is horn-coloured, the legs and feet brown. The female is mainly warm brown, slightly paler below; her upper surfaces have darker streakings.

Range Eastern areas of North America; it migrates to Central America, Colombia, Venezuela, Trinidad.

Status Widespread and abundant in many parts of the range; it has achieved a major recovery through adapting to changed habitat.

Avicultural rating Very rarely available; I have owned only a single pair obtained during the 1980s.

Feeding A varied diet is enjoyed; use a mix of Diet A and C with seeding grasses, abundant green food and a varied selection of live food. Grit is essential.

Breeding There is no information about captive breeding. A bulky, cup-shaped nest of plant fibres is made, usually low-down and concealed in dense cover. Clutch 4; incubation 12–13 days. Abundant live food is needed for rearing.

General management My pair proved extremely nervous and did not settle down until in a planted aviary. They proved extremely hardy, but both birds succumbed within 18 months following the appearance of carbuncle-like growths on the sides of the face.

Cardinalis cardinalis [PLATE 113]
Virginian Cardinal, Common Cardinal

PLATE 113 *Cardinalis cardinalis*, Virginian Cardinal

7–9 in (18–23 cm). The entire plumage is red, apart from a black 'mask' at the base of the bill. It has an upstanding crest. The bill is orange-red. the legs and feet light-brown. The female is a dull light brown or buff-brown with paler underparts.

Range Much of the United States except northwestern and central areas, eastern Canada, Mexico and Central America.

Note 18 subspecies are described and there is some variation in size; the nominate subspecies is *C. c. cardinalis* (described) from eastern USA.

Status It adapts well to changing habitat and is still abundant in many parts of the range.

Avicultural rating This is a popular avicultural species, although it is now only occasionally available. It is hardy and easily managed; but loses its brilliant colour except in a large natural aviary and with a varied diet.

Feeding Use diets A and C, with seeding grasses, green food and some live food. Many individual birds enjoy fruit. Modern colour-food would arrest the colour-fading but I have not tried this. Grit is essential.

Breeding This bird builds a cup-shaped nest of plant fibres lined with fine grass etc, usually among natural or artificial cover; it will also use a woven basket or an open box. Clutch 3–4; incubation 12–13 days. Abundant live food is essential for successful rearing. Breeding pairs need a separate aviary.

General management They are usually extremely hardy and thrive in an outdoor shelter and flight without heat. They usually mix well with similar-sized (but not related) companions. They are excellent songsters, although probably not of the quality to merit the occasional alternative name of 'Virginian Nightingale'!

Passerina cyanea [PLATE 114]
Indigo Bunting

$5\frac{1}{2}$ in (14 cm). Most of the plumage is bright blue, darker on the crown and with some purple-blue reflections on the sides of the head; the wings and tail are black and chestnut. The bill is lead-grey, the legs and feet black. The female is mainly grey-brown, paler below; the wing feathers have darker margins. The male in non-breeding dress resembles the female, but is usually darker; the females, out-of-colour adult males and juveniles often have splashes of blue on the shoulders or tail.

Range South-eastern areas of Canada, eastern USA. It migrates to Central America, Cuba,

PLATE 114 *Passerina cyanea*, Indigo Bunting

Colombia, Venezuela and various West Indian islands.

Status There are good numbers in many parts of the range, and they are even recovering in some areas of deforestation which have now been abandoned.

Avicultural rating A popular avicultural subject, although they have been available only occasionally within recent years.

Feeding Diet A, with seeding grasses, green food and live food. Some birds learn to take small amounts of insectile mixture. Soaked seeds are enjoyed but this species does not make rapid diet adjustments. Grit is essential.

Breeding It is best if a breeding pair are the sole occupants of a secluded shelter and flight with ample natural cover. They build a cup-shaped nest of small twigs, plant stems, leaves etc, with some fine grasses and rootlets as lining. Clutch 3–4; incubation 12–13 days. Abundant live food is essential throughout the nestling period.

General management Acclimatised birds are hardy but do best with a frost-free shelter for occupation during severe weather. Non-breeding birds mix with other species, but should preferably be kept in spacious quarters; the males become aggressive as breeding condition is achieved and are at their 'safest' when out of colour.

Passerina amoena
Lazuli Bunting

5 in (13 cm). The upper surfaces including the head, cheeks, chin and throat are bright blue, darker on the back and tail; the wings have white bars and darker margins to the feathers. The breast is pale chestnut, the lower underparts buff-white. The tail is grey or blue, the underside grey. The bill is lead-grey, the legs and feet black. The female is mainly grey-brown, paler below, she has white wing-bars.

Range Western USA; it winters in California (Baja), western Mexico.

Status As abundant in many areas of the range as the Indigo Bunting is in eastern North America.

Avicultural rating An attractive species but only occasionally available at present.

Feeding/Breeding/General management Similar to Indigo Bunting but this species is an infrequent breeder. Unlike some of its relatives, it is usually very conspicuous in a suitable aviary. The males sing from the highest visible perch.

Passerina versicolor
Versicolor Bunting, Varied Bunting

5 in (13 cm). The forehead and chin are black, the nape and throat red. The crown, cheeks, sides of the neck and rump are purple-blue. The rest of the plumage is mainly purple-chestnut with grey-black wings and tail. The bill, legs and feet are horn-coloured. The female is mainly grey-brown with paler underparts.

Range Southern USA, Mexico, Guatemala.

Status Generally uncommon.

Avicultural rating An attractive species which is now rarely available.

Feeding/Breeding/General management Similar to Indigo Bunting, but this species is more difficult to establish successfully; it is not as hardy, and does best in slightly heated quarters during winter. Cold or damp conditions are not tolerated.

Passerina ciris [PLATE 115]
Painted Bunting, Nonpareil Bunting

PLATE 115 *Passerina ciris*, Painted Bunting

5 in (13 cm). The head, nape and shoulders are blue, the mantle and back yellow-green. The lower back, rump and under surfaces are red. There is an area of moss-green at the bend of the wing; the wings and tail are yellow-green and brown. The bill, legs and feet are horn-coloured. The female's upper surfaces are grey-green, her underparts yellowish-green.

Range South and south-eastern areas of the USA; it winters in south-eastern Mexico, Central America, Bahamas.

Status Variable, but abundant in some areas of suitable habitat.

Avicultural rating A popular avicultural subject, although it is by no means easy to maintain it in good condition for long periods; there is a tendency for the red in the plumage to fade unless kept in a large natural aviary with abundant live food etc.

Feeding Diet A, with seeding grasses, germinated seed, green food and ample small live food. Grit is essential. Offer some Diet F.

Breeding Only occasional breeding successes have been recorded. It is best if they are the sole occupants of a secluded, planted shelter and flight; the cocks are belligerent with companions when in breeding condition. They build a cup-shaped nest of plant fibres etc, usually low down among vegetation. Clutch 3–4; incubation 12–13 days. Abundant live food is essential for rearing.

General management Imported birds need time, and sympathetic handling, to establish properly. Eventually they are quite hardy, but should not be exposed to winter cold or wet. They mix fairly well, except when coming into breeding condition. Loss of colour is a problem, but use of artificial colour-food would probably solve this.

Passerina lechlancherii [PLATE 116]
Rainbow Bunting, Orange-breasted Bunting

$5\frac{1}{2}$ in (14 cm). The forehead and crown are yellow-green; the rest of the upper surfaces and tail are sky-blue. An area around the eyes, the chin, throat and the whole of the underparts are yellow, suffused with orange on the breast. The bill is horn-coloured, the legs and feet grey. The female's upper surfaces are yellowish-green, her underparts paler and slightly more yellow.

Range South-western areas of Mexico.

Status Not abundant.

PLATE 116 *Passerina lechlancherii*, Rainbow Bunting

Avicultural rating A popular avicultural species, but it is often difficult to establish and does best in experienced hands only. It is rarely available.

Feeding/Breeding/General management Similar to Painted Bunting. Many Rainbow Buntings have been carefully acclimatised, only to die soon afterwards; fits and seizures are frequent causes of death. This suggests the possibility of environmental or dietary problems; these birds are nervous and easily stressed, but many are or were acquired by enthusiastic exhibitors. Efforts to 'steady' the birds for the showbench almost certainly included offering a surfeit of mealworms etc, whilst the birds were housed in limited-space quarters with eventual fatal consequences.

Emberizidae—Thraupinae (Tanagers)

This subfamily comprises more than 200 species which are confined to the New World. A few are North American, but the greatest concentration of species is in South America. This subfamily includes some of the most brilliantly-coloured of all birds; they range in size from small ($3\frac{1}{2}$-in, 9-cm) euphonias to the Magpie Tanager, measuring more than 10 in (25 cm) long. Many are multi-coloured with areas of metallic, iridescent feathers. They were formerly among the most frequently imported of all South American softbills; now supplies to aviculture are much reduced and there is an urgent need for captive-breeding programmes to be established with such stock as is available. Despite the opinions of some, tanagers are by no means impossible to breed in controlled conditions; several species have successfully raised young, but in the past perhaps too much emphasis on these birds' value on the showbench has inhibited interest in breeding. Some species become hardy and can even winter outside provided that they have a frost-free (and preferably slightly warm) shelter; many of the South American species are ideal occupants of either tropical house or conservatory accommodation where breeding is much more likely than if birds are regularly transferred between indoor (winter) and outdoor (summer) quarters.

Most tanagers are good mixers even when breeding, although in the latter circumstances some aggression between related or closely-related species is likely; they usually mix well with other small softbills, especially honeycreepers, zosterops etc. The Magpie Tanager is not trustworthy with other, weaker birds and some of the mountain tanagers are also bullies. The latter birds do not thrive in the warm humid conditions of a tropical house; after acclimatisation they are better in spacious outdoor aviaries throughout the year with access to a comfortable shelter. Extra hours of artificial light (to enable late feeding) are more important than high winter temperatures. Whilst many tanagers consume substantial quantities of fruit, a more varied diet is essential to maintain health; watch their consumption of liquid foods, which some will utilise as a staple item of diet if allowed to.

Cissopis leveriana
Magpie Tanager

$10\frac{1}{2}$ in (26.5 cm). The head, mantle, chin, throat, breast and lower breast, wings and tail are black;

[278]

the rest of the plumage is white. There is a narrow bar of white on the secondaries and white tips to the tail feathers. The eye is bright yellow. The bill, legs and feet are black. The sexes are alike.

Range Eastern Colombia and Venezuela, north-eastern Bolivia, western, southern and south-east Brazil, Paraguay, northern Argentina, the Guianas.

Status Variable, but almost certainly adversely affected by habitat loss.

Avicultural rating Rarely available, they are not among the most popular avicultural subjects.

Feeding Use a mix of Diets E and G with various items of live food; crickets or locusts and even pinky mice should be offered.

Breeding It builds an open cup of small twigs, grass, rootlets etc, usually among vegetation. Clutch 2; incubation 14 days. Offer a variety of fruit and live food for rearing.

General management This tanager is fairly hardy when acclimatised and probably capable of wintering outside with access to a frost-free shelter. Opinions vary about the wisdom of mixing these birds with other species; I have seen an example living over a long period in a tropical house with smaller tanagers, sunbirds, hummingbirds and honeycreepers as companions without problems arising, but on balance I do not recommend this kind of association.

Ramphocelus bresilius
Scarlet Tanager, Brazilian Tanager

---

7 in (18 cm). The entire plumage, except for the black wings and tail, is scarlet, darker on the mantle and back. The base of the lower mandible is silvery-white. The bill, legs and feet are black. The female's upper surfaces are brown, with more rufous on the back and with darker wings and tail; her underparts are red-brown, and the rump dull crimson.

Range Eastern Brazil.

Status Variable and much affected by habitat loss.

Avicultural rating Very popular avicultural subjects and easy to manage, but now very rarely available.

Feeding Use a mix of Diets E and F with additional berries (both wild, as available, and cultivated) and live food. The brilliant colour fades fairly rapidly in controlled conditions and careful use of a colouring agent is needed to help retain or regain colour; loss appears to be slowed down in large aviaries with abundant plant and insect life which aid the production of pigments.

Breeding They make an open cup-shaped nest of plant fibres, usually among vegetation, but at varying heights. Clutch 2; incubation 12–13 days. Provide abundant small live food during the nestling period. Several breeding successes have been recorded.

General management Properly acclimatised birds are usually easy to maintain in good health over a long period; they are hardy but should not be exposed to winter cold or wet etc. A temperate bird-house provides the ideal environment and offers the best chance of breeding. They generally mix well with other species, although some with a conspicuous area of red in their plumage often incite attack. Loss of this bright colour is the main problem associated with captive management of this species, which is now, sadly, almost unobtainable.

Anisognathus flavinuchus [PLATE 117]
Blue-winged Mountain Tanager

---

8 in (20 cm). The forehead, sides of the head and neck, mantle and upper back are black; the centre of the crown to the nape and the entire underparts are yellow. The back is moss-green, olive and black. The wings and tail are black, the wings with some shining cobalt or violet-blue on the shoulders, and turquoise-blue in the flight

PLATE 117 *Anisognathus flavinuchus*, Blue-winged Mountain Tanager

feathers. The bill is black, the legs and feet dark grey. The sexes are alike.

Range Northern Venezuela southwards through Colombia, Ecuador, Peru, Bolivia; mainly found in subtropical zones especially the slopes of the Andean range.

Note Some plumage variation exists among the 9 described subspecies.

Status They are believed to be reasonably abundant in some areas of the range but are affected by habitat loss in several places.

Avicultural rating A good avicultural subject, which acclimatises well and is eventually hardy.

Feeding Use a mix of Diets E and G, with the addition of wild and cultivated berries, and various items of live food; crickets, locusts and other large-bodied insects are accepted.

Breeding They build a cup nest of twigs, plant fibres etc. Clutch 2; incubation 13–14 days. Abundant live food is important if rearing is to be successfully accomplished in the aviary; pinky mice are probably an acceptable supplementary item of diet as the nestlings grow. Breeding pairs should be housed alone.

General management These are robust birds which appear to travel well and come through quarantine in good health and usually good plumage. Acclimatised birds become hardy and can be housed in a shelter and flight throughout the year; a frost-free shelter is needed, although the birds may not use it. They are generally aggressive and not suitable companions for other than birds of similar size. They do not flourish in the humid conditions of a tropical house or heated conservatory.

Anisognathus igniventris
Scarlet-bellied Mountain Tanager

8 in (20 cm). The head, chin, throat, breast and upper surfaces are black, the ear-coverts, lower breast and underparts red and black. The wing coverts and rump are an iridescent blue. The under tail coverts are red or black-red. The bill, legs and feet are black. The sexes are alike.

Range Western Venezuela, Ecuador, western and south-eastern areas of Peru, western and central Bolivia; mainly the subtropical to temperate zones of the Andean foothills.

Notes Some variation in plumage exists between the four described subspecies.

Status See Blue-winged Mountain Tanager.

Avicultural rating Strikingly beautiful, but rarely available.

Feeding/Breeding/General management Similar to the Blue-winged Mountain Tanager, but in my experience a more difficult species to establish and manage.

Anisognathus notabilis
Black-chinned Mountain Tanager

8 in (20 cm). The head, chin and sides of the neck are black; the narrow band on the nape and the entire underparts are yellow. The mantle and rest

of the upper surfaces are moss-green. The margins of the wing-feathers have some blue, but this is less conspicuous than in the Blue-winged Mountain Tanager.

Range South-west Colombia, north-west Ecuador.

Status See Blue-winged Mountain Tanager.

Avicultural rating Rarely available.

Feeding/Breeding/General management Similar to Blue-winged Mountain Tanager. They need care after importation and quarantine but eventually are not difficult to manage. They are hardy; confident and aggressive, they are not good mixers.

Euphonia saturata
Orange-crowned Euphonia

---

4 in (10 cm). The forehead and crown are yellow-orange; the face, chin, throat, nape and sides of the neck, the upper surfaces and tail are glossy purple/blue. The breast and underparts are yellow-orange, lighter on the flanks. The bill, legs and feet are black. The female's upper surfaces are yellow-olive, lighter below with an area of yellow in the centre of the abdomen.

Range Western areas of Colombia, Ecuador and Peru.

Status Not known.

Avicultural rating Occasionally available, these are attractive and relatively easily-managed little birds.

Feeding Use mainly Diet E, but also offer Diet F mixed with a *small* amount of diced fruit in a separate feeding vessel; persevere if the birds ignore the combination at first. Various wild and cultivated berries are much enjoyed. Provide small live food. This is one species that will consume considerable amounts of liquid nectar or nectar and sponge cake mixtures; this is not recommended and birds thus 'addicted' are best not kept in the company of true nectar-feeding species.

Breeding This species constructs a domed nest of fine plant fibres, usually among vegetation. Clutch 3–4; incubation 13–14 days. Although the adults are mainly fruit and berry-eaters, small live food is necessary for rearing the nestlings.

General management These attractive and interesting small tanagers are ideal occupants of a tropical house or conservatory. They usually mix well with other small softbills. They are fairly hardy; the related Violet Tanager has been housed successfully in an outdoor shelter and flight throughout the year.

Tangara chilensis [PLATE 118]
Paradise Tanager

$5\frac{1}{2}$ (14 cm). The head is a shining golden-green; the nape, shoulders, mantle, back and tail are black, with a narrow area of black down the centre of the belly and abdomen. The lower back and rump are scarlet and orange-yellow. There is a black 'mask' at the base of the bill; the throat and upper breast are cobalt-purple. The breast and underparts are bright blue. There is also some blue in the mainly black wings. The bill, legs and feet are black. The sexes are alike.

Range Northern and central Peru, Ecuador, southern and eastern Colombia, southern Venezuela, the Guianas, north and north-west Brazil.

Status Reasonably abundant in some parts of the range, but loss of habitat through deforestation is having an adverse effect on populations.

Avicultural rating A popular avicultural subject and one of the most striking and colourful members of the family. It is reasonably easy to manage, but recommended for experienced aviculturists only.

Feeding Use a mix of Diets E and F, with wild

PLATE 118 *Tangara chilensis*, Paradise Tanager

and cultivated berries as available and some live food. Persevere with Diet F if the birds are, at first, unwilling to investigate, and 'bait' the receptacle with other sought-after items of diet. Watch their consumption of liquid food, if housed with nectar-feeders; too much will lead to trouble.

Breeding This species builds a cup-shaped nest of plant fibres etc in vegetation. Clutch 2; incubation 13–15 days. Breeding is likely to be a difficult proposition; apart from the problem of sexing, the environment may be a big factor in success, or the lack of it. They are probably best kept in a tropical house, or a planted conservatory, if a potential breeding pair can be the sole occupants. Small live food is likely to be an important part of the rearing diet for nestlings but artificially-produced items may not by themselves prevent deficiency problems with youngsters.

General management They are fairly long-lived in the correct environment, but many succumb after a few months, mainly due to incorrect handling. The Paradise Tanager is a particularly nervous species; many imported birds are destined for the show bench, but (even though individual specimens may appear to become steady in a show-cage) stress often takes its toll. Despite their eye-catching appearance, they are generally not good exhibits because of their nervous disposition. They need care until acclimatised, then are reasonably easy to manage but never hardy.

Tangara schrankii [PLATE 119]
Schrank's Tanager, Green and Gold Tanager

5 in (13 cm). The forehead, chin and sides of the head are black; an area between the base of the bill and the eye is green. The crown, nape, centre of the breast and the rump are golden-green or yellow; the rest of the upper surfaces are emerald-green with prominent black streaking. The throat, sides of the neck and flanks are green;

[282]

there is some blue in the wing coverts. The bill is black, the legs and feet dark grey. The sexes are similar but the female has a green crown.

Range Southern Venezuela, southern areas of Colombia, northern and eastern Peru, northern Bolivia.

Status An abundant species in many parts of the range, but loss of habitat is having some adverse effects on populations.

Avicultural rating A popular and easily managed species which is occasionally available.

Feeding/Breeding/General management Similar to Paradise Tanager, but it is easier to establish and not nearly as nervous and highly-strung. This species thrives on a fruit and insectile mix diet with some live food. It becomes fairly hardy and can spend the summer months in a sheltered outdoor aviary, but must be wintered inside with slight heat. This is an excellent tropical house or conservatory species which usually associates well with other small softbills.

PLATE 119 *Tangara schrankii*, Schrank's Tanager

Tangara arthus [PLATE 120]
Black-eared Golden Tanager, Golden Tanager

---

5 in (13 cm). Most of the plumage is golden-yellow with pronounced darker lacing and streaking on the mantle and wings. The lores and ear-coverts are black. The bill is black, the legs and feet dark grey. The sexes are alike.

Range Northern Venezuela, western Colombia, Ecuador, western and central Peru and Bolivia.

Note There is some plumage variation among the 9 described subspecies; one of the best known to aviculturists is *T. a. aequatorialis*, the Gleaming Gold Tanager, from parts of Ecuador and Peru.

PLATE 120 *Tangara arthus*, Black-eared Golden Tanager

Status There are good populations in many parts of the range.

Avicultural rating A popular and widely-kept Tanager; acclimatised birds are easily managed.

Feeding/Breeding/General management Similar to the Paradise Tanager but it is not difficult to establish or maintain. It becomes hardy, but must not be over-wintered outdoors. It mixes well and is a good tropical house occupant.

Tangara icterocephala [PLATE 121]
Silver-throated Tanager

5 in (13 cm). The head and nape are yellow, separated from the silvery-white of the throat and sides of the neck by a narrow black line from the base of the bill; the mantle and back are yellow with black streaks. The under surfaces are yellow, the wings and tail black and green. The bill is black, the legs and feet grey. The sexes are similar but the female is usually slightly duller.

Range Costa Rica and Panama to western Colombia and Ecuador.

Status There are believed to be good numbers in some parts of the range.

PLATE 121 *Tangara icterocephala*, Silver-throated Tanager

Avicultural rating See Black-eared Gold Tanager.

Feeding/Breeding/General management Similar Paradise Tanager but easier to establish a maintain. It becomes hardy but is not resistant prolonged cold or wet and therefore should over-wintered indoors or in a tropical house. has good potential for captive-breeding pr grammes.

Tangara parzudakii [PLATE 122]
Flame-faced Tanager

6 in (15 cm). The forehead, crown and pat behind and below the eye are golden-orang

[284]

PLATE 122 *Tangara parzudakii*, Flame-faced Tanager

there is a band of black from the base of the bill, through the eye of the black area behind the ear-coverts. The rear crown and nape are yellow; the mantle, wings and tail are black. The under surfaces and wing coverts are an iridescent green. The bill is black, the legs and feet grey. The sexes are similar.

Range Western Venezuela, Colombia, Ecuador and northern/central areas of Peru.

Status There are believed to be fairly good populations, but habitat loss is probably affecting the species in some areas.

Avicultural rating See Black-eared Golden Tanager.

Feeding/Breeding/General management Similar to the Paradise Tanager, this is a reasonably robust species which appears to travel and quarantine rather better than related species. In my experience the birds settle down quickly and well; they are usually confident and speedily develop a good bond with their owner. They enjoy live food, which is a good inducement towards a lasting friendship, but beware of giving an excess of mealworms—crickets and small locusts are preferable. Despite the difficulty of sexing, this species might also provide a good subject for captive-breeding programmes.

Tangara gyrola
Bay-headed Tanager, Chestnut-headed Tanager

$5\frac{1}{2}$ in (14 cm). The head is a dull brick-red, sometimes bordered with yellow on the neck. The upper surfaces are grass-green, the underparts similar but lighter. The thighs are chestnut. There is some yellow in the wing coverts; the tail is dark green. The bill is black, the legs and feet grey. The sexes are alike.

Range Costa Rica and Panama to Colombia, Ecuador, north and north-eastern areas of Peru, northern Bolivia, north-west, northern and central Brazil, the Guianas, Trinidad, Venezuela.

Note There are considerable plumage variations between the 9 described subspecies.

Status There are good populations in many parts of the range.

Avicultural rating An easily managed and popular species, this is usually among the more inexpensive and freely available of tanagers.

Feeding/Breeding/General management Similar to Paradise Tanager, but it is among the easiest of tanagers to establish and manage. It can be housed outdoors during the summer but does not tolerate exposure to cold or damp conditions, and must be over-wintered indoors in slight heat. A good tropical house or conservatory species.

Tangara cyanicollis
Blue-necked Tanager

5 in (13 cm). The whole of the head, neck, chin and throat are turquoise-blue; there is a black streak from the base of the bill to the eye. The mantle, wings, tail and breast are black, the rump blue-green. The lower underparts are blue. There is some blue and bronze in the wings. The bill, legs and feet are black. The sexes are similar.

Range North-west Venezuela, western Colombia, Ecuador, western Peru, northern Bolivia.

Status There are fairly good populations in several areas and they are numerous in some parts of the range.

Avicultural rating See Black-eared Golden Tanager.

Feeding/Breeding/General management Similar to Paradise Tanager, but it is easily managed after acclimatisation. This is a good tropical house or conservatory species; it can be housed outdoors during the summer months but needs an aviary which offers good shelter from winds and driving rain.

Tangara nigroviridis
Beryl-spangled Tanager

5 in (13 cm). The forehead, face, chin and mantle are black; the crown, back and rump are blue. The under surfaces are black, heavily spotted with silvery-blue. There is some white in the centre of the abdomen; the wings and tail are black. The bill is black, the legs and feet grey. The sexes are similar.

Range North-western Venezuela, central Colombia, Ecuador, eastern Peru, Bolivia.

Status Variable, but there are good populations in some parts of the range.

Avicultural rating Less freely available than related species.

Feeding/Breeding/General management Similar to Paradise Tanager; this species does not appear either to travel or to quarantine well and is often in need of considerable care when purchased. It is not easy to establish but eventually prospers with careful handling.

Dacnis cayana [PLATE 123]
Blue Dacnis, Blue Sugarbird

PLATE 123 *Dacnis cayana*, Blue Dacnis

4½ in (11.5 cm). The chin, throat, a streak through the eye, the back, wings and tail are black; the remainder of the plumage is blue, including some blue in the wings and tail. The bill is black, the legs and feet dark grey. The female's head and cheeks are blue, the chin and throat grey, the rest of her plumage light green with darker wings and tail.

Range Nicaragua, Costa Rica, Panama southwards to Ecuador, Colombia, Venezuela, the Guianas, Brazil, northern and eastern Peru, northern Bolivia, Paraguay, northern Argentina.

Status Reasonably abundant in many areas.

Avicultural rating A good subject for experienced aviculturists, but now infrequently available.

Feeding Diet L, with fruit and some live food; fruit (pears, grapes, tomato etc) should be given in large (half) sections and not diced. Live food is very important for these birds; *Drosophila* and house flies should be offered daily and other small flying insects, spiders, tiny smooth caterpillars etc will also be enjoyed. Both wild and cultivated berries are useful additions to the diet. A small quantity of sponge cake soaked in the nectar mixture can be offered each day.

Breeding It builds a deep cup-shaped nest of plant fibres etc. Clutch 2; incubation 11–12 days. Various wild berries and 'seeds' are reported to be the main rearing food in the wild.

General management These are attractive birds but not easy to establish. Once properly acclimatised and accepting a varied diet they often prove to be long-lived and present few problems to their owners. Some individual birds show no interest in live food. They are not hardy and do best in a tropical house or conservatory, or similar sheltered accommodation.

Chlorophanes spiza
Black-headed Sugarbird, Green Honeycreeper

$5\frac{1}{2}$ in (14 cm). The head and sides of the neck are black; the remainder of the plumage is a shining blue-green, darker on the wings and tail. The upper mandible is black, the lower mandible yellow; the legs and feet are grey. The female is mainly grass-green, her chin and throat yellow and lower underparts suffused yellow.

Range Southern Mexico, Honduras, Ecuador, Colombia, Venezuela, the Guianas, northern Peru, north-western and northern areas of Brazil, northern Bolivia, eastern Brazil.

Status There are fairly good populations in some parts of the range.

Avicultural rating This is a hardy and easily-managed species when properly acclimatised; but its aggressive temperament is a problem.

Feeding Use a mix of Diets E, F and L, with the addition of various wild and cultivated berries, as available, and regular live food.

Breeding These birds build a cup-shaped nest among vegetation or into a tree-fork. Clutch 2; incubation 12–13 days. Abundant supplies of small insects are essential for rearing (*Drosophila*, house flies etc). The breeding pair should be the sole occupants of their accommodation.

General management These are attractive and easily-established birds if they are obtained in good health following importation and quarantine. The species is one of the most hardy of Central/South American softbills and acclimatised birds thrive in planted flights and shelters outside; they are best brought indoors before the winter, although some have remained outdoors throughout the year but with a warm shelter available. They are also good for a tropical house or conservatory, but not good mixers; they are very spiteful with similar-sized birds and especially aggressive as breeding condition is assumed.

Cyanerpes caeruleus
Purple Honeycreeper, Purple Sugarbird

$4\frac{1}{4}$ in (11 cm). The entire plumage is blue except for the black eye-streak, chin and throat, wings and tail. The bill is black, the legs and feet yellow. The female is mainly grass-green, paler on the underparts and with some grey streaking. There is some gold-buff around the eyes and on the throat, and a blue moustachial streak.

Range Ecuador, Colombia, Venezuela, the Guianas, Trinidad, western and northern areas of Brazil.

Status There are good populations, but numbers are declining in some areas following loss of habitat.

Avicultural rating Very popular avicultural subjects, they need care but can be long-lived. They are now only occasionally available.

Feeding Diets E and L, with regular small live food such as *Drosophila*, house flies, spiders etc. Birds in poor condition after importation and quarantine can be offered a small quantity of sponge cake soaked in nectar mixture once a day.

Breeding A cup-shaped nest of fine rootlets etc is built. Clutch 2; incubation 11–12 days. Provide plenty of small live food when the young hatch.

General management Established birds thrive best in a tropical house or conservatory; they can also be housed in sheltered outdoor accommodation during the summer months only, but are not sufficiently hardy to winter outside. They are generally good mixers.

Cyanerpes cyaneus [PLATE 124]
Red-legged Honeycreeper, Yellow-winged Sugarbird

4½ inches (11.5 cm). There is a black streak from the base of the bill through the eye; the mantle, upper back, wings and tail are black. The crown is an iridescent turquoise-blue. The rest of the plumage is royal-blue. The bill is black, the legs and feet red. The female's upper surfaces are dark grey-green, greyer on the sides of the head, chin and throat; her underparts are paler. The bill is black, the legs light brown. The males in eclipse plumage resemble the females but may be easily distinguished by their red legs and feet.

Range Eastern and southern Mexico to northern Colombia, Venezuela, the Guianas, Trinidad; Ecuador, northern Peru, northern, central and eastern Brazil, western Colombia.

Note 12 subspecies are described.

Status There are generally good populations where deforestation has not destroyed their habitat.

PLATE 124 *Cyanerpes cyaneus*, Red-legged Honeycreeper

Avicultural rating They are only occasionally available at present, but are a long-standing avicultural favourite. Easily managed, they have bred in a cage.

Feeding Diets E and L, with regular small live food.

Breeding Established pairs are often willing breeders. Clutch 2; incubation 12–13 days. A cup-shaped nest of fine grasses, rootlets etc is built. They may use a basket or box. Soft fruit and small insects are the main rearing food; *Drosophila*, house flies etc are valuable.

General management Acclimatised birds are hardy, but are best not exposed to winter cold or damp. They are ideal occupants of a tropical house, they mix well with other small softbills, but should be housed with non-aggressive companions.

Diglossa cyanea [PLATE 125]
Masked Flower-piercer

---

6 in (15 cm). The forehead, crown, sides of head, chin and throat are black; the rest of the plumage

PLATE 125 *Diglossa cyanea*, Masked Flower-piercer

is slate-blue. The bill, legs and feet are black. The female is similar but duller.

Range Northern and western Venezuela, Colombia, Ecuador, Peru, northern Bolivia.

Status There are good populations in some parts of the range, but numbers are being affected by habitat loss.

Avicultural rating Occasionally available; this species needs care but is generally a good avicultural subject.

Feeding Diets E and L, with small flies (*Drosophila*, house flies etc). Fruit should be provided in large (half) sections; pears, grapes and tomato are the favourites.

Breeding A cup-shaped nest is usually built in the depths of a shrub or similar vegetation. Clutch 2; incubation 13–14 days. Small insects form the bulk of the rearing diet.

General management Careful management is needed during the immediate post-importation and quarantine period, but properly established birds are not difficult. They are ideal tropical house occupants and mix well with other small softbills. They are not suitable for an outdoor shelter and flight except during the warmest period of the summer.

Tersina viridis [PLATE 126]
Swallow Tanager

PLATE 126 *Tersina viridis*, Swallow Tanager

$5\frac{3}{4}$ in (14.5 cm). The forehead, face, chin and throat are black; the rest of the plumage is turquoise-blue, slightly lighter on the crown. The abdomen and under tail coverts are white; there is some black in the flight feathers. The tail is turquoise and black. The bill is black, the legs and feet grey. The female is mainly green or grey-green, with a grey facial mask.

Range Eastern Panama, Colombia, Ecuador, Peru, Bolivia, northern Argentina, Paraguay, Brazil, the Guianas.

Status Local and nowhere abundant throughout the extensive range.

Avicultural rating Occasionally available but difficult to establish and for experienced aviculturists only.

Feeding Diet F with abundant live food; also fruit (Diet E) with various soft berries as available.

Flying insects are particularly important and this species is often difficult to switch to an inanimate menu after importation and quarantine. A high-protein diet is essential.

Breeding This species is unusual among tanagers in being a hole-nester; it excavates a nest-chamber in an earth-bank, or may use a tree-cavity. Clutch 3; incubation 14–15 days. Fruit and insects, mainly the latter, provide the rearing diet.

General management Patience is needed to establish these birds on the necessary high-protein diet, but provide flies hatching at regular intervals so Swallow Tanagers can hawk. They are excellent tropical house inhabitants; they can also be housed in shelter and flight outdoors during the summer months but the species does not tolerate cold and wet conditions well. It can be aggressive in mixed company, so keep in separate accommodation, or choose companions carefully.

Coereba flaveola [PLATE 127]
Bananaquit

4½ in (11.5 cm). The crown and stripe from the base of the bill, through the eye, to the sides of the neck, are black; the rest of the upper surfaces, wings and tail are dark grey. The chin and throat are grey, the rest of the underparts pale yellow-orange shading to grey-white on the abdomen. There is a conspicuous white eyebrow stripe. The bill is black, the legs and feet dark grey. The female is similar but duller.

Range Mexico and Central America, Colombia, Ecuador, northern areas of Peru, northern Bolivia, western and northern Brazil, Venezuela, the Guianas, Trinidad, Tobago and most Caribbean islands except Cuba.

Note There is some plumage variation among the 41 described subspecies.

Status Abundant in many parts of the range, including gardens and cultivated areas.

PLATE 127 *Coereba flaveola*, Bananaquit

Avicultural rating These are attractive and interesting avicultural subjects which are not difficult to manage.

Feeding Diets E, F and L with the addition of small live food. Items of fruit are best offered in large (half) sections. *Drosophila* and other small flies are the best live food. Watch that the birds do not consume artificial nectar to the exclusion of other foods. Despite the bird's common name, use of banana in the diet is not recommended.

Breeding They make a globular nest of grasses and plant fibres lined with fine, soft stems, usually in bushes, or low-growing trees. Clutch 2–3; incubation 12–13 days. The young are said to be reared mainly on nectar with some small flying insects.

General management These are excellent tropical house birds, but they can be housed in an outdoor shelter and flight during the summer months when acclimatised; they do not tolerate wet and cold conditions in the winter. They usually mix well, but pairs approaching breeding condition are aggressive and best separated from companions unless in spacious quarters.

Icteridae (New World Blackbirds)

This New World family contains more than 80 species, including grackles, cowbirds, orioles, blackbirds, oropendolas, caciques and troupials. They are mainly tropical, but some are from temperate zones. They are sparrow to crow-sized; with mainly black or brown plumage with yellow, orange and red highlights. Some are polygamous, others parasitic. Many are familiar avicultural subjects, although never among the most popular species; they are now less frequently available. They are mainly omnivorous, but can exist on a simple seed and live food diet, although this is not recommended.

Agelaius icterocephalus [PLATE 128]
Yellow-hooded Blackbird, Yellow-headed Marshbird

PLATE 128 *Agelaius icterocephalus*, Yellow-hooded Blackbird

7½ in (19 cm). The head, neck, chin, throat and breast are yellow; the rest of the plumage, including a small area from the base of the bill to the eye, is black. The bill, legs and feet are black. The female's head is olive, her upper surfaces olive-brown, eyebrow stripe and throat yellow and the underparts grey-olive.

Range Northern Peru, Colombia, Venezuela, the Guianas, northern Brazil.

Status Abundant in many areas.

Avicultural rating They are hardy, easily managed and occasionally available.

Feeding Use a mix of Diets A and G with some fruit and live food. Grit is essential.

Breeding This species is a social breeder, and there is some polygamy. A cup-shaped nest is built low among cover. Clutch 3–4; incubation 12–13 days. Insects provide the bulk of the rearing diet.

General management They are hardy and easily managed after acclimatisation; they can be housed in an outdoor shelter and flight throughout the year, but there should be a frost-free, dry shelter. They are usually good mixers. They do best in spacious accommodation, if possible, as they are nervous and highly-strung in close confinement. Hens are only rarely available; most shipments are made up of brighter-coloured males.

Molothrus bonariensis
Common Cowbird, Shiny Cowbird, Glossy Cowbird

8 in (20 cm). The entire plumage is black with purple reflections. The bill, legs and feet are black. The female's upper surfaces are grey-brown, paler below, with some light streaking on the breast.

Range Panama, South America and Lesser Antilles.

Status Extremely abundant in many parts of the range.

Avicultural rating This is not a popular avicultural subject, although it was once freely available.

Feeding Use a mix of Diets A and G, with regular live food. Some fruit may be taken. Grit is essential.

Breeding This bird is parasitic on many host species including mockingbirds, kingbirds, tyrant flycatchers etc. Some females occasionally attempt nest construction, but this is not typical. They are very prolific, with each female laying 60–100 eggs during the breeding season; incubation 11–12 days.

General management These are easily managed and hardy birds. Acclimatised specimens will over-winter in an outdoor shelter and flight. They are sociable and are best in small groups. They usually mix well, but watch for aggressive behaviour if a mixed-sex collection.

Fringillidae (Finches)

There are more than 120 species of finch in 2 subfamilies: Fringillinae (chaffinches and brambling) and Carduelinae (serins, siskins, goldfinches, hawfinches, etc). They have a wide distribution, mainly in Europe and Asia, but there are several species in Africa, and the New World; they are absent from Australasia, Madagascar and the Antarctic regions. Seeds of various kinds are the main diet, although some are also highly insectivorous when breeding. Many are kept in aviculture; they are generally easily-managed, long-lived birds. They are mainly small or medium-sized; brown plumage is general but many species have brighter markings. Most build cup-shaped nests in vegetation; nest-construction, incubation and brooding are mainly carried out by the female, although the cock assists with feeding the young.

Serinus atrogularis
Yellow-rumped Serin, Yellow-rumped Seedeater, Black-throated Canary

$4\frac{1}{2}$ in (11.5 cm). The upper surfaces are grey-brown streaked and spotted with dark brown; the feathers of the wings (and tail) have lighter margins. The underparts are grey, darker on the breast and with some streaking; the rump is yellow. The bill is horn-coloured, the legs and feet flesh-brown. The sexes are alike.

Range Southern Arabia, Sudan, Ethiopia southwards to Kenya, Tanzania, Uganda, Zaire to Angola, Namibia and South Africa.

Note There is some plumage variation among 11 described subspecies.

Status Numbers fluctuate, but they are abundant in many parts of the range, including areas of cultivation.

Avicultural rating A fairly popular avicultural species, mainly for its prowess as a songster; it is easily managed and hardy after acclimatisation.

Feeding Diet B, with millet sprays, seeding grasses and regular green food. Grit is essential.

Breeding True pairs are fairly willing breeders in suitable accommodation. They are aggressive and active custodians of their territory when nesting; breeding pairs are best kept alone in limited-space accommodation, but there are few problems in a spacious mixed aviary. A cup-shaped nest of plant fibres etc with soft lining is built, frequently in open boxes, baskets etc placed among cover. Clutch 4; incubation 13–14 days. Small live food is important while the nestlings are small; also supply germinated seed, canary rearing food, insectile mixture.

General management The lack of bright colours or distinctive markings is the probable reason why this species is not more widely kept. Sexing is difficult, and both sexes sing, although the females not as persistently as the cocks. Acclimatised birds can over-winter in an outdoor shelter and flight. They are not as belligerent as related species, but watch their behaviour with smaller or weaker companions (waxbills etc).

Serinus canaria
Wild Canary

5½ in (14 cm). The upper surfaces are dark olive-yellow with some darker streaking. The underparts are yellow-green. The wing and tail feathers are black with yellow-green edges. The bill is lead-grey, the feet horn-coloured. The sexes are similar.

Range The Canary Islands, Madeira and the Azores.

Status Reasonable numbers in some areas.

Avicultural rating Infrequently available.

Feeding Diet C with various wild seeds and screenings. Also provide both wild and cultivated green food (groundsel, chickweed, shepherd's purse and sow thistles are all acceptable—together with invaluable cress when natural green food is in short supply). Germinated seeds are also enjoyed—but are not essential except when the birds are breeding. Grit is essential.

Breeding These nervous and easily-disturbed birds are by no means easy to breed until they have had the opportunity to settle down in a suitable aviary. Wicker baskets are frequently used as a base for the nest. Provide plenty of dried grass, moss and some soft material for lining. Clutch 4–5; incubation 13–14 days. A breeding pair should be provided with a variety of rearing foods including germinated seeds, insectile mixture and egg food.

General management They are generally adaptable and hardy after acclimatisation—although a little highly-strung. They thrive in a small aviary of which they are the sole occupants. If possible it should provide good natural or artifical cover.

Canary (domesticated)

4½–7 in (11.5–17.5 cm) Selective breeding has led to the establishment of many varieties, of varying shape and size, as well as an increasing number of colours. Among the best known and most popular domestic canary breeds are Borders, Norwich, Rollers, Glosters, Yorkshires and Lizards. New Colours are rapidly increasing in popularity—and many of the older breeds such as the Lancashire (both Copy and Plainhead), Scotch Fancy and others are making a comeback. Several Continental varieties (Belgian, Parisian Frilled, etc) are also back in fashion.

Avicultural rating One of the most popular of all cage birds.

Feeding A staple Diet C should be supplemented with various wild seeds and green food similar to those recommended for the Wild Canary.

Breeding Most are bred in cages provided with a nest-pan and felt lining. It is usual to remove each egg as it is laid, store carefully—preferably stood on end in bran or some other soft material and turned each day—in a cool room; replace the removed eggs with artificial (plastic) ones which are readily available from most pet shops. When the clutch is judged to be complete return all real eggs to the nest (removing the artificial ones, of course) so that incubation and hatching are synchronised. Clutch 4–5; incubation about 13 days. Provide eggfood and proprietary rearing foods. Canaries housed in aviaries hatch and rear young without artificial management—just leave them to it to get on with things.

General management Adaptable and hardy, Canaries are among the most popular and easily-managed of birds. They make excellent pets—but remember only the cocks sing.

Fringilla coelebs
Chaffinch

6 in (15 cm). The head, nape and sides of the neck are blue-grey, the face, cheeks, mantle, chin, throat and underparts vinous-pink, paler on the lower breast and shading to grey-white on the

abdomen. The lower back and rump are moss-green. The wings are brown, with a large area of white on the shoulder, a white wing bar and some yellow-green on the outer webs of the flights; the tail is grey-black with white outer feathers. The bill is horn-coloured (lead-blue in the breeding season); the legs and feet are brown. The female is similar but browner above and with buff-coloured underparts.

Range Europe, North Africa, Central Asia, Siberia.

Status They are generally abundant in many parts of the range.

Avicultural rating A popular and easily-kept species, but aggressive in mixed company.

Feeding Diet C, with green food and regular small live food. Wild seed mixtures and ripe weed seeds are beneficial. Grit is essential.

Breeding They are not as free-breeding as some European finches. A pair need a separate aviary with adequate natural or artificial cover. Clutch 3–6; incubation 13–14 days They are very insectivorous when breeding; abundant small live food is needed to rear the young, plus germinated seeds, extra green food etc. Courtship is often turbulent, so ensure that the female has plenty of refuges. This species builds a beautiful nest using lichens, cobwebs, moss etc.

General management They are hardy and easily managed, doing best in an outdoor shelter and flight, but can be pugnacious even outside the breeding season. They tend to be nervous when closely confined.

Note In the UK only aviary-bred and closed-ringed birds of this species can be offered for sale under the provisions of the Wildlife and Countryside Act, 1981 (Schedule 3, Part 1).

Serinus leucopygius (III)
Grey Singing Finch, White-rumped Seedeater

---

4 in (10 cm). The head is light-grey, the rest of the upper surfaces grey-brown with some brown streaking. The underparts are grey, whiter towards the abdomen; the rump is white. The bill, legs and feet are horn-coloured. The sexes are alike.

Range Senegal eastwards to Nigeria, Chad, Central African Republic, Sudan, northern Ethiopia.

Status Variable, but abundant in some parts of the range including areas of cultivation, villages etc.

Avicultural rating It is popular as a songster. Acclimatised birds are hardy and usually long-lived.

Feeding/Breeding/General management See Yellow-rumped Serin. Birds of this species seem not to travel as well as some relatives and are often difficult to establish, although eventually they are very hardy. They can be housed with other birds in spacious accommodation but their aggressive tendencies can be a problem in a restricted space. They are difficult to sex but true pairs are fairly willing breeders. They are excellent songsters.

Serinus mozambicus (III)
Green Singing Finch, Yellow-fronted Canary

4½ in (11.5 cm). The forehead, eyebrow-streak, cheeks, chin, throat, underparts and rump are yellow. The top of the head, a small area from the base of the bill to eyes, ear-coverts and sides of the neck are grey; the rest of the upper surfaces are olive-grey with darker striations. There are conspicuous grey-black moustachial streaks. The bill, legs and feet are horn-coloured. The females are similar but can usually be distinguished by dark grey-green 'necklace' markings across the throat; immature birds of both sexes also exhibit indistinct spotting on the throat and upper breast.

Range Senegal, Gambia, Guinea eastwards

through southern Chad, Central African Republic to Sudan, Ethiopia, Kenya, south to Angola, Namibia in west, Zimbabwe, Mozambique, South Africa in east.

Note There is some variation in plumage among the 11 described subspecies.

Status Abundant in many parts of the range.

Avicultural rating This is a long-established avicultural favourite, which combines an attractive appearance with a pleasant song. They are hardy and long-lived, but aggressive in mixed company.

Feeding/Breeding/General management Similar to Yellow-rumped Serin. Ease of sexing makes breeding opportunities more frequent, but they are best kept in separate small accommodation (shelter and flight) because of their aggressive disposition and disruptive behaviour with companions; they can also be bred in roomy cages.

Carduelis chloris
Greenfinch

$5\frac{3}{4}$ in (14.5 cm). The upper surfaces are olive-green, the underparts more yellow. The wings are darker, with a yellow wing-bar and outer tail feathers. The bill is yellow-brown, the legs and feet brown. The female's upper surfaces are grey-green, the underparts less yellow.

Range Europe, North Africa eastwards to Caucasus, Siberia.

Status Abundant in many parts of the range.

Avicultural rating A popular and widely-kept species which is easily managed and a free-breeder.

Feeding Diet C with wild seed mixtures, green food etc. Grit is essential.

Breeding This species is generally a free breeder, although the young birds are often difficult to rear. They can be bred in cages, but are best in a small outdoor shelter and flight. Provide artificial cover unless in a planted flight; put a wicker basket among the vegetation. Clutch 4–6; incubation 13 days. Provide germinated seed, live food, extra green food etc for rearing.

General management It is hardy and thrives in suitable outdoor accommodation; provide a sheltered area for roosting. It is usually a good mixer with other species.

Note In the UK only aviary-bred and closed-ringed birds of this species can be offered for sale under the provisions of the Wildlife and Countryside Act, 1981 (Schedule 3, Part 1).

Carduelis spinoides
Himalayan Greenfinch, Black-headed Greenfinch

5 in (13 cm). The forehead, eyebrow stripe, neck collar, rump and underparts are yellow; the crown, mantle and back are greenish-brown. The wings are brown with a conspicuous yellow patch; the tail is brown and yellow. The bill is horn-coloured, the legs flesh-brown. The female is similar but duller; she has more green on the upper surfaces, and a smaller yellow wing patch.

Range Pakistan eastwards across northern India, Himalayas to Assam, western Burma.

Status There are believed to be good populations in many areas.

Avicultural rating Occasionally available; established birds are hardy and will over-winter in an outdoor shelter and flight. They mix well with similar-sized species, but there may be some aggression when breeding condition is assumed.

Feeding Use a mix of Diets A and C, with seeding grasses, abundant green food, some live food etc. Grit is essential.

Breeding This bird builds a cup-shaped nest of plant fibres, moss etc in trees and shrubs. Clutch

3–4; incubation 13–14 days. Provide canary rearing food, germinated and soaked seeds, extra green food and live food for rearing.

General management It is generally easy to manage, and acclimatised birds will successfully overwinter in an unheated outdoor shelter and flight. They are generally peaceful, but I have noted some fighting among a small group as individual birds improved in condition after importation and quarantine.

Carduelis spinus
Siskin

$4\frac{3}{4}$ in (12 cm). The forehead, crown and chin are black, the upper surfaces olive green with darker streaking. The underparts are yellow-green, becoming paler towards the abdomen; the wings are black with two irregular yellow bars. The rump is yellow-green, the tail green or black. The bill is flesh-coloured with a darker tip; the legs and feet are flesh-brown. The female is generally paler and more grey; she lacks the black head and chin.

Range Northern Europe and Asia.

Status There are good populations, and they are increasing in some areas.

Avicultural rating Excellent avicultural subjects, they are easily managed and good breeders.

Feeding Diet C, with a variety of green food etc. Wild seed mixtures are enjoyed. Grit is essential.

Breeding They are usually willing breeders. Cover in the breeding aviary should be either growing conifers or bunches of spruce wired into place. They will build a neat cup-shaped nest of plant fibres, moss, lichens etc; they also utilise wicker baskets placed among cover. Clutch 3–5; incubation 11–12 days. Provide germinated seed, live food etc for rearing.

General management These delightful little birds are usually tame and confiding from the outset. They mix well with other species, but it is best if the breeding pair occupy their own small shelter and flight; colony breeding is also feasible. Tameness sometimes leads to lethargy, and obesity; so ensure that there is plenty of flying space in the accommodation and that food and water containers are not too conveniently close to the perches. They are hardy but provide a sheltered area for roosting.

Note In UK only aviary-bred and closed-ringed birds of this species can be offered for sale under the provisions of the Wildlife and Countryside Act, 1981 (Schedule 3, Part 1).

Acanthis flammea
Redpoll, Lesser Redpoll

$4\frac{3}{4}$ in (12 cm). The upper surfaces are brown with darker streaking; the forehead is red, the chin black. The underparts are buff-white, rose-red in the breeding season. There is an indistinct dark brown eye-stripe. There is some pink on the rump during the breeding season. The bill is yellow, the legs and feet brown. The female is similar, but lacks the red on the breast.

Range Northern Europe, Asia, North America, Iceland.

Status There are good populations in many parts of the range.

Avicultural rating They are cheerful, popular avicultural subjects, which are easy to manage and good breeders.

Feeding Diet C, with only small quantities of hemp—or none at all. Wild seed mixtures are enjoyed. Also provide green food (wild and cultivated) and seeding grasses or weeds etc as available. Grit is essential.

Breeding This bird is usually a free breeder, either maintained as single pairs or when several pairs share a larger aviary. Provide plenty of natural or artificial cover; wicker nest baskets are used. Clutch 4–6; incubation 11 days. Provide abundant small live food, germinated seed etc for rearing.

General management These active and sociable little birds are very hardy and thrive in an outdoor shelter and flight, but need dry, sheltered roosting areas. They usually mix well with other finches.

Note In the UK only aviary-bred and closed-ringed birds of this species can be offered for sale under the provisions of the Wildlife and Countryside Act, 1981 (Schedule 3, Part 1).

Acanthis cannabina
Linnet

5¼ in (13.5 cm). The head and neck are grey-brown with darker streaks; the mantle, back and wings are chestnut-brown. The chin and throat are grey-white, the rest of the underparts grey-brown, paler towards the abdomen (the forehead and breast are crimson in the breeding season). The rump is whitish, the tail brown with white outer feathers. The bill is horn-coloured (lead-grey in summer); the legs and feet are brown. The female is generally duller and with only a trace of red on the forehead; she has pronounced streaking above and below.

Range Europe, North Africa eastwards to Asia Minor, south-west Asia.

Status There are generally good populations in many parts of the range.

Avicultural rating A long-established favourite which is easily managed and a good breeder. It is a fluent songster.

Feeding/Breeding/General management Similar to Redpoll. These birds are often shy and nervous; they are free breeders, but do best in a well-planted aviary in a quiet situation.

Note In the UK only aviary-bred and closed-ringed birds of this species can be offered for sale under the provisions of the Wildlife and Countryside Act, 1981 (Schedule 3, Part 1).

Carpodacus mexicanus
House Finch

5¾ in (13.5 cm). The forehead, eyebrow stripe, rump, throat and breast are red, the rest of the upper surfaces brown with darker streaking. The flanks and abdomen are pink or grey-white. The wings are brown with indistinct white wing-bars; the tail is dark brown. The bill is horn-coloured, the legs and feet flesh-brown. The female is mainly grey-brown, with paler underparts and some streaking.

Range South-western areas of Canada, west and west-central areas of the USA to Mexico.

Status Abundant in many parts of the range. It is increasing its range eastwards.

Avicultural rating Infrequently available at present, but a good avicultural subject and a willing breeder; it loses the red colour in close confinement, but this may be retained in a spacious, natural environment.

Feeding Diet C with seeding grasses, soaked seed, green food etc. Grit is essential. Some live food should be offered.

Breeding They are fairly willing breeders in a good environment. They do best in a large, secluded shelter and flight with ample natural cover. They have been bred on a 'colony' system with several pairs sharing the same aviary; there is some bickering and competition for prime nest sites among the females. The nest is cup-shaped, usually low down in vegetation. They will also use an open wicker basket or box. Clutch 4–6 (occasionally 7); incubation 13–14 days. Provide extra green food, soaked seed, some canary rearing food and live food for rearing.

General management These are nervous little

birds, and they are not suited to close confinement. They are generally well-behaved and mix with other species. They are hardy when acclimatised.

Pyrrhula pyrrhula
Bullfinch

5¾ in (14.5 cm). The face, crown, wings and tail are black, the nape, mantle and back grey. The sides of the head and the underparts are rose-pink, paler towards the abdomen; the rump is white. There is a white bar on the wings. The bill is black, the legs and feet brown. The female is duller and of a much browner shade but has similar, though less clearly defined, markings to the cock.

Range Europe eastwards across northern Asia to Manchuria, Korea.

Status There are generally good populations, despite persecution in many fruit-growing districts.

Avicultural rating A well-established avicultural subject, it is easily managed and often a good breeder, but is often shy and secretive.

Feeding Diet C with extra sunflower seed; hemp is greatly enjoyed but is best fed in modest amounts. Provide wild seed mixtures, green food etc. Grit is essential. Fruit buds and berries should be offered.

Breeding A secluded aviary suits these birds best; planting, although desirable, is difficult because of the species' habit of consuming buds from a wide range of trees and shrubs. Provide shallow wicker baskets among the natural or wired-in artificial cover; they make a nest of rootlets etc, often with abundant twigs in the outer construction. Clutch 4–5; incubation 13 days. Breeding pairs should not be housed with companions. Provide live food, germinated seeds etc.

General management When not breeding, these birds can be housed with other species in spacious accommodation; there is a strong pair bond. They do best in an outdoor shelter and flight; they are hardy but need dry, frost-free sleeping accommodation.

Note In the UK only aviary-bred and closed-ringed birds of this species can be offered for sale under the provisions of the Wildlife and Countryside Act, 1981 (Schedule 3, Part 1).

Coccothraustes migratorius
Black-tailed Hawfinch, Chinese Hawfinch, Yellow-billed Grosbeak

7½ in (19 cm). The whole of the head, chin, throat, wings and tail are glossy black, the rest of the upper surfaces fawn-grey, lighter on the nape to form a collar. The underparts are fawn-grey or beige, becoming a paler grey-white on the abdomen. The flanks have a rufous wash. The rump and tips of the flight feathers are white. The bill is massive and yellow, grey at the base and tip; the legs and feet are flesh-coloured. The female has grey-brown upper surfaces, paler below; she lacks the distinctive glossy-black feathers of the cock's head, chin and throat.

Range Eastern China, North Korea, Taiwan.

Status There is little information available, but there are believed to be good populations in some parts of the range.

Avicultural rating Occasionally available, they are hardy and easily managed after acclimatisation.

Feeding Diet C, with extra hemp, sunflower, safflower and buckwheat. They are fond of berries, including hawthorn, rowan, elder, blackberry, blackcurrants, raspberries. Like the European Hawfinch this species enjoys fresh peas. Also offer green food and some live food. Grit is essential.

Breeding It has proved difficult to breed successfully. The nest is usually a substantial cup of small twigs, rootlets and other plant fibres lined with soft, fine grasses etc. Clutch 3–4; incubation 14 days. Live food is likely to be an important part of the rearing diet while the nestlings are small. The breeding pair should occupy a secluded shelter and flight with ample natural or artificial cover; the cock is dominant and aggressive and thickets of vegetation will provide some shelter for the female during courtship.

General management Established pairs are best kept in roomy shelter and flight accommodation throughout the year; they are very hardy. They are not recommended for housing with other species; they are powerfully-built with a bill capable of exerting 150-lb (68 kg) pressure to crack cherry stones. The cocks outnumber the hens in most importations.

Coccothraustes personatus
Masked Hawfinch, Japanese Hawfinch, Japanese Grosbeak

8½ in (21.5 cm). Similar to Black-tailed Hawfinch, but this bird has the crown, cheeks and chin black, with no rufous on the flanks, and it is larger in size. The bill is yellow with blue-grey at the base and a black tip; the legs and feet are flesh-coloured. The female lacks the black markings on the head and chin. The upper parts are grey-brown; the under surfaces are paler.

Range North-east Asia, Japan; it winters in southern Japan, eastern China.

Status Little information is available, but there are believed to be good populations in some areas.

Avicultural rating See Black-tailed Hawfinch.

Feeding/Breeding/General management Similar to Black-tailed Hawfinch. They are infrequently available, and females are always in short supply. They are very hardy and weather-resistant. Both this and the previous species are sometimes called 'Black-headed Hawfinches', leading to some confusion.

Estrildidae (Waxbills, etc)

There more are than 130 species of the Estrildidae distributed throughout the Old World tropics. It is probably the most important seedeating family for aviculturists, embracing many popular and widely-kept species. Most are small and many are colourful. They are usually easy to manage after acclimatisation, although some with more specialist habits are less so; a few species (Zebra Finch, various other Australian finches, Cut-throat etc) are well on the way to the type of domestication that has made the canary and budgerigar so popular and available in various colours etc. They are usually willing breeders in a suitable environment, and it is vital that purchasers do not 'waste' imported stock by not giving them every encouragement to nest; already some familiar species are no longer available because of export/import restrictions, and this is a trend likely to continue and embrace other species.

Pytilia phoenicoptera [PLATE 129] (III)
Red-winged Pytilia, Aurora Finch

PLATE 129 *Pytilia phoenicoptera*, Red-winged Pytilia

4¾ in (12 cm). The whole of head, mantle and upper surfaces are grey; the underparts are similar but with fine white barring from the lower breast to the abdomen. The wings are grey-brown with a conspicuous area of red on the coverts and primaries. The upper tail coverts are red; the tail is black and red. The bill is black, the legs and feet flesh-brown. The female is similar, but readily distinguished by her slightly browner plumage and smaller and paler areas of red.

Range Senegal and Gambia eastwards to Central African Republic, northern Zaire, Uganda, southern Sudan, Ethiopia.

Status Distribution is somewhat patchy, and nowhere are they abundant, although populations in the western part of the range are generally larger than in the east.

Avicultural rating This is a popular and easily managed species; it is also a very willing breeder. Birds are regularly available, but in small numbers.

Feeding Use Diet B, with millet sprays, seeding grasses, live food; the latter is essential daily and can include housefly or blowfly pupae, tiny mealworms etc. Mealworms are best cut up if large and thick-skinned. Individual birds vary in their liking for green food, but offer items such as groundsel, chickweed and cress on a regular basis. Grit is essential.

Breeding Established pairs are usually free breeders and may be multi-brooded. They prefer globular nest baskets with a side entrance; the nest itself is of grass stems and plant fibres with a softer lining, although I have had successful hatching and rearing in a rudimentary nest of strips of torn paper inside a basket, despite the availability of more suitable materials. Clutch 3–5; incubation 13 days. Both cock and hen may sit together on the nest, and the cock usually sleeps alongside the hen. Provide abundant small live food, seeding grasses, soaked millet sprays and germinated seed for rearing. This species is very willing to attempt nesting in a spacious cage, but the lack of some *natural* live food often results in failure although some pairs succeed in raising chicks on little other than tiny gentles or mealworms and spray millet.

General management Acclimatised birds are eventually fairly hardy, but do not appreciate cold and wet conditions and short winter days; it is best if they are housed in a bird-room during the winter. They can be housed with other small birds, and even breeding pairs seem to retain their gentle disposition, although species with red in their plumage may induce threatening behaviour. These birds have an odd habit of turning completely about when disturbed or upset; the rear end faces the object of alarm, with the tail cocked like that of a wren.

Pytilia hypogrammica (III)
Red-faced Pytilia, Yellow-winged Pytilia

4½ in (11.5 cm). The forehead, face, cheeks, chin and throat, rump, upper tail coverts and tail are red, the rest of the upper surfaces grey-brown. The underparts are grey with some white barring on the sides and abdomen; the wings are brown with yellow margins to the flights. The bill is black, the legs and feet flesh-coloured. The female lacks the red head and is generally browner.

Range Sierra Leone eastwards to Cameroon, Chad and Central African Republic.

Status They are locally distributed and nowhere abundant.

Avicultural rating Only occasionally available, they need similar treatment to the Red-winged Pytilia.

Feeding/Breeding/General management Similar to Red-winged Pytilia. They are sometimes delicate and difficult to establish, probably depending on the quality of care and diet during quarantine; eventually they are fairly hardy and easily managed, but not weather-resistant, and they do best in heated accommodation during the colder period of the year. They are much less willing breeders than the Red-winged Pytilia.

Pytilia afra
Orange-winged Pytilia, Red-faced Waxbill

4¾ in (12 cm). The forehead, face, chin, upper tail coverts and tail are red, the rest of the upper surfaces grey-olive. The underparts are grey, more olive towards the abdomen, with a faint orange suffusion. The wings are olive, the greater wing coverts and margins of the flights orange. The bill is red, the legs and feet flesh-coloured. The female lacks the red head, and is generally duller and paler.

Range Sudan and Ethiopia southwards to Kenya, Tanzania, Zambia and through the Central African Republic, Zaire, to northern Angola.

Status Variable throughout the range; abundant in certain parts, they are rarely seen in some areas.

Avicultural rating Now less often available but a good avicultural subject; they have frequently bred.

Feeding/Breeding/General management Similar to Red-winged Pytilia. They are not weather-resistant and are best in heated indoor quarters during the winter. They appear slightly more insectivorous than the Red-winged Pytilia and should have abundant small live food plus germinated seeds when rearing young.

Pytilia melba
Melba Finch, Green-winged Pytilia

5 in (13 cm). The forehead, cheeks, chin, throat, upper tail coverts and tail are red; the rest of the head, crown and nape are grey. The upper surfaces are olive-brown. The breast is orange, the rest of the underparts and flanks grey with white barring. The bill is red, the legs and feet flesh-brown. The female lacks the red on the head; she has paler upper surfaces and the grey underparts are flecked with white.

Range Senegal eastwards to Sudan, Ethiopia, then southwards to Kenya, Uganda, eastern Zaire, Tanzania, Malawi, Zambia, Mozambique, South Africa.

Note There is some plumage variation among the 9 described subspecies.

Status Variable. Abundant in a few areas, they are scarce elsewhere.

Avicultural rating This long-established avicultural favourite is still available in reasonable numbers. They are easily managed when established but are best in experienced hands only. They need careful handling until acclimatised.

Feeding Diet B, with millet sprays, seeding grasses and regular supplies of small live food; the latter is essential if the birds are to remain in good health. Individual birds vary in their interest in green food; offer groundsel, cress etc. Grit is essential.

Breeding Established pairs are often willing breeders, but need abundant live food to succeed. They do best in a separate shelter and flight when nesting; provide ample natural or artificial cover. A domed nest of plant fibres is made; the birds will use woven globular baskets, even cylinders of welded-mesh offcuts stuffed with dried grass etc. Clutch 3–5; incubation 12–13 days. The nestlings are fed exclusively on small live food at first; provide *Drosophila*, house flies, small crickets and as much natural live food as possible. The nestlings will probably be ejected or abandoned if the supply of insects is not maintained.

General management They are probably more delicate (in terms of their ability to withstand cold and wet conditions) than their relatives. Keep in an outdoor shelter and flight only in the summer months, then in warm, indoor quarters for the winter. This species mixes fairly well with other birds; but there are occasional displays of aggression, usually involving similar-sized species with some red in their plumage.

Mandingoa nitidula (III)
Green-backed Twinspot, Green Twinspot

4 in (10 cm). The upper surfaces, throat, upper breast and under tail coverts are olive-green, the face, cheeks and chin red. The lower breast to the abdomen is black heavily spotted with white. The wings are olive-brown, some feathers with brighter olive-green margins; the tail is olive-green and black. The bill is black, the legs and feet flesh-coloured. The female is similar but with the red area on the face replaced by yellow or orange.

Range Fernando Po, Sierra Leone and Cameroon eastwards to Zaire, southern Sudan and Ethiopia and southwards through Kenya, Tanzania, Zambia, Mozambique to eastern areas of South Africa.

Note Of the 4 described subspecies, Schlegel's Twinspot (*M. n. schlegeli*), which occupies a western part of the range, is the most frequently available at the present time.

Status Variable, and rarely seen throughout much of the range; but this may be more a reflection of their secretive behaviour and the heavy vegetation of their habitat than an indication of increasing rarity.

Avicultural rating They need care when first imported, then are fairly easy to manage, but are not for inexperienced aviculturists.

Feeding Diet B, with millet sprays, seeding grasses and regular items of small live food; the latter must be included in the daily diet if the birds are to thrive. Some individual birds enjoy live food; cress appears to be a favourite. Grit is essential.

Breeding They are rarely bred. This species builds a large (for the size of the bird) domed nest or may use a globular basket. Clutch 3–5; incubation 12–13 days. Abundant small live food is essential if rearing is to be successfully accomplished; the breeding pair are best kept in separate small shelter and flight unit, so that competition for insects is reduced. *Drosophila*, house flies (and pupae), tiny crickets and as much natural live food as can be collected, is vital until the chicks fledge.

General management They are often delicate until properly established, then successful in outdoor accommodation only during the warmest part of the summer and autumn. Comfortable, indoor accommodation is needed at other periods of the year, and this species does not tolerate cold and wet conditions at any time. It may be a good tropical house species, but is fairly unobtrusive and likely to remain mainly unseen among dense vegetation. Schlegel's Twinspot, although best regarded as fairly delicate, appears easier to manage than the nominate subspecies (*M. n. nitidula*) from south-eastern areas of Africa, which was occasionally exported some years ago; I have kept *M. n. nitidula*, *M. n. schlegeli* and *M. n. chubbi* at various times, and regard the first-mentioned as the most demanding of the three.

Pyrenestes ostrinus (III)
Black-bellied Seedcracker

5½ in (14 cm). The whole of the head, neck, breast and upper tail coverts are red; the rest of plumage is black or brown. The tail is a dull red. The bill is blue-black, the legs and feet brown. The female is similar, but the black of her back and underparts is replaced with brown. Birds from the extreme west of the range have brown back and underparts in both sexes.

Range Gambia, Guinea, Ghana eastwards to Cameroon, northern Zaire and south to Angola.

Status Fairly scarce in most of the range.

Avicultural rating They are spectacular, but usually extremely difficult to maintain in good health.

Feeding Diet B, with spray millet and ample small live food. Little interest is shown in green food, although some authorities recommend it. Grit is essential.

Breeding This species builds a domed nest low down in vegetation; plant fibres and stems are used in construction with a finer, softer lining. Clutch 3–4; incubation 12–13 days. Abundant live food is needed for successful rearing.

General management These birds should be acquired only by experienced aviculturists accustomed to dealing with other delicate seedeaters. The period immediately following importation and quarantine is usually the most testing. They are fairly hardy when acclimatised, but not resistant to cold and wet conditions, and are suitable for outdoor accommodation only during the warmest part of the summer. This bird is an excellent occupant of a tropical house, and this type of environment may eliminate some of the present question marks about the species' longevity in controlled conditions, when few live for any length of time.

Spermophaga haematina (III)
Western Bluebill, Blue-billed Weaver

5½ in (14 cm). The plumage is mainly glossy black, except for the chin, throat, breast, flanks and upper tail coverts, which are red. The bill is silvery-blue, the legs and feet black. The female's upper surfaces are dark slate, her face and upper tail coverts red; her under surfaces are black, heavily spotted white.

Range Gambia, Guinea eastwards to Central African Republic; south to Gabon, Congo, Zaire, northern Angola.

Status Abundant in many parts of the range.

Avicultural rating This species is often difficult to establish after importation but improves with careful management. It is not for inexperienced aviculturists.

Feeding Diet B, with spray millet, seeding grasses, live food and items of green food (offer groundsel or cress). Some birds quickly adapt to taking fly pupae. Provide grit.

Breeding They build an untidy domed nest with a side entrance. Clutch 3; incubation 12–14 days. The nestlings appear to be fed on large quantities of insects.

General management Although somewhat similar in appearance, in my experience they are relatively easier to maintain than Seedcrackers. Much depends on the ability of the importers to handle them correctly through quarantine; if the birds are not debilitated afterwards there is a better prospect of establishing them successfully. Their wild habitat suggests that these birds, like Seedcrackers, might be more successful in the moist, warm conditions of a tropical house or conservatory.

Hypargos niveoguttatus
Peters' Twinspot, Peters' Spotted Firefinch

---

5 in (13 cm). The top of the head and nape are olive-brown; the mantle, back and wings are brown. The face, sides of the neck, chin, throat, breast, upper tail coverts and rump are red, the lower underparts black with multiple white spots. The tail is black and red. The bill is blue-grey, the legs and feet grey. In the female the sides of the head are grey-brown, with the red areas paler and the underparts dark grey, spotted white.

Range Eastern Zaire, Tanzania, Kenya, Malawi, Zambia, Zimbabwe, Mozambique.

Status Widespread and abundant in some parts of the range.

Avicultural rating This species needs care after importation but eventually is a good avicultural subject. It is a fairly willing nester.

Feeding Diet B, with spray millet, seeding grasses, some green food and regular small live food. Individual birds may take small amounts of Diet F. Grit is essential.

Breeding This bird builds a domed nest, usually low down among vegetation; globular wicker baskets or suitable open-fronted boxes may attract them. Clutch 3; incubation 12 days. Abundant small live food, germinated seeds, extra millet sprays and seeding grasses should all be available for rearing. Successful breeding probably depends to a considerable degree on the availability of suitable small insects in quantity during the nestling period.

General management They are not weather-resistant, and fare best in a secluded shelter and flight only during the summer; winter them indoors with slight heat. They are good mixers with other similar-sized birds.

Euschistospiza dybowskii
Dybowski's Twinspot, Dusky Twinspot

4½ in (11.5 cm). The head, neck, chin, throat and breast are slate-grey, the mantle, back, rump and upper tail coverts red. The wings are brown. The lower breast, abdomen and under tail coverts are black with white spots on the sides of the breast and flanks. The tail is black. The bill, legs and feet are black. The female is similar but the whole of her underparts are grey with white spots on the breast and flanks.

Range Sierra Leone eastwards to Nigeria, northern Cameroon, Chad, Central African Republic, northern Zaire and south-west Sudan.

Status Uncommon in most areas.

Avicultural rating They are rare in aviculture, but fairly hardy when established.

Feeding Diet B, with millet sprays, seeding grasses and regular small live food; they are adept at capturing small flies.

Breeding Little is recorded. A domed nest is usually built in low-growing vegetation. Clutch 3–5; incubation probably 11–13 days. Abundant small live food, germinated seeds and spray millet are likely to be favoured for rearing. They have successfully reared young in aviaries.

General management After acclimatisation they are not difficult to manage and appear slightly hardier than many species from West or Central Africa. They seem happiest in a shelter and flight with plenty of natural cover; they are very active but mix well with other small species. It is likely that they would prosper in this type of accommodation if an enclosed, heated shelter was available for severe weather housing. They are probably not a good tropical house species.

Lagonosticta senegala (III)
Red-billed Firefinch, Common Firefinch, Senegal Firefinch

---

4 in (10 cm). The whole of the head, neck, chin, throat and breast and upper tail coverts are red; the mantle and back are brown suffused with red. The lower underparts are red shading to buff or pink on the abdomen and under tail coverts. The wings are brown, the tail dark brown and red. There is a narrow yellow eye-ring, and some fine white spots on the sides of the breast. The bill is red, the legs and feet flesh-coloured. The female's upper surfaces are grey-brown or earth-brown, with paler underparts; her lores and upper tail coverts are red.

Range Senegal eastwards to Nigeria, Chad, Central African Republic, Zaire, south-west Sudan, Ethiopia, Kenya, Uganda; southwards to Angola, Namibia, Zambia, South Africa.

Note There is slight plumage variation among the 9 described subspecies.

Status Abundant in many areas, including gardens, villages, cultivated land etc.

Avicultural rating A long-established avicultural favourite which is easily managed, but it is by no means the easiest species to establish successfully.

Feeding Diet B with spray millet, seeding grasses, various items of green food including cress, groundsel and occasional small live food. Grit is essential.

Breeding They are usually very willing breeders. A domed nest of fine plant fibres is built, usually lined with feathers. They will use a variety of receptacles for nesting: boxes, baskets, wire cylinders etc. Clutch 3–5; incubation 12 days. Abundant small live food (*Drosophila*, house flies and pupae, ant pupae, wasp grubs etc) is needed for successful rearing; also supply some germinated seeds and extra millet sprays or seeding grasses. A shovelful of garden compost or manure will attract small flies to the breeding aviary.

General management The period immediately after importation and quarantine is a critical one for these birds; they are best described as 'delicate' until the necessary period of rehabilitation is complete, and the females are especially vulnerable. Imported birds are best housed in spacious box-pattern cages until properly acclimatised, which may take weeks or months according to the season. Provide a varied diet and occasional artificial nectar and sponge cake mixture, if the birds are especially debilitated. Eventually hardy, they are not recommended for occupation of unheated outdoor quarters throughout the year; cold and damp conditions are damaging to these and other small exotics. They mix with other small birds and will breed in some community groupings. They are territorial when breeding and will pursue intruders; other species with red in their plumage may be persecuted, if small enough.

Lagonosticta nitidula
Brown Firefinch

$4\frac{1}{4}$ in (11 cm). The head and upper surfaces, including the rump and upper tail coverts, are grey-brown; the sides of the head and neck are a dull red. The underparts are vinaceous-red with fine white spotting; the abdomen and under tail coverts are grey-brown to buff-white. The wings are brown, the tail black. The bill is red, the legs and feet flesh-brown. The female is similar but her cheeks are more grey and the red under surfaces are less bright.

Range Eastern Angola, southern Zaire, Zambia, Botswana.

Status Uncommon throughout the range.

Avicultural rating Rarely available and difficult to establish.

Feeding/Breeding/General management Similar to Red-billed Firefinch. This species is decidedly difficult to establish, even if properly shipped and quarantined. Stress may be part of the problem, as the birds can appear to be in good condition after the short journey from retailer to aviculturist but then go into decline. Both sexes are delicate in the early stages; properly-acclimatised birds should be treated as recommended for other related species. They do best in a secluded, densely-planted aviary, but are not weather-resistant and must be housed indoors during the colder months of the year. It has been bred.

Lagonosticta rhodopareia
Jameson's Firefinch, Jameson's Ruddy Waxbill, Abyssinian Firefinch

$4\frac{1}{4}$ in (11 cm). The lores and chin are deep rose-pink; the rest of the upper surfaces are grey or fawn-brown suffused with rose-pink. The upper tail coverts are red. The sides of the head, neck and underparts are vinaceous pink, darker towards the abdomen. There are a few white spots on the flanks. The bill is blue-grey, the legs and feet light brown. The female is similar but duller. Many imports are juvenile birds, which may give rise to identification problems; Jameson's Firefinch differs from the similar African Firefinch (*L. rubricata*) in lacking the notch on the second primary.

Range Southern Chad eastwards to Ethiopia, Kenya and south to Tanzania, Zimbabwe and eastern areas of South Africa.

Status There is local distribution, but they are reasonably abundant in many parts of the range.

Avicultural rating Available from time to time, and a good avicultural subject.

Feeding/Breeding/General management Similar to Red-billed Firefinch. They need care during acclimatisation but then are reasonably easy to manage; they seem slightly more robust than other firefinches but do not tolerate cold and wet conditions. Breeding successes have been recorded.

Uraeginthus angolensis
Blue-breasted Waxbill, Cordon Bleu

4¾ in (12 cm). The upper surfaces are beige-fawn, the face, cheeks, chin, throat, underparts, rump and tail blue. There is some pale grey-buff on the lower underparts. The tail is dark blue. The bill is pearl-grey, the legs and feet pale brown. The female is paler and less blue, but the sexes are often difficult to distinguish because of plumage variations between the 4 subspecies.

Range Northern Angola eastwards to southern Zaire, Tanzania, Kenya and south to Zimbabwe, eastern areas of South Africa.

Status There is wide distribution and they are abundant in many parts of the range; they are frequently found in villages, gardens and cultivated areas.

Avicultural rating A popular and widely-kept species, although availability fluctuates. They are good breeders.

Feeding Diet B, with spray millet, seeding grasses, various ripe weed seeds, green food and regular live food. Grit is essential. A varied diet is important for these birds, including daily live insects.

Breeding They are usually very willing breeders if true pairs are obtained. A domed nest of grass stems and other plant materials, sparsely lined with feathers, is built, with a side entrance. A variety of sites are used, but the nest is usually fairly low down and often in a bush. Baskets or boxes are also used. Clutch 4–6; incubation 11–12 days. They are almost completely insectivorous when rearing and abundant small live food is necessary; also offer extra sprays of millet, seeding grasses and some germinated seed.

General management They are delicate when first imported but eventually quite hardy; but they should not be housed in outdoor accommodation during the winter. They usually mix well with other small birds. Despite the need for care after import and quarantine, this is arguably the easiest of the 'blue' waxbills to care for in aviaries.

Uraeginthus bengalus [PLATE 130] **(III)**
Red-cheeked Cordon Bleu, Cordon Bleu

4¾ in (12 cm). The forehead, top of the head and upper surfaces, including the wings, are beige-fawn; the face, sides of the head, chin, throat, breast and flanks are blue. The abdomen and under tail coverts are grey-fawn. There is a prominent red patch on the cheeks. The bill is pearl-pink, the legs and feet flesh-brown. The female is similar, but paler and lacks the red cheek patch.

Range Mauritania and Senegal to Chad, Central African Republic, western Sudan, Uganda, Kenya, northern Tanzania, Zambia.

Status Generally abundant in suitable areas of habitat.

Avicultural rating This is one of the most widely-kept and popular of all exotic seedeaters, but it is delicate and difficult to establish.

Feeding/Breeding/General management Similar to Blue-breasted Waxbill, but the Red-cheeked Cordon Bleu needs great care after importation and the females are particularly difficult to establish properly. Provide a good, varied diet and remember the importance at all times of small live food; do not attempt to hurry the acclimatisation process and ensure that the birds are not subjected to upset or a rapid change of scene which can lead to stress. They are generally good breeders, but need plenty of small live food to rear young successfully.

PLATE 130 *Uraeginthus bengalus*, Red-cheeked Cordon Bleu

Uraeginthus cyanocephala [PLATE 131]
Blue-capped Cordon Bleu, Blue-headed Waxbill

5 in (13 cm). The whole of the head, chin, throat, breast and flanks are blue, the mantle, back and wings a warm beige-fawn. The abdomen and under tail coverts are beige-buff. The bill is red, the legs and feet flesh-brown. The female lacks the blue top of the head (the forehead is blue in some specimens) and is generally paler.

Range Kenya, southern Somaliland, Tanzania.

Status Locally distributed and nowhere abundant.

Avicultural rating Compared with the Red-cheeked Cordon Bleu, this is a relatively new introduction to aviculture. It has been scarce for many years, but is now more readily available. The birds need care when first received, but are eventually easy to manage and good breeders.

Feeding/Breeding/General management Similar to Red-cheeked Cordon Bleu. Like its relatives, it will not tolerate cold and wet conditions and should be housed indoors during the winter.

PLATE 131 *Uraeginthus cyanocephala*, Blue-capped Cordon Bleu

These birds are usually very free breeders once established.

Uraeginthus granatina
Violet-eared Waxbill, Common Grenadier

5½ in (14 cm). The lores and chin are black, the forehead cobalt-blue; the forecrown and sides of the head are violet-mauve. The upper surfaces are chestnut-brown, darker on the wings; the rump and upper tail coverts are blue. The underparts are a warm brown, darker on the abdomen; the under tail coverts are dark blue. The tail is black. The bill is red, the legs and feet grey-brown. The female has similar markings, but they are much paler.

Range Angola eastwards to Mozambique and south to Natal and Cape Province.

Status Variable. There are generally good populations but nowhere are they abundant.

Avicultural rating A popular avicultural subject, which is now less frequently available; it is delicate when first imported and always sensitive to damp conditions and low temperatures.

Feeding Diet B, with spray millet, seeding grasses, abundant small live food; few take any interest in green food. Grit is essential. Some birds of this species appear to take little water (although it must always be available, of course) and refrain from bathing; this is probably linked with their area of origin, which may be naturally arid and waterless.

Breeding They build an untidy and fairly loose domed nest of plant stems and grasses with a feather lining, usually in vegetation. Wicker baskets or open-fronted boxes may be used. Clutch 3–4; incubation 12–13 days. Abundant small live food is the key to successful rearing; flies (including *Drosophila*), fly pupae, live ant pupae, tiny crickets and ultra-clean 'feeder' (house-fly) gentles are among the most easily obtained, and these should be supplemented with natural live food collected from safe (untreated) areas of gardens and the countryside. Extra millet sprays and some germinated seeds should also be provided.

General management This is a species best acquired only by experienced aviculturists. The birds are reasonably hardy after acclimatisation but will not tolerate cold and damp; they are good occupants of a shelter and flight during the summer, but need comfortable indoor accommodation in winter. Single birds usually mix fairly well with other waxbill species, but pairs are best provided with separate quarters. Many individual birds become tame and practically all of those I have owned over the years have been what might be described as 'characters'.

Uraeginthus ianthinogaster
Purple Grenadier

5½ in (14 cm). The lores, forehead and cheeks are blue; the head, neck, mantle and back are chestnut-brown with the upper tail coverts blue. The chin and throat are a warm brown, the rest of the under surfaces violet-blue with some areas of chestnut-brown. The wings are earth-brown, the tail black. The bill is red, the legs and feet dark grey. The female is mainly brown, paler below; there is an area of whitish-mauve around her eyes. She has faint but regular white barring across her breast and underparts; her rump is blue.

Range Uganda, Kenya, Somaliland, northern Tanzania.

Status Abundant in some parts of the range.

Avicultural rating See Violet-eared Waxbill.

Feeding/Breeding/General management Similar to Violet-eared Waxbill. They are delicate when first imported and never completely resistant to periods of cold and wet weather. Abundant live food is essential, especially if breeding. Several successful breedings have been recorded. They are intelligent and soon become bold and confident when accustomed to their new surroundings.

Estrilda caerulescens (III)
Lavender Finch

4½ in (11.5 cm). Most of the plumage is blue-grey, except the black lores, and the rump, tail coverts and tail which are red. The bill is red, the legs and feet black. The sexes are similar.

Range Senegal and Gambia eastwards to Chad, Central African Republic.

Status Variable, but generally not as abundant as other 'common' Estrildine species from West Africa.

Avicultural rating This is a long-established avicultural favourite, but it is delicate when first imported and needs care to establish successfully.

Feeding Diet B, with spray millet, seeding grasses, various ripe weed seeds, items of green food and some live food. Grit is essential.

Breeding Established true pairs are usually fairly willing breeders but need ample live food to rear youngsters successfully. They build a typical domed nest; it may have a 'tunnel' entrance. Globular wicker nests or open boxes are also favoured; site these in natural or artificial cover. Clutch 4; incubation 11–12 days. Supply flies and fly pupae, 'feeder' gentles, tiny cut-up mealworms etc for rearing; substantial quantities are required or the adults will abandon the young. Also offer germinated seed, some insectile mixture (Diet F), extra millet sprays.

General management Acclimatised birds can occupy an outdoor shelter and flight during the summer months only; they are not weather-resistant and cold and wet conditions and shorter hours of daylight lead to trouble. They are best kept indoors and with slight heat during the winter, but need to be fully occupied if caged, for bored Lavender Finches are enthusiastic feather-pluckers.

Estrilda perreini
Black-tailed Lavender Finch, Black-tailed Waxbill

4½ (11.5 cm). Similar to Lavender Finch but the plumage is darker, the chin and under tail coverts black; the tail is a darker red and with some black. The bill, legs and feet are black. The sexes are similar.

Range Gabon south to northern Angola and eastwards to Tanzania, Mozambique, Zimbabwe, eastern areas of South Africa.

Status There are good populations, but these are local and nowhere abundant.

Avicultural rating Only occasionally available.

Feeding/Breeding/General management Similar to Lavender Finch. It is a very active bird and, like its relative, given to feather-plucking if confined in an unfurnished cage for any length of time.

Estrilda melanotis
Dufresne's Waxbill, Swee Waxbill

4 in (10 cm). The forehead and crown are blue-grey, the mantle, back and wings grey-green with darker vermiculations; the rump and upper tail coverts are red. The face, cheeks, chin and tail are black; the breast and underparts grey—grey-green on the flanks, buff-yellow towards the abdomen. The upper mandible is black, the lower mandible red; the legs and feet are black. The female is similar, but lacks the black mask.

Range South Africa.

Status Variable, but there are good populations in areas of suitable habitat.

Avicultural rating They are rarely available at the present time, but are delightful little birds and good avicultural subjects.

Feeding Diet B, with spray millet, seeding grasses and some small live food. Various ripe weed seeds and soaked or germinated seed are also enjoyed. Few of these birds take much interest in green food but I have had some take small amounts of cress after much coaxing.

Breeding It builds a domed nest of fine grass stems and other plant fibres, usually in a bush or vegetation. It may use a globular wicker basket. Clutch 4; incubation 11–12 days. Provide a varied mixed diet, including plenty of small live food for rearing.

General management These birds need care when first imported; they are eventually not especially delicate, but cannot stand cold and damp conditions. They are active and tit-like in habit; they enjoy a planted shelter and flight in the summer months. Some of the birds of this species that I have previously owned have been loath to bathe.

Estrilda melanotis quartinia
Yellow-bellied Waxbill

4 in (10 cm). The forehead, crown and nape are blue-grey, the rest of the upper surfaces and wings grey-green. The upper tail coverts and rump are red; the tail is black. The chin, throat and breast are grey-white merging with the yellow-buff of the lower underparts. The upper mandible is black, the lower mandible red; the legs and feet are black. The sexes are alike.

Range Southern Sudan, Ethiopia, Eritrea.

Note This is the best known in aviculture of 5 subspecies of Dufresne's Waxbill (*E. m. melanotis*).

Status Local distribution; there are good populations, but it is not abundant.

Avicultural rating Rarely available at the present time, these are attractive and interesting waxbills that are good avicultural subjects.

Feeding/Breeding/General management Similar to Dufresne's Waxbill. Like the previous species, they are very intolerant of cold and wet conditions.

Estrilda paludicola
Fawn-breasted Waxbill

4 in (10 cm). The top and sides of the head and the sides of the neck are grey; the upper parts are a pale rufous-brown with darker vermiculations. The upper tail coverts and rump are red, the tail black. The under surfaces are buff-white with some pink on the flanks and abdomen. The bill is red, the legs and feet flesh-brown. The sexes are alike.

Range Eastern Zaire, Uganda, Sudan, Kenya, Ethiopia, western Tanzania, Zambia, Angola.

Note There is some plumage variation among the 6 described subspecies.

Status Abundant in some parts of the range.

Avicultural rating Only occasionally available, they are not easy to acclimatise.

Feeding/Breeding/General management Similar to Dufresne's Waxbill. They are delicate when first received and need care until established; they do not tolerate cold and wet conditions. This species has been bred.

Estrilda melpoda (III)
Orange-cheeked Waxbill

4¼ in (11 cm). The head, nape, chin, throat and upper breast are grey-fawn; there is a variable area of orange on the lores and cheeks. The upper surfaces, wings and flanks are fawn-brown; the underparts are grey, with some orange-buff on the abdomen. The tail is black. The bill is red, the legs and feet horn-coloured. The sexes are similar, although some females are paler and have smaller or less bright orange areas on the cheeks; this is not an infallible method of sexing as the variation of colour (particularly the orange cheek-patches and orange-buff on the abdomen) among individual populations of these birds can be confusing.

Range Senegal and Gambia eastwards to Cameroon, Chad, northern Zaire and south to northern Angola, Zambia.

Status Abundant in most parts of the range, it is a crop pest in some areas and is destroyed in substantial numbers.

Avicultural rating This is a long-established and popular avicultural subject, which is not difficult to acclimatise and is then easy to manage.

Feeding Diet B, with spray millet, seeding grasses, green food and occasional small live food. Grit is essential.

Breeding Established true pairs are often free breeders; a large untidy domed nest is built, sometimes with a tunnel entrance. Boxes, woven globular nests, wire-mesh tubes etc are also used; the nest is usually in a bush or other cover, but other sites are used. Clutch 5–6; incubation 11–12 days. Provide small live food, germinated seed, extra green food (groundsel, chickweed, cress etc) for rearing.

General management They are easily managed and fairly hardy, although they do not appreciate cold and wet conditions; they are best kept indoors during the winter. This species is an excellent occupant of an outdoor shelter and flight from spring to autumn; it mixes well with other small birds. Active and colourful, it is gregarious and does well in small groups.

Estrilda rhodopyga
Sundevall's Waxbill, Crimson-rumped Waxbill, Rosy-rumped Waxbill

---

4½ in (11.5 cm). The head, upper surfaces, wings and tail are fawn-brown with fine vermiculations; the chin, throat and lower cheeks are grey-white. The underparts are grey-buff, suffused red on the flanks and a deeper shade on the abdomen, with fine, darker vermiculations; the rump and upper tail coverts are red. There is a red eye streak, and some red in the wings. The bill is black, the legs and feet dark brown. The sexes are alike.

Range Sudan, Ethiopia, Uganda, Kenya, Tanzania, Malawi.

Status Variable, but abundant in some parts of the range.

Avicultural rating Available from time to time, this is a good avicultural subject.

Feeding Diet B, with spray millet, seeding grasses, various ripe weed seeds and items of green food, occasional small live food.

Breeding Established true pairs are often good breeders; they make a typical Estrildine nest, usually well hidden low down in vegetation. They will use globular nest baskets, or open-fronted boxes. Clutch 3–5; incubation 11–12 days. Provide abundant small live food, extra millet sprays, seeding grasses, germinated seeds, green food for rearing; some pairs will take nectar-mixture and sponge cage, bread and milk etc.

General management These attractive, vivacious little birds live well in pairs or small groups. They are reasonably hardy when properly acclimatised but not suitable for all-year occupation of an outdoor shelter and flight. They mix well with other birds, but are not assertive and may be bullied by more robust species.

Estrilda troglodytes (III)
Red-eared Waxbill, Pink-cheeked Waxbill, Grey Waxbill, Black-rumped Waxbill

---

3¾ in (9.5 cm). The upper surfaces are fawn-brown, darker on the wings and with very indistinct fine vermiculations; the underparts are grey-fawn, lighter on the chin and throat. The rump, upper tail coverts and tail are black. There is some rosy-pink on the lower breast and abdomen, and a red eye streak. The bill is red, the legs and feet flesh-brown. The sexes are alike.

Range Senegal eastwards across equatorial Africa to Ethiopia.

Status Patchy distribution, but abundant in many areas.

Avicultural rating An inexpensive and widely-kept species, which is fairly easy to maintain after acclimatisation.

Feeding/Breeding/General management Similar to Sundevall's Waxbill. True pairs are often free breeders. They are active and vivacious, but easily bullied by more aggressive species. They are fairly easy to acclimatise and then easily managed; they are invariably in impeccable plumage and robust health if cared for properly.

Estrilda astrild (III)
St Helena Waxbill, Common Waxbill

3¾–4¼ inches (9.5–10.5 cm). The upper surfaces are fawn-brown with fine but distinctive darker vermiculations; the sides of the face, the chin and throat are grey-white. The breast and lower underparts are grey-brown, finely barred, especially on the flanks, and with a deepening rosy hue from the centre of the breast to the abdomen (red in some subspecies). The upper tail coverts are red or rose; the under tail coverts and tail black. There is a red eye streak. The bill is red, the legs and feet flesh-brown. The female is similar but often paler and usually with less extensive areas of red or rose and black on her under surfaces.

Range Sierra Leone, Liberia eastwards to Sudan, Ethiopia and south to Angola, Namibia, Zaire, Zambia, Mozambique, etc to South Africa. Introduced to Cape Verde Islands, Portugal.

Note There is some variation in size and colour among the 17 described subspecies.

Status Variable throughout the extensive range, but there are substantial populations in some areas including forests, savannah, cultivated areas, villages etc.

Avicultural rating A long-established, popular and widely-kept avicultural subject.

Feeding/Breeding/General management Similar to Sundevall's Waxbill. This is an especially active species which is always investigating its territory. It is fairly easy to establish and acclimatise, and management is then straightforward. It thrives in an outdoor, planted shelter and flight during the summer but is not weather-resistant and should be housed indoors from autumn to spring. They are usually good breeders, if true pairs are obtained.

Estrilda nonnula
Black-crowned Waxbill, Black-capped Waxbill, White-breasted Waxbill

---

4 in (10 cm). The lores, head and nape are black; the mantle, back and wings are a finely barred grey-black. The rump and upper tail coverts are red. The sides of the face, the chin, throat and breast are white; the lower underparts are grey-white, darker on the under tail coverts and with pale red on the flanks. The tail is black. The bill is black and red; the legs and feet are black. The female is similar, but usually a paler grey on the mantle and wings.

Range Fernando Po, Cameroon eastwards to Central African Republic, northern Zaire, southern Sudan, Uganda, western Kenya, north-west Tanzania.

Status Abundant in many parts of the range in both West and East Africa, but locally distributed.

Avicultural rating This is an attractive but rarely available species, which is not difficult to manage, and they have bred.

Feeding Diet B, with spray millet, seeding

grasses, green food, some germinated seed, ripe weed seeds and small live food; fresh buds of various kinds are enjoyed by these birds. Grit is essential.

Breeding A domed nest of plant fibres with softer lining is made, with a side entrance. Clutch 3–6; incubation 11–12 days. Provide abundant small live food for rearing: *Drosophila*, house-flies and pupae, ant pupae etc. Also provide additional seeding grasses and germinated seed.

General management Although it is some years since I had these birds in my collection, they are remembered as no more difficult to establish and care for than other Estrildine waxbills. They are not weather-resistant and are best kept in indoor quarters during autumn to spring. They are good occupants of an outdoor shelter and flight during the summer. They are vivacious, active, and mix well with other small species.

Estrilda atricapilla
Black-headed Waxbill

---

4 in (10 cm). Very similar to Black-crowned Waxbill but the under surfaces are grey-white, and it is much darker on the abdomen and under tail coverts; there is more red on the flanks. The female has browner upper parts. There is less red on the bill in both sexes.

Range Cameroon, Gabon, Congo, northern Zaire, Uganda, western areas of Kenya.

Status Locally distributed, but there are good populations in many parts of the range.

Avicultural rating See Black-crowned Waxbill.

Feeding/Breeding/General management Similar to Black-crowned Waxbill. This species has been bred. Provide a variety of seeding grasses, germinated seeds and soaked millet sprays as well as other items recommended for the preceding species.

Estrilda erythronotos
Black-cheeked Waxbill, Black-faced Waxbill

---

5 in (13 cm). The face, ear-coverts and chin are black, the forehead grey. The crown, nape, mantle, throat and breast are mauve-grey, the rump and upper tail coverts red, the under tail coverts black. The lower underparts are grey suffused with vinous, the flanks washed with dull red and the abdomen black. The wings are finely barred black and white; the tail is black. There are fine vermiculations on the mantle. The bill, legs and feet are black. The female is similar but sometimes paler and lacks the black on the abdomen and under tail coverts.

Range Uganda, southern Kenya, Tanzania, Zambia, Angola, Namibia, Zimbabwe, eastern areas of South Africa.

Status There are good populations in many parts of the range.

Avicultural rating A popular avicultural subject, but it is not easy to establish; although it is eventually relatively easy to manage after acclimatisation has been successfully accomplished.

Feeding Diet B, with spray millet, seeding grasses and small live food; the latter should be provided daily. Some birds accept green food but others refuse it; this may be an indication of their country or area of origin e.g. whether from arid or wooded habitat. Grit is essential.

Breeding They build a large domed nest, with a side entrance, usually well hidden among vegetation. Clutch 4–5; incubation 11–12 days. Abundant small live food is essential during the nestling period; it is best if the breeding pair are the sole occupants of a suitable shelter and flight to eliminate competition for live food.

General management They need great care during the immediate post-importation and quarantine period; house in a spacious box-pattern cage, and keep them warm and quiet. They are eventually fairly hardy but *very* intolerant of cold and wet conditions; they are not suitable for outdoor accommodation except during the warmest months of the year, and then it is best if their aviary, preferably planted, is well sheltered. They usually mix well with other small birds.

Amandava amandava [PLATE 132]
Red Avadavat, Tiger Finch, Bombay Avadavat, Red Munia

PLATE 132 *Amandava amandava*, Red Avadavat

4 in (10 cm). The lores are black; the head, mantle, upper tail coverts and underparts are red. The wings and tail are dark brown, the under tail coverts ochre-yellow, brown or blackish. There is a narrow white streak below the eye and white spots on the wings and flanks. The bill is red, the legs and feet brown. The female (and male in non-breeding plumage) has grey-brown upper surfaces, darker on the wings, paler on the underparts; the rump and upper tail coverts are red, and there are white spots on the wings.

Range Pakistan, India, Burma eastwards to south-west China; Thailand, Indochina, Java and Lesser Sundas. Introduced to Singapore, Malaysia, Sumatra, Philippines, Fiji, Mauritius etc.

Note There is some variation in size and plumage between the 3 described subspecies. The best known (though now rarely available) is *A. a. punicea* (Indochina, Java, Bali) which is aviculture's 'Strawberry Finch'. It is smaller, more brightly coloured and with heavier white spotting than either the nominate sub-species, *A. a. amandava* (Pakistan and India), or *A. a. flavidiventris* (Burma, China, Sundas).

Status Abundant in many parts of the range.

Avicultural rating A popular favourite which is easily managed and a good breeder.

Feeding Diet B, with spray millet, seeding grasses, green food and some small live food.

Breeding Established pairs are usually ready breeders. They build a domed nest, with a side entrance, of plant fibres, feathers etc, usually low down in vegetation. They will use globular baskets or open-fronted boxes etc. Clutch 4–6; incubation 11–12 days. Provide small insects, extra millet sprays and green food, seeding grasses and some germinated seed for rearing. They are aggressive when nesting and do best in separate accommodation.

General management These are among the easiest of waxbills to establish and manage, but are not as weather-resistant as some authorities claim; birds of this species *may* survive a mild European winter without a heated shelter but it is entirely possible that exposure to prolonged cold and damp, plus reduced hours of daylight, will have a permanently debilitating effect. They usually mix fairly well, but are antagonistic to other birds with red in their plumage; they are pugnacious when breeding condition is reached. This is one of the most pleasant songsters of the waxbill family. It has an eclipse plumage when the male loses its red colour. Individual males frequently assume a form of melanistic plumage if confined in cages for long periods and fed an unvarying seed diet; this is probably a deficiency problem which can usually be rectified by turning the affected bird into a naturally-planted outdoor aviary.

Amandava formosa
Green Avadavat

4¼ in (11 cm). The upper surfaces and wings are olive-green; the chin and throat a pale grey-olive, the underparts yellow-olive. The flanks are dark olive with white vertical stripes; the abdomen is grey-white. The upper tail coverts are green, the tail black. The bill is red, the legs and feet flesh-coloured. The female is usually paler and with more indistinct barring on the flanks.

Range Central India; a population also exists near Lahore (East Pakistan).

Status There are good populations in some parts of the range, but habitat loss is reducing numbers elsewhere.

Avicultural rating A popular avicultural subject, but now only rarely available; it is not easy to establish and is less robust than its relatives.

Feeding/Breeding/General management Similar to the Red Avadavat, with the proviso that this species needs extra care during the post-importation and quarantine period and is generally a little more delicate and 'difficult' than its cousin. Several breeding successes have been recorded, but it is not as reliable as the Red Avadavat. It does well in small groups.

Amandava subflava [PLATE 133] (III)
Golden-breasted Waxbill, Zebra Waxbill, Orange-breasted Waxbill

3½ in (9 cm). The head and upper surfaces are a dark grey-green; the upper tail coverts and rump are orange-red. The underparts are yellow, yellow-orange or orange, barred grey on the flanks; the under tail coverts are orange-yellow or orange. The eyebrow stripe is red. The bill is red, the legs and feet flesh-brown. The female is similar, but the barring on her flanks is less distinct, and she usually has less bright underparts.

PLATE 133 *Amandava subflava*, Golden-breasted Waxbill

Range Senegal, Gambia, Guinea eastwards to northern Cameroon, Chad, Central African Republic, southern Sudan, Uganda, Kenya; south (in the west) to Gabon, Congo, Angola and (in the east) to Mozambique, South Africa.

Note There is some variation in size and plumage between the 2 described subspecies. *A. s. clarkei* (Angola to Kenya and south to Mozambique, South Africa) is slightly larger than the nominate subspecies and often has brighter (orange) underparts; it is sometimes offered by dealers as 'Orange-breasted Waxbill', 'Giant Orange-breasted Waxbill' or 'South African Orange-breasted Waxbill'.

Status Abundant in many parts of the range.

Avicultural rating One of the tiniest seedeaters

and an excellent avicultural subject. It is a good breeder.

Feeding Diet B, with spray millet, seeding grasses, various items of green food, ripe weed seeds, occasional live food. Grit is essential.

Breeding A domed nest with the entrance near the top is built; in the wild disused nests of other small birds are used. They usually build close to the ground among vegetation. Clutch 4–5; incubation 11–12 days. Provide small live food, germinated seeds, extra spray millet or seeding grasses etc when rearing young.

General management Despite their tiny size this is one of the most robust of waxbills, but it is not recommended for all-year occupation of a shelter and flight; it is best kept in indoor quarters from autumn to spring. It mixes well with other species that will not bully it, and it is a very good subject for the sort of breeding programmes that are helping to establish aviary-bred strains of many Australian finches and a few other species.

Ortygospiza atricollis [PLATE 134] (**III**)
African Quail Finch

PLATE 134 *Ortygospiza atricollis,* African Quail Finch

4 in (10 cm). The forehead, face and chin are black, the rest of the upper surfaces a dark grey-brown. The breast and flanks are dark brown with white barring; there is a large area on the lower breast which is rufous-brown, and the lower underparts are grey-white. The tail is blackish tipped white. The bill is red and black, the legs and feet flesh-brown. The female is similar but her forehead, face and chin are earth-brown.

Range Senegal eastwards to Chad, west and south-west Sudan, northern Uganda.

Status Widespread, but local throughout the range.

Avicultural rating These interesting little birds have evolved with the habits and appearance of a Partridge or Quail; appropriate accommodation is needed.

Feeding Diet B, with spray millet, seeding grasses, ripe weed seeds, some germinated seeds, small live food. Provide adequate grit.

Breeding Some breeding successes have been recorded, but are an infrequent occurrence. They build an untidy domed nest of grass stems and other plant materials, lined with feathers or soft grasses, usually at ground level among grass tussocks. Clutch 4–6; incubation 11–13 days. Provide abundant small insects for rearing together with germinated seeds, soaked spray millet etc.

General management They are almost entirely terrestrial in habit and do not thrive in some conventional captive environments. They do best in an outdoor shelter and flight during the summer months with access to a natural area of grass, but beware of chilling if the grass becomes saturated through heavy rain; indoor accommodation in the winter needs careful thought—grass is easily grown in shallow trays of compost and can be rotated in indoor flights. Peat is good if kept moist; sand and grit are too abrasive, and newspaper pads are too dry and will give rise to cracked toes. A tropical house or conservatory environment is good for year-round occupation and may be more conducive to successful breeding. If caged for any length of time (especially whilst nervous after import and quarantine), provide a padded top (or fabric false 'ceiling') to the cage as the birds take off vertically (like game birds) when upset or disturbed.

Ortygospiza atricollis muelleri (III)
East African Quail Finch

4 in (10 cm). The forehead and lores are black; the prominent eye-streak and chin are white. The upper surfaces are grey-brown. The white chin is bordered black, including the black throat. The breast and flanks are a paler grey-brown with black and white barring; there is more rufous-brown in some cases and deeper rufous-chestnut on the abdomen. The lower tail coverts are grey-white; the tail is black with a white tip. The upper mandible is black, the lower mandible red (all red in non-breeding plumage); the legs and feet are flesh-brown. The female is similar but has more muted patterns of plumage.

Range Southern Kenya, Tanzania, Malawi.

Status Local distribution throughout the range.

Avicultural rating Occasionally available, and popular because of the more clear-cut markings.

Feeding/Breeding/General management Similar to African Quail Finch.

Aegintha temporalis
Sydney Waxbill, Red-browed Waxbill

4½ in (11.5 cm). The top of the head is grey; the nape, mantle, back and wings grey-green. The sides of the head and the under surfaces are light grey, suffused with yellow on the abdomen. The prominent eyebrow stripe, rump and upper tail coverts are red. The bill is red, the legs and feet yellow-brown. The sexes are similar.

Range Eastern and south-eastern Australia, northern Queensland.

Status Good populations, and the range is being slightly extended following escapes from aviaries.

Avicultural rating Only limited numbers have been imported in recent years. This is an attractive and interesting species, but it needs care.

Feeding Diet B, with spray millet, seeding grasses, green food. Grit is essential.

Breeding They are not as free-breeding as other familiar Australian seedeaters. They will use boxes, wicker baskets or globular nests. Clutch 4–6; incubation 12 days. Provide germinated seeds, extra spray millet or seeding grasses, artificial nectar or honey water and sponge cake or bread and milk, a variety of green foods and occasional small live food.

General management They are usually delicate during acclimatisation; they need careful treatment at all times but are excellent in a shelter and flight during the summer months.

Emblema guttata
Diamond Sparrow, Diamond Firetail Finch

5 in (13 cm). The top of the head and nape are silver-grey, the mantle, back and wings brown. The rump and upper tail coverts are red, the tail black. The sides of the head are grey-white, the underparts white with a black band across the breast. The sides and flanks are black with large white spots. The bill is red, the legs and feet brown. The female is similar but often smaller; the sexes are difficult to distinguish with certainty, except by behaviour.

Range Eastern and south-eastern Australia.

Status Populations have been reduced following habitat loss and other disturbance.

Avicultural rating A popular and widely-kept Australian finch, which is easily managed and a good breeder.

Feeding Diet B, with spray millet, seeding grasses, green food etc. Grit is essential and many Australian finch breeders provide mineralised and oystershell together with vitamin and mineral supplements.

Breeding They are generally good breeders and will use a box or basket. Clutch 4–6; incubation 11–13 days. They occasionally build a large, untidy nest, if housed in an aviary, in a bush or among cover. Provide extra spray millet, seeding grasses, green food, germinated seed and live food when rearing.

General management The majority of these birds are kept and bred in cages or indoor flights, but they are good occupants of an outdoor shelter and flight during the summer months; like their relatives, they do not thrive in cold and wet conditions. If closely confined in an unfurnished cage they can prove persistent feather-pluckers; they are also inclined to obesity in such situations. They are frequently aggressive, so choose companions with care, and breeding pairs are best isolated.

Emblema picta [PLATE 135]
Painted Finch, Painted Firetail

PLATE 135 *Emblema picta*, Painted Finch

4½ in (11.5 cm). The lores, forehead, face, chin, throat, upper tail coverts and an irregular band across the centre of the breast are red; the upper surfaces and wings are brown. The rest of the underparts are black with white spots on the sides and flanks; the tail is blackish-brown. The bill is black and red, the legs and feet flesh-coloured. The female is similar, but has less red on the face; the white spotting is more pronounced and extends up the sides of her breast.

Range Central Australia.

Status Variable; distribution is local and there are good populations in many parts, although nowhere are they abundant.

Avicultural rating This is one of the most desirable of Australian finches. It is rare and expensive, but is now being captive-bred in increasing numbers.

Feeding Diet B, with spray millet, seeding grasses, green food. Some germinated seed is enjoyed; individual birds may not eat green food. Live food is taken by a few. Grit is essential.

Breeding They will use an open-fronted box or a wicker basket. Clutch 3–5; incubation 12–13 days. Provide germinated seed, extra spray millet, seeding grasses, green food and some live food when young are being reared.

General management Control-bred stock is much more reliable than the wild birds previously imported from Australia. The majority of these and other Australian finches are now cage-bred, but most thrive in an outdoor shelter and flight through the summer. Captive-bred stock is hardy but should not be exposed to prolonged cold and wet conditions. They usually mix well with other birds.

Neochmia phaeton
Crimson Finch, Blood Finch

5 in (13 cm). The crown and nape are grey-brown, the mantle, lower back and wings brown. The forehead, face and underparts are red; the back is red and brown. The tail is red with some dark brown. There are small white spots on the flanks. The bill is red, the legs and feet flesh-brown. The female is generally duller; she has red

on the face, chin and throat, but the rest of her underparts are a light grey-brown.

Range Coastal areas of northern and north-eastern Australia.

Note 1 of the 3 described subspecies (*N. p. evangelinae*) is from southern New Guinea.

Status Abundant in some parts of the range.

Avicultural rating They are rarely available in the UK at present, although they are being captive-bred in increasing numbers in Continental Europe and elsewhere.

Feeding/Breeding/General management Similar to Diamond Sparrow; it has a reputation for aggression, particularly when breeding. It takes more live food than related species, especially when rearing young.

Neochmia ruficauda
Star Finch, Ruficauda

$4\frac{1}{2}$ in (11.5 cm). The face, forehead, forecrown, cheeks and chin are red, the upper surfaces a pale olive-green. The underparts are grey-green with more yellow towards the abdomen and under tail coverts. The upper tail coverts are dull red, the tail brown and red. The sides of the face, throat, sides of the neck and flanks are spotted white. The bill is red, the legs and feet flesh-brown. The female's plumage is generally duller and she has less red on the head and face.

Range North-western, northern and north-eastern areas of Australia.

Status Variable. Numbers are decreasing in some areas, increasing in others; some land reclamation schemes linked with agricultural development have had a beneficial effect on this species.

Avicultural rating A popular and widely-kept Australian finch which is not difficult to manage and usually a good breeder.

Feeding/Breeding/General management Similar to Diamond Sparrow; it is one of the hardier grass-finches, but needs heat in winter.

Poephila guttata [PLATE 136]
Zebra Finch, Chestnut-eared Finch

PLATE 136 *Poephila guttata*, Zebra Finch

$4\frac{1}{2}$ in (11.5 cm). The top of the head and the nape are grey, the mantle, back and wings fawn-grey. There is a vertical black stripe at the base of the bill and under the eye, with a white area in between; the cheek patches are chestnut. The chin, throat and sides of the neck are barred black and white; the flanks are chestnut with white spots. A band across the breast is black; the rest of the underparts, rump and tail coverts are grey-white. The bill, legs and feet are red. The female is similar but lacks all the distinctive markings except for some black and white on the sides of the face.

Note Zebra Finches are now available in many different colours and patterns which have been produced by selective breeding; among the most popular are White, Fawn, Cream, Silver, Chestnut-flanked White, Pied and Penguin. More will undoubtedly follow in the future.

Range There is local distribution throughout the Australian interior.

Status This varies from abundant to scarce.

Avicultural rating One of the most popular and widely-kept of all exotic seedeaters; an ideal beginner's species.

Feeding Diet B, with millet sprays, green food. It appreciates seeding grasses, fruit etc. Grit is essential.

Breeding They are very free breeders in a cage or aviary. Use a variety of nesting receptacles, but a box with a half-front is preferred. Some pairs develop a bad habit of 'sandwiching'—half-filling the box with material, laying a clutch of eggs, and then repeating the process *ad infinitum* until the entire box area is occupied. Clutch 3–6; incubation 12–13 days. Some germinated seeds, extra millet sprays or seeding grasses are recommended for rearing, but are not essential. They are very prolific and will breed throughout the year if permitted; 3 broods is the maximum recommended.

General management These hardy and adaptable little birds will succeed in a cage or aviary; most serious breeders prefer the former which allows greater control of matings etc. They are hardy and mix well with other birds, but are intolerable busybodies and will keep other small species from their nests and even attempt to take over their eggs or nestlings, or eject them so that they can install their own families.

Poephila bichenovii [PLATE 137]
Bicheno's Finch, Double-barred Finch, Owl Finch

4¼ in (11.5 cm). The face, cheeks, chin, throat and breast are white; the forehead and a narrow border encircling the cheeks and chin, and a similar band across the breast, are black. The crown, nape, mantle and back are brown; the wings are dark brown with a very small black and white chequered pattern. The rump is white, the under

PLATE 137 *Poephila bichenovii*, Bicheno's Finch

tail coverts black; the tail is blackish. The bill is silver-grey, the legs and feet dark grey. The female is said by some authorities to have narrower black bands encircling her cheeks and chin and across the breast; sexing is difficult by visual means.

Range Northern and eastern Australia; also north-western Australia, Northern Territory and some offshore islands.

Status Abundant in many suitable areas of habitat.

Avicultural rating Popular and widely-kept, active, vivacious birds, which are fairly free breeders.

Feeding Diet B, with the addition of spray millet, seeding grasses, green food, occasional live food. Mineral and vitamin supplements should be offered together with mineralised and oyster-shell grit.

Breeding True pairs are fairly willing breeders; most birds are bred in large cages or indoor flight units. They use a box or basket; in aviaries they may build an untidy, spherical nest of plant stems lined with feathers etc. Clutch 4–5; incubation 11–12 days. Provide germinated seeds, live food, extra millet sprays, seeding grasses and green food when rearing, and especially during the first 10 days.

General management Formerly this species was among the more difficult of Australian finches to establish, but specialist breeders are now producing much improved stock. They are excellent in an outdoor shelter and flight during the summer months, but need a temperature of approximately 55°F (13°C) during the winter, and are best housed indoors. They are usually good mixers.

Poephila personata
Masked Grassfinch

5¾ in (15 cm). The lores, face, forehead, chin and throat are black, the upper surfaces cinnamon-brown, darker on the head and wings. The underparts are similar but slightly paler; the abdomen and rump are white with a broad black band across the top and down the sides of the rump. The tail is black. The bill is yellow-orange, the legs and feet flesh-coloured. The female is similar but usually smaller; sexing is difficult, except by using behaviour and song as a guide.

Range Northern Australia.

Status Variable, but there are good populations in some areas.

Avicultural rating Not as widely kept as other Australian finches, they are fairly good breeders.

Feeding Diet B, with the addition of spray millet, seeding grasses, green food, some germinated seed, live food. Also provide vitamin and mineral supplements, and mineralised and oystershell grit.

Breeding They are fairly willing breeders if a true pair is obtained. They will use a box or basket; if in vegetation a bulky, domed nest of plant fibres with the entrance at the side is built. Clutch 4–6; incubation 12–13 days. Provide germinated seeds, extra millet sprays, seeding grasses, green food and live food when nestlings are present.

General management They are excellent in an outdoor shelter and flight during the summer months, but need moderate heat (55°F or 13°C) during the winter months. They are gregarious and small non-breeding groups can occupy suitable accommodation; they are good mixers, but aggressive with other closely-related species, and breeding pairs need separate accommodation.

Poephila acuticauda
Long-tailed Grassfinch

6–6½ in (15–16.5 cm). The forehead, crown, nape, sides of the head and neck are a pale blue-grey; the lores and a bib extending over the chin, throat and upper breast are black. The rest of the upper and under surfaces are fawn-brown, darker on the wings. The rump and upper tail coverts are white; there is a black band across the top and down the sides of the rump. The bill is yellow, the legs and feet flesh-coloured. The female is similar but may be slightly smaller and have a smaller bib.

Range Northern Australia.

Status Good populations in many areas.

Avicultural rating One of the most popular and widely-kept of Australian grassfinches, it is easy to manage and a fairly free breeder, but can be aggressive.

Feeding Diet B, with the addition of millet sprays, seeding grasses, green food and live food. Also provide vitamin and mineral supplements, as well as mineralised and oystershell grit.

Breeding True pairs are usually free breeders; most are bred in cages or indoor flights. They use a box or basket; in an aviary they occasionally construct a domed nest among vegetation. Clutch 5–6; incubation 11–12 days. Provide germinated seed, extra spray millet or seeding grasses and a generous supply of small live food when nestlings are present.

General management This species is among the

easiest of grassfinches to maintain but, like its relatives, it does not enjoy cold and wet conditions or shorter hours of daylight, so is best housed indoors at around 55°F (13°C) during the winter. It can be housed with other species during the non-breeding season, but is usually belligerent, and pairs are especially truculent.

Poephila cincta [PLATE 138] **(II)**
Parson Finch, Black-throated Finch

PLATE 138 *Poephila cincta*, Parson Finch

4½ in (11.5 cm). Very similar to Long-tailed Grassfinch, but this species has a black bill and a shorter tail (with central, elongated feathers).

Range North-east Australia.

Status Abundant in many parts of the range.

Avicultural rating A popular bird, but it is not as widely-kept as the Long-tailed Grassfinch; it is easily managed and a fairly reliable breeder.

Feeding/Breeding/General management Similar to the Long-tailed Grassfinch, it is invariably aggressive in mixed company.

Erythrura prasina
Pin-tailed Parrot Finch, Pin-tailed Nonpareil

5½ in (14 cm). The lores are black; the forehead, sides of the head and throat are blue. The crown, nape, mantle, back and wings are grass-green. The upper tail coverts are red. The underparts are a golden-buff; there is red on the abdomen. The tail is black and red; the elongated central tail feathers are red. The bill is black, the legs and feet flesh-coloured. The female is similar, but the blue on the forehead and throat is either absent or indistinct; the central tail feathers are shorter and her underparts less bright.

Range Burma, Thailand, south-eastwards to Malaysia, Sumatra, Java and Borneo.

Status Variable, but there are good populations in some parts of the range.

Avicultural rating A popular avicultural subject, mainly because it is the most inexpensive of parrot finches, but now is less frequently available.

Feeding They fare best on a varied seed diet; use a combination of Diets A and C, with the addition of millet sprays, seeding grasses, some germinated seed, green food, diced apple and some live food. Newly imported birds are usually accustomed to a staple diet of paddy rice, but responsible importers begin the 'weaning' process during quarantine; it is best if some paddy rice is available when recently-imported birds are acquired in case they are still not accustomed to a more varied diet.

Breeding They are very infrequently bred, especially considering the huge numbers shipped some years ago; an abundance of wild-caught birds and a shortage of true pairs may have been factors towards this situation. A domed nest of plant materials is built, usually fairly high; they will use boxes and other receptacles. Clutch 3–5; incubation 13–14 days. Provide eggfood, germinated seed, extra millet sprays or seeding grasses and some live food for rearing.

General management They are by no means difficult to keep, except when permanently caged, when all kinds of problems, from obesity to fits, seem to arise. They are initially nervous and inclined to take fright easily; it will help if recent imports are provided with some seclusion, even a lightweight cover over part of the cage-front will help reduce stress situations. It is best if a potential breeding pair is housed in a separate small shelter and flight; otherwise they mix well with species not inclined to bully them but not with other members of the same family. They are eventually hardy, but do best in moderately warm indoor quarters during the winter months.

Erythrura trichroa [PLATE 139]
Blue-faced Parrot Finch

PLATE 139 *Erythrura trichroa*, Blue-faced Parrot Finch

$4\frac{1}{2}$–5 in (11.5–13 cm). The lores are black, the forehead, face and cheeks blue. The crown, mantle, back and wings are grass-green, darker on the flight feathers; the rump and upper tail coverts are red, the tail brown and red. The underparts are bright grass-green. A faint orange suffusion is sometimes present on the nape. The bill is black; the legs and feet are light brown. The female is similar but duller and with less blue on her face.

Range Celebes, Moluccas, New Guinea, New Britain, New Ireland, New Hebrides; now rare or absent from north-eastern Australia.

Note 10 subspecies are described.

Status There are good populations in some parts of the range, but habitat loss is having an adverse effect in several areas.

Avicultural rating This is an increasingly popular and widely-kept species, but it is nervous and best kept in spacious quarters. It can be prolific.

Feeding Diet A, with millet sprays, seeding grasses, green food and some live food. Also supply vitamin and mineral supplements, mineralised and oystershell grit.

Breeding Compatible pairs can be free breeders; they use boxes or, less often, baskets. Clutch 3–6; incubation 13–14 days. Provide abundant live food, germinated seed, extra millet sprays or seeding grasses, eggfood when eggs hatch. They fare best in roomy compartments indoors or, better still, a small aviary.

General management The best environment is a sheltered, planted aviary with a shelter offering moderate warmth (50–55°F or 10–13°C) in the winter; they are eventually hardy but do not thrive if forced to endure a combination of cold, wet and short winter days without a comfortable shelter. If housed in indoor cages or flights provide some form of seclusion or cover.

Erythrura psittacea
Red-headed Parrot Finch, Red-throated Parrot Finch

$4\frac{3}{4}$ in (12 cm). The lores are black; the forehead, top and sides of the head, chin, throat, rump, upper tail coverts and central tail feathers are red; the rest of the plumage is grass-green with some brown in the wings. The bill is black, the legs and feet brown. The female is similar, with possibly less red on the head and generally she may be slightly paler, but visual sexing is not easy.

Range New Caledonia.

Status One of the best known species, but it is by no means abundant in the wild; increased numbers are being bred in control.

Avicultural rating This is one of the most strikingly-coloured of all finches; it is reasonably easy to maintain and a fairly free breeder.

Feeding/Breeding/General management Similar to Blue-faced Parrot Finch. They are best kept in secluded aviaries (shelter and flight) in the summer, or a spacious indoor compartment; they are active and nervous. Eggshells (first baked in an oven and then finely crumbled) are a valuable additive for these and other Parrot Finches. They are moderately hardy but need some warmth (50–55°F, 10–13°C) in winter.

Chloebia gouldiae [PLATE 140]
Gouldian Finch, Lady Gould's Finch

---

5 in (13 cm). The forehead, top of head, cheeks and throat are black; there is a narrow light-blue border which is wider on the sides of the neck. The neck, mantle, back and wings are grass-green; the rump and upper tail coverts are turquoise, the tail black. There is a broad band of purple across the breast; the lower breast and underparts are yellow to buff-white. The bill is pink-white, with a red tip; the legs and feet are flesh-coloured. The female is similar but her breast is more mauve and her underparts paler.

Range Northern Australia.

Status Populations are decreasing in many areas.

Avicultural rating This is the most colourful of

PLATE 140 *Chloebia gouldiae*, Gouldian Finch

finches; it is now captive-bred in substantial numbers. It is widely kept.

Feeding Diet B, with millet sprays, seeding grasses, green food. Provide vitamin and mineral supplements, mineralised and oystershell grit, granulated charcoal.

Breeding They are fairly willing breeders. Like many Australian birds, they approach breeding condition in mid to late summer in the northern hemisphere; breeding thus takes place mainly during the autumn and winter. They will use a box or basket; the former is most often used, with a half-open front. Clutch 4–8; incubation 12–13 days. Provide germinated seeds, eggfood, extra millet sprays and seeding grasses etc for rearing. Some breeders prefer to entrust the hatching and rearing to Bengalese finches, who usually accomplish the task of fostering without mishap; but many authorities believe that parent-reared Gouldians are important for the future captive-propagation of the species. Because of the inclination to breed out of season, most are housed in spacious breeding cages in a bird-room. Ensure that such units allow space for flying exercise, or egg-binding may be a problem.

General management They are excellent in an outdoor shelter and flight during the summer months; they usually mix well with other similar-sized birds or can be kept as a single-species group with little trouble. It is best to move them to their breeding quarters by late summer. They are reasonably hardy but do not tolerate cold and wet conditions, and need moderate heat (55°F, 13°C) in winter; extra artificial lighting is essential at this period of the year, and especially when young birds are in the nest.

Note In addition to Black-headed Gouldians, both Red-headed and Yellow-headed occur naturally in the wild and are well established in aviculture. Now new mutations are being produced; one which is readily available at present is a white-breasted form and others will undoubtedly follow.

Aidemosyne modesta
Cherry Finch, Plum-headed Finch

4½ in (11.5 cm). The lores are dark red, the forehead and chin dark purplish-red. The upper surfaces are brown with white spots on the wings and tail; the cheeks and throat are grey-white. The underparts are barred brown and white. The bill is black, the legs and feet brown. The female is similar but has a smaller purple-red area on her forehead, and a white streak in front of her eye; the chin is grey-white.

Range Inland areas of western Queensland, western New South Wales.

Status Variable, but reasonably abundant in a few areas.

Avicultural rating These attractive but quietly-coloured little birds are not among the most widely-kept of Australian finches.

Feeding Diet B, with spray millet, seeding grasses, green food, etc. Provide vitamin and mineral supplements, mineralised and oystershell grit.

Breeding They are fairly willing breeders; but do best in separate accommodation. Clutch 3–7; incubation 12–13 days. Provide eggfood, germinated seed, extra millet sprays or seeding grasses, live food for rearing. They will use a box or basket for nesting; some seclusion or cover is necessary as the adults are often nervous parents.

General management Captive-bred stock is much easier to handle than previously imported wild birds, but these are not among the most robust of birds. They need some care when first received and until settled into their new accommodation; like related species, they do not tolerate cold or wet and need modest heat (50–55°F, 10–13°C) during the winter months.

Lepidopygia nana
Bib-Finch, African Parson Finch

3½ in (9 cm). The lores, forehead and chin are black, the crown grey and black. The upper surfaces are grey-fawn, grey-green on the rump and upper tail coverts. The cheeks are grey, the throat and underparts light fawn-brown, darker towards the abdomen and with black under tail coverts. The tail is black and white. The upper mandible is black, the lower mandible grey; the legs and feet are horn-coloured. The sexes are alike.

Range Madagascar.

Status No information is available, but they are probably declining because of habitat loss.

Avicultural rating Extremely rare, and never especially popular with aviculturists. They are hardy after acclimatisation, but aggressive.

Feeding Diet B with spray millet, seeding grasses, green food, ripe weed seeds and germinated seeds. Grit is essential.

Breeding Established pairs are fairly willing breeders and can be prolific. A domed nest of plant fibres is made; they will use a box or globular basket. Clutch 3–8; incubation 11 days.

Provide germinated seed, extra millet sprays or seeding grasses, green food, eggfood etc for rearing; some live food may be taken when the nestlings are small.

General management These dapper little birds are agile and tit-like in habit. They have a reputation for pugnacious behaviour, but my birds lived with various waxbills and were themselves bullied by the latter; breeding pairs *are* aggressive. They are hardy after acclimatisation and good occupants of an outdoor shelter and flight, but are not happy in the cold and wet and do best if provided with slight heat during the winter.

Lonchura cantans
African Silverbill

4½ in (11.5 cm). The upper surfaces are a pale biscuit-brown with very fine vermiculations; the wings are darker, and the upper tail coverts and tail black. The chin and throat are grey-white, the underparts creamy-buff, palest on the abdomen and under tail coverts. The bill is blue-grey, the legs and feet flesh-coloured. The sexes are alike.

Range Mauritania, Senegal, Gambia, Mali, eastwards through Niger, Chad to Sudan, Ethiopia, Kenya, north-west Tanzania; South Yemen.

Status Abundant in many parts of the range.

Avicultural rating This inexpensive and widely-kept species is easily managed and hardy when acclimatised.

Feeding Diet B, with spray millet, seeding grasses, green food; some live food is enjoyed. Grit is essential.

Breeding Established true pairs can prove free breeders. They will use a box or basket; occasionally they build their own nest among vegetation in the aviary, or attempt to evict owners of other nests, causing some disruption in mixed groups. Clutch 3–8; incubation 11 days. Provide eggfood, soaked seed, extra millet spays and live food for rearing, although the nestlings may be raised successfully without the latter. They will also breed fairly readily in a spacious cage or indoor flight with minimal cover, but some seclusion is necessary to ensure success.

General management This is a good beginner's species; it is hardy when acclimatised but is best kept in frost-free accommodation in winter. It mixes well with other species, and is usually completely inoffensive except occasionally when nesting. Sexing is difficult and the best guide is the cock's song and display.

Lonchura malabarica
Indian Silverbill

4¾ in (12 cm). The upper surfaces are creamy-brown, the rump and upper tail coverts white. The tail and flight feathers are black. The cheeks and underparts are pale fawn. The bill is blue-grey, the legs and feet flesh-coloured. The sexes are alike.

Range Saudi Arabia, southern Iran, Afghanistan, Pakistan, India, Sri Lanka.

Status Variable, but there are good populations in some parts of the range.

Avicultural rating See African Silverbill; these are now less freely available than that species.

Feeding/Breeding/General management Similar to African Silverbill. This species can be equally prolific; hybrids have been produced on many occasions between the two species, but it is important nowadays to ensure the purity of captive-bred stock of both species.

Lonchura griseicapilla
Pearl-headed Silverbill, Grey-headed Silverbill

4¼ in (11 cm). The lores and chin are black, the head, cheeks and throat grey, the cheeks and chin with numerous tiny white spots. The nape, mantle and back are biscuit-brown, the rump and tail coverts white. The underparts are a pale biscuit-brown. The upper mandible is black, the lower mandible grey; the legs and feet are horn-coloured. The female is similar, but usually with paler underparts.

Range Southern Ethiopia, Kenya, Uganda, to central Tanzania.

Status Generally scarce in the northern parts of the range; more abundant in Kenya, Tanzania.

Avicultural rating Available at intervals, it is not as easy to establish as African or Indian Silverbills.

Feeding Diet B, with spray millet, seeding grasses, green food and regular live food. Grit is essential.

Breeding Established true pairs are often good breeders. They build an untidy domed nest of plant fibres; they will use a globular basket or box. Clutch 4–5; incubation 13–14 days. Provide abundant live food, germinated seeds, eggfood or insectile mixture, millet sprays etc for rearing.

General management They need care during acclimatisation. They are good occupants of an outdoor shelter and flight during the summer months but are not winter-hardy; they are best in indoor quarters with a minimum 45°F (7°C) temperature if acclimatised.

Lonchura cucullata (III)
Bronze Mannikin, Bronze-winged Mannikin

3¾ in (9.5 cm). The forehead, crown, sides of the head, bend of the wing and an area on the flanks are black with iridescent green reflections; the chin, throat and upper breast are black with bronzy reflections suffused with green. The nape, mantle, back and wings are grey-brown, the rump and upper tail coverts grey and black, the tail black. The underparts are white, the flanks and under tail coverts barred black. The bill is grey-black, the legs and feet black. The sexes are alike.

Range Islands of Principé and São Thomé; Senegal and west coast of Africa eastwards to Sudan, Ethiopia, Uganda, and south to Angola, South Africa.

Status Abundant in many parts of the range.

Avicultural rating A popular and widely kept species. Acclimatised birds are easily managed and often good breeders.

Feeding Diet B, with millet sprays, seeding grasses, green food, occasional live food. Grit is essential.

Breeding Sexing is the biggest obstacle to more frequent breeding successes with this species; a small group can occupy an aviary and pairs will be formed on a self-selection basis, provided that birds of both sexes are present. They will breed successfully on a colony system. They are aggressive with other birds when breeding and, despite their small size, can be very disruptive in a mixed group of other similar-sized species; this behaviour is upsetting but rarely damaging to their neighbours. They will use a variety of nest receptacles, such as a box or basket. Clutch 4–8; incubation 11–12 days. Provide some live food, germinated seed, extra millet sprays, seeding grasses or green food when rearing.

General management These attractive and amusing little birds are best described as 'assertive' in mixed company. They like to sleep in a box or basket, and if several are housed together they will crowd into a communal dormitory; they roost very close together, even perching on the backs of companions. Acclimatised birds are hardy but do best in frost-free quarters during the winter.

Lonchura bicolor nigriceps (III)
Rufous-backed Mannikin

3¾ in (9.5 cm). The head, nape, chin, throat and breast are black with green reflections; the mantle, back and wing coverts are chestnut, the flights darker. The underparts are white, the flanks barred black; the tail is black. The bill is lead-grey, the legs and feet black. The sexes are alike.

Range Kenya south to Mozambique and Natal.

Status Abundant in some parts of the range.

Avicultural rating Occasionally available, they are fairly hardy and not as aggressive in mixed company (except when nesting) as some related species.

Feeding Diet B, with spray millet, seeding grasses, green food and some live food. Grit is essential.

Breeding It builds a globular nest with a side entrance and will also use a box or basket. Clutch 4–6; incubation 11–12 days. Breeding pairs are aggressive and best kept in separate quarters, unless sharing substantial-sized accommodation. Provide extra live food.

General management They are hardy after acclimatisation, but do best in frost-free winter accommodation. They usually mix well with other species, although individual birds are occasionally aggressive, especially towards related subspecies.

Lonchura bicolor poensis (III)
Fernando Pó Mannikin, Black and White Mannikin, Black-breasted Mannikin

---

3¾ in (9.5 cm). The head and upper surfaces, including the wings and tail, are glossy black with a bottle-green sheen; there is some white on the primaries. The underparts are white, the flanks black and white. The rump, tail coverts and tail are black. The bill is lead-grey, the legs and feet black. The sexes are alike.

Range Fernando Pó; Cameroon south to northern Angola and east to Congo, Uganda, southwestern Sudan and Ethiopia, Kenya and northwest Tanzania.

Status Abundant in some parts of the range.

Avicultural rating They are fairly frequently available, and hardy after acclimatisation. They have an uncertain temperament in mixed groups.

Feeding/Breeding/General management Similar to Rufous-backed Mannikin. They need care during acclimatisation, then are best kept in a shelter and flight; they are very nervous in close confinement and this can lead to stress problems.

Lonchura fringilloides (III)
Magpie Mannikin

4½ in (11.5 cm). The whole of the head, neck, chin, throat, upper breast, rump and upper tail coverts are glossy black; the mantle, back and wings are dark brown. There are some black and warm brown markings on the flanks, and a small black area on the sides of the breast. The upper mandible is black, the lower mandible lead-grey; the legs and feet are grey. The sexes are alike.

Range Senegal eastwards across Africa to southern Sudan, Uganda, Kenya and south to Mozambique, Zimbabwe, eastern areas of South Africa; Zanzibar.

Status There are good populations in the western parts of the range, but they are less abundant in East Africa.

Avicultural rating Available from time to time, they are easily managed when acclimatised, although they are belligerent in mixed company.

Feeding Diet A, with millet sprays, seeding grasses, ripe weed seeds, green food, occasionally live food. Grit is essential.

Breeding True pairs are often prolific breeders; they will nest in cages as well as aviaries. They build a large nest of plant fibres etc and will use a box or basket readily. Clutch 3–6; incubation 11–12 days. Provide eggfood, live food, extra millet sprays or seeding grasses, germinated seed for rearing. They are vigorous defenders of territory and are best (and safest) kept in separate accommodation; they need a secluded nest site.

General management They are fairly straightforward when acclimatised, but choose companions carefully in a mixed aviary; the stout bill can inflict substantial damage on small or weaker companions. They are hardy but best in frost-free winter quarters; they sleep in a nest box or basket if available.

Bengalese Finch (domesticated)

---

5 in (13 cm). There is still some debate as to the precise origins of this invaluable little bird, but it is generally regarded as a domesticated form of the White-backed or Striated Munia (*Lonchura striata*) of South-east Asia with—possibly—one or two other anonymous species included in its 'pedigree'. Available in fawn, white and chocolate, as well as combinations of those colours; there is also a crested form.

Feeding Diet B, with the addition of millet sprays, seeding grasses, green food etc. Grit is essential.

Breeding They are very free breeders, which probably do better in cages or indoor units than conventional outdoor aviaries, although they are completely at home in the latter. Clutch 3–8; incubation 13–14 days. Little extra is needed in the way of rearing foods, but some germinated seeds may be given. These birds are excellent foster parents for other similar-sized grain-feeding birds' eggs and chicks.

General management This species probably just about edges the Zebra Finch into second place as the ideal bird for the beginner. It is adaptable and hardy, a good mixer and a free breeder; if it has a fault it is in being just *too* maternal. In a mixed group, it will often attempt to take over the paternal and maternal roles of other breeding pairs with eggs or nestlings.

Lonchura punctulata [PLATE 141]
Spice Bird, Nutmeg Finch, Spotted Munia, Scaly-breasted Finch

PLATE 141 *Lonchura punctulata*, Spice Bird

$4\frac{1}{2}$ in (11.5 cm). The head, neck, mantle, back, wings, chin, throat and upper breast are a bright chocolate-brown; the rump, upper tail coverts and tail are slightly paler. The lower breast and underparts are white, the feathers with brown margins giving a scale-like effect; the centre of the abdomen and the under tail-coverts are buff-white. The bill is lead-grey, the legs and feet dark grey. The sexes are alike.

Range India, Sri Lanka, southern Nepal, Bhutan, Bangladesh, Burma, eastwards to southern China, Hainan, Taiwan and south-east to Thailand, Indochina, Malaysia, Sumatra, Java, Bali, Lesser Sundas, Celebes, Philippines; introduced to Australia, Mauritius, Seychelles etc.

Note 11 subspecies are described.

Status Abundant in many parts of the range.

Avicultural rating A long-established and inexpensive avicultural favourite; a good beginner's bird.

Feeding Diet B, with millet sprays, seeding grasses, green food and some live food. Grit is essential.

Breeding Established true pairs breed fairly readily; they use a box or wicker nest, or build a nest among vegetation. They are usually peaceful in mixed company when nesting. Clutch 3–6; incubation 11–13 days. Provide live food, germinated seeds, eggfood, extra millet sprays or seeding grasses etc when eggs hatch.

General management They are hardy after acclimatisation, but should have at least frost-free winter accommodation. They are excellent occupants of a shelter and flight with other small birds; they are usually quiet and retiring. They can also be kept in small flocks.

Lonchura malacca
Tri-coloured Mannikin, Tri-coloured Nun, Three-coloured Nun, Chestnut Mannikin

---

4½ in (11.5 cm). The head, neck, chin, throat, upper breast, abdomen and under tail coverts are black, the rest of the upper surfaces chestnut-brown, brighter on the rump and tail coverts. The lower breast and underparts are white. The bill is light grey, the legs and feet dark grey. The sexes are alike.

Range India, Sri Lanka, Bangladesh, Assam, Burma, south-west China, Thailand, Indochina, Malaysia, Sumatra, Java, Greater Sundas, Celebes, Philippines, Hainan, Taiwan; introduced to Moluccas.

Status Abundant in many parts of the range.

Avicultural rating A popular and widely-kept species which is easily managed and an occasional breeder.

Feeding Diet B, with millet sprays, seeding grasses, green food, occasional live food. Grit is essential.

Breeding They build an open nest of plant fibres, and may use an open box or basket placed among cover. Clutch 3–7; incubation 12–13 days. Provide germinated seed, live food, extra millet sprays or seeding grass, eggfood etc for rearing.

General management They are fairly hardy when acclimatised, but frost-free winter accommodation is recommended. They usually mix well with other similar-sized species, but breeding successes are more likely if pairs are housed separately in a secluded and planted shelter and flight.

Lonchura malacca atricapilla
Black-headed Mannikin, Black-headed Nun, Black-headed Munia

4¼ in (11 cm). The head, neck, chin, throat, upper breast, abdomen, thighs and under tail coverts are black; the mantle, back and wings are chestnut-brown, the underparts similar but slightly brighter. The tail is dark brown. The bill is light grey, the legs and feet dark grey. The sexes are alike.

Range North-east India, Bangladesh, Assam, Burma. In aviculture this is the best known of the 11 described subspecies of Tri-coloured Mannikin.

Status Abundant in many areas.

Avicultural rating See Tri-coloured Mannikin.

Feeding/Breeding/General management Similar to Tri-coloured Mannikin. As in related species, the claws grow extremely quickly in cages or indoor flights; they wear more readily in natural aviaries. Ensure that the claws are not overgrown before the birds are released into an aviary, as they can prove a major hazard catching in wire-mesh, among vegetation, in rough bark etc; trim them carefully if necessary using good-quality nail-clippers and ensuring that the cut is not close to the blood-vessel running through the nail (visible if held up to the light).

Lonchura maja
White-headed Mannikin, Pale-headed Mannikin, White-headed Nun, Maja Munia

4½ in (11.5 cm). The head, nape, chin and throat are white; the rest of the plumage is chestnut-brown, darker on the wings. The centre of the belly, the abdomen and under tail coverts are black. The bill is light grey, the legs and feet grey-brown. The sexes are alike.

Range Southern Thailand, Malaysia, Sumatra, Java, Bali.

Status Abundant in much of the range.

Avicultural rating This species is popular and widely-kept and an occasional breeder.

Feeding/Breeding/General management Similar to Tri-coloured Mannikin. Apart from the difficulty of obtaining true pairs (because of the similarity in appearance of male and female) these birds appear shy and secretive when nesting; they are best kept in a reedy aviary which offers good natural nesting sites; they may also use an open box or basket.

Lonchura castaneothorax [PLATE 142]
Chestnut-breasted Mannikin, Chestnut-breasted Finch

4½ in (11.5 cm). The forehead, crown and nape are cinnamon, the feathers with grey margins; the mantle and back are chestnut-brown, the rump, upper tail coverts and tail rufous-yellow. The face, cheeks, chin and throat are black; there is a broad band of chestnut across the breast, a narrow black border below. The lower underparts are white; the flanks have some black barring. The thighs and under tail coverts are black. The bill is blue-grey, the legs and feet grey-brown. The female is similar but paler or duller, and with less clear-cut markings.

Range Northern and eastern Australia; New Guinea. Introduced to Tahiti, New Caledonia.

PLATE 142 *Lonchura castaneothorax*, Chestnut-breasted Mannikin

Status Abundant in some parts of the range; scarce in northern area.

Avicultural rating These handsome birds are not as widely kept as other Australian seedeaters. They are easily managed and can be good breeders.

Feeding Diet B, with millet sprays, seeding grasses, some green food, live food. Grit is essential.

Breeding Some pairs construct a natural nest among vegetation; it is globular, built of grass stems etc with a softer lining. They will also use boxes or globular baskets. Clutch 5–6; incubation 12–13 days. Provide extra live food, germinated seed, millet sprays etc for rearing.

General management They do best in spacious quarters, but will breed in a cage or indoor flight; if housed in the latter, ensure that the birds obtain adequate flying exercise or obesity may be a problem. They are hardy but do best in frost-free quarters in winter. Individual birds may bully smaller companions; they generally mix well, except when coming into breeding condition when they may become aggressive.

Lonchura pectoralis
Pictorella Mannikin, Pictorella Finch

4¾ in (12 cm). The upper surfaces are silvery-fawn and grey, darker on the wings and tail; the face, chin and throat are black with a narrow yellow-orange border. A band of black feathers with very broad white margins gives a scalloped effect across the upper breast; the lower underparts are fawn-pink. The bill is blue-grey, the legs and feet flesh-brown. The female is similar but duller and paler.

Range Northern Australia.

Status Variable; abundant in some areas, but scarce or absent in others.

Avicultural rating Available in limited numbers, but being bred in the UK, Continental Europe and elsewhere.

Feeding Diet B, with millet sprays, seeding grasses, some green food. Grit is essential.

Breeding This species builds a spherical nest of plant fibres; it will also use a box or basket. Clutch 4–6; incubation 12–13 days. Provide germinated seed, extra millet sprays or seeding grasses and live food for rearing.

General management They need care until established, although captive-bred stock is now easier to manage than the previous wild exports from Australia; they seem never completely hardy and need some warmth (50°F, 10°C) in winter. The combination of short winter days and low temperatures can have a markedly debilitating effect on this species.

Padda oryzivora
Java Sparrow, Java Rice Bird, Paddy Bird

5¾ in (14.5 cm). The whole of the head, the nape and chin are black with a large white area on the cheeks. The mantle, back, wings and breast are blue-grey; the rump, upper tail coverts, tail and flight feathers are black. The lower underparts are beige-pink. The bill is pink, the legs and feet flesh-coloured. The sexes are alike.

Note White, fawn and pied are among the various mutations which are now available.

Range Islands of Java and Bali; also introduced to southern China, Malaysia, Sumatra, Borneo, Burma, parts of Thailand, Taiwan, Philippines, Zanzibar, St Helena, coastal areas of East Africa.

Status There are still good populations in the original natural range, and they are also now well established in several other locations.

Avicultural rating A long established avicultural favourite, this species is hardy and easily managed. The grey form (wild type) is often not a free breeder, but the mutations reproduce freely.

Feeding Diet A, with millet sprays, green food; they also enjoy unhusked rice. Grit is essential.

Breeding There is some evidence that the sight and sound of birds of the same species acts as a stimulus to breeding activity, although only one breeding pair to each unit of accommodation is the rule. They will use a large box (a budgerigar nest-box or even slightly bigger with a half-open front); the nest is bulky and made of grass stems, plant fibres etc. Clutch 3–6; incubation 11–13 days. Provide germinated seed, extra millet sprays, seeding grasses, green food, live food and eggfood for rearing. They may also breed in a spacious cage or indoor flight.

General management Except when sick, this is one of the sleekest, most well-groomed birds imaginable. Acclimatised birds are very hardy and can be wintered in an outdoor shelter and flight, although frost-free sleeping quarters are desirable. They mix with other birds of comparable size (weavers etc) but are not safe with small finches, waxbills etc.

Amadina erythrocephala [PLATE 143]
Red-headed Finch, Paradise Sparrow

PLATE 143 *Amadina erythrocephala*, Red-headed Finch

5 in (13 cm). The lores are buff-white, the head and chin red. The upper surfaces are grey-brown; there are some buff-white spots on the wing coverts and secondaries. The tail is dark brown, the feathers with white tips. The underparts are creamy-white with dark-brown edgings to the feathers, giving a scaly appearance; the abdomen is light brown, the under tail coverts grey-brown. The bill is horn-coloured, the legs and feet flesh-coloured. The female is similar but lacks the red head, and her markings are less distinctive; she has a buff-white throat.

Range Angola, Zimbabwe, South Africa.

Status Good populations in some areas of the range.

Avicultural rating Occasionally available, this species is fairly hardy but needs winter warmth.

Feeding Diet A, with millet sprays, seeding grasses, green food, some live food. Grit is essential.

Breeding Some pairs prove very prolific, others less so. Provide a box or basket as a nest receptacle. Clutch 3–6; incubation 11–12 days. Offer egg-food, germinated seeds, seeding grasses, live food for rearing. If only in limited-size accommodation, they are best segregated when breeding; if sharing a spacious aviary disputes are usually limited to bickering and defence of territory.

General management This bird needs care during acclimatisation and is never as adaptable as the related Cut-throat. It does best in indoor accommodation in winter; in a temperature of around 50°F (10°C). This is probably a species which could be established in aviculture through captive breeding.

Amadina fasciata (III)
Cut-throat, Cut-throat Finch, Ribbon Finch

5 in (13 cm). The upper surfaces are beige-brown with fine barring and speckling; the rump and upper tail coverts are light brown. The wings are grey-brown, the tail black with white tips to the outer feathers. The underparts are yellow-fawn, the darker margins giving a partial scale effect; the abdomen is rufous. There is a conspicuous red band across the throat. The bill is horn-coloured, the legs and feet flesh-brown. The female lacks the red band but is otherwise similar.

Range Senegal, northern Nigeria eastwards to west and south-west Sudan.

Note Three subspecies are described. Alexander's Cut-throat (*A. f. alexanderi*), from Abyssinia, Somaliland, south-east Sudan, Kenya and eastern parts of Tanzania is occasionally available and is usually listed as 'East African Cut-throat'. It differs from the nominate subspecies mainly in having more prominent barring on the upper and under surfaces.

Status Abundant in most parts of the range.

Avicultural rating These long established avicultural favourites are easily managed and free breeders.

Feeding Diet A, with millet sprays, seeding grasses, green food. Grit is essential.

Breeding They are usually free breeders and will nest in cages. They sometimes build a spherical nest in a bush; they regularly use boxes or baskets and other receptacles. Clutch 4–7; incubation 11–12 days. Provide eggfood, germinated seed, extra millet sprays or seeding grasses, live food. They have been known to rear young successfully on a much more spartan diet. Hens may be subject to egg-binding if in small cages offering insufficient exercise.

General management This is among the easiest species to establish after importation and quarantine, but, like other exotic species, they *must* be acclimatised properly before being transferred to an outdoor shelter and flight. Even though hardy acclimatised birds are able to over-winter with minimum heat, they will not thrive in cold and wet conditions. They are best not associated with small waxbill-sized species.

Ploceidae (Weavers, Sparrows)

This Old World family includes many brilliantly-coloured species and others with remarkable feather development (weavers and whydahs) as well as one of the world's best-known and most ubiquitous species, the House Sparrow. Some are builders of ornate nests; others are parasitic and make no nests of any kind. One, the Quelea, which ranges over much of Africa, is regarded by some as the world's most destructive bird, to be exterminated by any and every means available in crop-growing districts. Included are many long-established avicultural favourites, especially among the weavers and whydahs; they are generally easily managed, hardy and (with few exceptions) mix well with similar-sized birds. Breeding successes with most are few and far between, mainly because of their parasitic habits which require aviculturists to have rare Estrildines nesting at the same time as whydahs come into breeding condition. Some weavers are more readily bred, but success with most is only intermittent.

Vidua chalybeata (III)
Village Indigo Bird, Senegal Combassou, Senegal Indigo Bird, Steel Finch

---

4½ in (11.5 cm). The entire plumage is a glossy blue-black, the primaries a dull black. The bill is pinkish-white, the legs and feet orange or light brown. The female (and male in non-breeding plumage), has brown upper surfaces with darker striations, the centre of the crown is grey-buff bordered dark brown. The underparts are buff.

Range Senegal, Gambia, Sierra Leone, eastwards to central and southern Sudan, Ethiopia, Uganda, Kenya, western Tanzania, Zambia, Angola, Botswana, Zimbabwe, Mozambique.

Status Abundant in many areas.

Avicultural rating Fairly regularly available and widely kept, they are easily managed and hardy.

Feeding Diet B, with millet sprays, seeding grasses, green food, some live food. Grit is essential.

Breeding They are mainly parasitic on the Red-billed Firefinch (*Lagonosticta senegala*) although there are some examples of captive birds hatching and rearing their own broods. Incubation 12 days.

General management Acclimatised birds are hardy but need frost-free winter accommodation. They mix well with other species; association with the Red-billed Firefinch in the aviary is clearly desirable if breeding is to be achieved.

Vidua fischeri
Fischer's Whydah, Straw-tailed Whydah

---

12 in (30 cm). The forehead, crown, lower underparts and four elongated, straw-like tail feathers

are golden-buff; the rest of the plumage is black with the rump and upper tail coverts buff, streaked black. The bill is red, the legs and feet orange. The female, and male in non-breeding plumage (5 in, 13 cm), is mainly deep buff with darker streaking; the crown, eyebrow streak and cheeks are tawny. The underparts are buff-white.

Range Eastern Ethiopia, Somaliland, Kenya, northern Tanzania.

Status Locally distributed and scarce in most areas.

Avicultural rating Occasionally available, this species is easily managed after acclimatisation and mixes with other small seedeaters.

Feeding Diet B, with millet sprays, seeding grasses. Individual birds enjoy green food, especially cress. Regular live food is important. Grit is essential.

Breeding They are parasitic on the Purple Grenadier (*Uraeginthus ianthinogaster*). Incubation 12–13 days.

General management They need care during the immediate post-importation and quarantine period; then are reasonably hardy, but need modest warmth (50°F, 10°C) during the winter and will not thrive in cold, damp conditions. They usually mix well with other species; individual males occasionally create minor problems by an exuberant display to the female.

Vidua regia
Queen Whydah, Shaft-tailed Whydah

13 in (33 cm). The nape, sides of the head and underparts are golden-buff; the remainder of the plumage, including four elongated, club-ended, wire-like tail feathers, is black. The bill, legs and feet are red. The female, and male in non-breeding plumage (5 in, 13 cm), is similar to Fischer's Whydah, but the upper parts are slightly darker.

Range Southern Angola and Mozambique, Botswana, South Africa.

Status Local distribution; nowhere abundant.

Avicultural rating A rare and much-prized avicultural subject, which is not difficult to care for after acclimatisation.

Feeding Diet B, with millet sprays, seeding grasses, live food. Grit is essential.

Breeding This bird is parasitic on the Violet-eared Waxbill (*Uraeginthus granatina*). It has been captive bred with both these and Red-cheeked Cordon Bleus (*Uraeginthus bengalus*) acting as the host species. Incubation 11–13 days.

General management They are elegant and beautiful when the male is in nuptial plumage; they are generally peaceful in mixed company, although the male may create some disruption among smaller species. One cock to two or three females is the best ratio with this and closely-related species. It needs care during acclimatisation and will not usually successfully over-winter in unheated accommodation.

Vidua macroura (III)
Pin-tailed Whydah

11 in (28 cm). The face, crown, mantle, back, wings and four elongated tail feathers are black; the rest of the plumage, including a large wing patch, is white. The bill is red, the legs and feet dark grey. The female, and male in non-breeding plumage (5 in, 13 cm), has brown upper surfaces with darker striations; the underparts are buff-white. The centre of the crown is rufous with a darker stripe on each side.

Range The islands of Fernando Pó and St Thomé; Senegal across Africa to southern Sudan, Ethiopia, and south to eastern areas of South Africa. Also Zanzibar and Mayotte (Comoros).

Status Abundant in many parts of the range.

Avicultural rating A popular, widely kept species which is among the more inexpensive of the Viduine Whydahs; it is easily managed.

Feeding Diet B, with millet sprays, seeding grasses, green food, regular live food. Grit is essential.

Breeding It is parasitic on several Estrildine species including Red-eared Waxbill (*Estrilda troglodytes*), Orange-cheeked Waxbill (*Estrilda melpoda*) and St Helena Waxbill (*Estrilda astrild*). Several captive breedings have been achieved over the years. Incubation 11–12 days.

General management This bird acclimatises easily and afterwards is hardy and easily managed; it is often long-lived up to 15 or more years. Males in breeding plumage and condition are often a handful in community aviaries; a frequent tactic is to dive on the food vessel, scattering any small birds around it. Sometimes more serious aggression occurs, but this is less likely if the birds are sharing spacious quarters. If breeding is to be achieved it is vital that a suitable host species shares the whydahs' territory. The best ratio is one male to two or three females; the latter are usually in short supply among imports.

Vidua paradisaea (III)
Paradise Whydah

---

16 in (41 cm). The head, chin, throat, upper breast, mantle, back, rump, tail coverts, wings and tail are black; a broad band on the nape is golden-buff. The lower breast is chestnut, the rest of the underparts buff, paler towards the abdomen. There is a spectacular development of the tail feathers in the nuptial dress: one pair is very wide and short with long, bare shafts; a further pair is also wide but greatly elongated (10–11 in, 25–28 cm). The female, and male in nonbreeding plumage (6 in, 15 cm), has yellow-brown upper surfaces, and grey-white underparts; the crown is black with a buff centre streak.

Range Angola and Namibia eastwards to Zambia, Zimbabwe, Mozambique; north Ethiopia, south-eastern Sudan, and south to Transvaal, Natal.

Status Abundant in many parts of the range.

Avicultural rating They are frequently available, widely kept and hardy after acclimatisation.

Feeding Diet B, with millet sprays, seeding grasses, green food, regular live food. Grit is essential.

Breeding They are parasitic on the Melba Finch (*Pytilia melba*) and its subspecies. Captive breedings have been achieved with appropriate host species. Incubation 12–13 days.

General management Although the Paradise Whydah is frequently presented by authors as an entirely virtuous species in mixed groupings, the reverse is often true; the nature of their display flights frequently leads to attacks on their companions. They are hardy after acclimatisation and a very spectacular aviary species but are unsuitable for cage accommodation when in full nuptial plumage.

Vidua orientalis (III)
Broad-tailed Paradise Whydah

Similar to Paradise Whydah (*V. paradisaea*) except it has a broader golden-buff or chestnut band on the nape, and the second elongated pair of tail feathers is wide throughout the length.

Range Senegal, Sierra Leone, eastwards to northern Nigeria and Cameroon, Chad, southern Sudan, Ethiopia, Kenya; also south to Angola, Mozambique.

Status Abundant in many parts of the range.

Avicultural rating See Paradise Whydah.

Feeding/Breeding/General management Similar to Paradise Whydah. 5–6 subspecies parasitise various *Pytilia* species and subspecies, as follows.

Subspecies	Host
Vidua o. orientalis (Chad to Ethiopia)	*Pytilia melba citerior* (Senegal to Ethiopia)
V. o. acupum (Senegal to northern Nigeria)	*P. m. citerior*
V. o. obtusa (Angola to Kenya/ Mozambique)	*Pytilia afra* (Sudan, Angola etc)
V. o. interjecta (Cameroon to Sudan)	*Pytilia phoenicoptera* (Senegal to Cameroon) and *P. p. emini* (Cameroon to Sudan)
V. o. togoensis (Sierra Leone to Togo)	*Pytilia hypogrammica* (Sierra Leone to Cameroon)

Auripasser luteus
Sudan Golden Sparrow, Golden Song Sparrow, Yellow Sparrow

5 in (13 cm). The head, nape, chin, throat and underparts are yellow, the mantle and back chestnut. The rump is yellow-buff, the wings black and chestnut. The tail is brown. The bill is horn-coloured (black during the breeding season), the legs and feet flesh-brown. The female's head, mantle back and rump are buff-brown.

Range Mauritania, Senegal, eastwards to Mali, Niger, northern Nigeria, Chad, northern Sudan, Ethiopia and Somaliland.

Status Abundant in many parts of the range.

Avicultural rating This inexpensive and widely-kept seedeater is a good aviary subject.

Feeding Diet B, with millet sprays, seeding grasses, green food, some live food. Grit is essential.

Breeding This is a fairly willing breeder if true pairs are obtained; the majority of imports are cocks with 'females' proving to be immature males. It is a good colony breeder, and some contact with others of the same species may stimulate breeding activity. They make a fairly bulky globular nest, and will also use a box or basket among cover. Clutch 3–4; incubation 11–12 days. Provide live food, germinated seed, extra millet sprays, seeding grasses and green food for rearing; some birds will take eggfood. They are generally peaceful when breeding.

General management Dealers who sell these little birds either as potential pets or songbirds are guilty either of a lack of knowledge or of sharp practice. Golden Sparrows have no song worthy of note; they are also nervous and easily stressed, doing better in aviaries. They are not good cage birds and often succumb quickly if closely confined. They need care until acclimatised, then are fairly hardy, but frost-free winter accommodation is needed. They mix well with other birds.

Sporopipes squamifrons [PLATE 144]
Scaly-crowned Weaver, Scaly Weaver

PLATE 144 *Sporopipes squamifrons*, Scaly-crowned Weaver

4 in (10 cm). The forehead and crown are black with white margins to the feathers giving a scaled appearance; the nape, sides of the head, mantle and back are grey. The wing and tail feathers are

black with distinctive white edges. The lores and moustachial stripe are black; there is a white eye-ring. The chin, throat and underparts are grey-white, paler towards the abdomen. The bill is pink, the legs and feet grey-brown. The sexes are alike.

Range South-western Angola and Namibia eastwards to Botswana, South Africa.

Status There are good populations in some parts of the range.

Avicultural rating This species is only occasionally available and is not as easy as *Ploceus/Euplectes* weavers, needing some heat during the winter.

Feeding Diet B, with millet sprays, seeding grasses, green food and regular small live food. Grit is essential.

Breeding They are only occasionally bred. They build a large, untidy globular nest of grasses and other plant fibres. Clutch 3–4; incubation 11–13 days. Provide abundant small live food for rearing plus germinated seeds, extra millet sprays or seeding grasses etc. They are aggressive when in breeding condition.

General management They need care during the immediate post-importation and quarantine period. Use sheltered accommodation: shelter and flight is best during the summer months, a warm (50°F, 10°C) birdroom during the winter. They do not tolerate cold and damp conditions. They are generally good mixers, except when coming into breeding condition.

Sporopipes frontalis (III)
Speckle-fronted Weaver

5 in (13 cm). The forehead and crown are black with white margins to the feathers giving a scaled appearance; the hind crown, nape and sides of the neck are rufous. The mantle, back, rump and wings are sandy-brown, the latter with lighter margins to the feathers; the tail is sandy and dark brown. The cheeks and underparts are grey-white; the moustachial stripe is black and white similar to the forehead and crown. The bill is pink, the legs and feet flesh-coloured. The sexes are alike.

Range Mauritania, Senegal, eastward to Niger, Chad, Sudan, Ethiopia, Somaliland, northern Uganda and Kenya.

Status There is local distribution, but they are abundant in some areas.

Avicultural rating See Scaly-crowned Weaver; this species is more often available.

Feeding/Breeding/General management Similar to Scaly-crowned Weaver. They are gregarious in the wild and sometimes nest in colonies.

Ploceus vitellinus (III)
Vitelline Masked Weaver, Half-masked Weaver

5 in (13 cm). The forehead is chestnut, the crown chestnut-orange, the nape yellow; mantle, back, wings and tail are greenish-yellow with darker markings. The face, cheeks and chin are black, the rest of the plumage yellow, suffused with chestnut on the breast. The bill is black, the legs and feet flesh-coloured. The female (and the male in non-breeding plumage) has brown upper surfaces with darker striations, and buff-white underparts suffused with yellow on the throat and breast. There is no mask.

Range Mauritania, Senegal, Gambia, eastwards to Chad, western Sudan, northern areas of Central African Republic.

Status Abundant in many parts of the range.

Avicultural rating This popular and widely-kept species is easily managed and hardy.

Feeding Diet A, with millet sprays, seeding grasses, green food and some live food. Grit is essential.

Breeding It is probably best if one cock and three to four hens are the sole occupants of an aviary. The cock builds a pear-shaped, pendulous nest of strips of grass and other plant materials, with a softer lining. Clutch 2–4; incubation 11–12 days. Provide abundant live food, germinated seed, extra spray millet or seeding grass etc for rearing. The male is constantly adding to the nest or building new ones, including 'cock's' nests.

General management After acclimatisation this is a hardy species which can over-winter in an outdoor shelter and flight; it is usually proof against all but the most severe weather conditions. It can be housed with other robust species in a large aviary and it is not aggressive when out of colour but becomes so when nuptial plumage is assumed.

Ploceus cucullatus (III)
Village Weaver, Rufous-necked Weaver, Black-headed Weaver, Spotted-backed Weaver

---

6½ (16.5 cm). The whole of the head, chin, throat and upper breast are black; there is a narrow area of chestnut on the neck and breast bordered with black. The mantle, back and rump are yellow with a darker 'V' marking; the wings are black and yellow. The breast and underparts are yellow, the tail olive and yellow. The bill is black, the legs and feet light brown. The female (and male in non-breeding plumage) has olive-green and brown upper surfaces, with darker wings; the underparts are a pale yellow-buff.

Range Island of Fernando Pó; Senegal eastwards across Africa to Sudan, Ethiopia; south to northern Angola, eastern areas of South Africa.

Note There are some size and plumage differences among the 7 subspecies.

Status Abundant in many areas of the range.

Avicultural rating Fairly frequently imported, it is hardy and easily managed.

Feeding Diet A, with millet sprays, seeding grasses, green food, some live food. Grit is essential.

Breeding This weaver is rarely bred in captivity; most successes appear to have occurred when the birds were housed in colonies. They build a pendulous nest with the entrance (sometimes a tube) at the base; it is built by the male, although the females usually attend to the interior 'furnishings'. Clutch 1–3; incubation 13–14 days. The males have been known to destroy aviary nests while the hen is part-way through incubation. Provide ample live food, extra spray millet and seeding grasses, germinated seed etc for rearing.

General management They are hardy when acclimatised and resistant to all but the most severe weather. They are aggressive and not trustworthy with smaller birds; they are best kept with cardinals etc, and isolated if breeding seems likely. Damage to plants will occur during nest-building activities.

Ploceus philippinus
Baya Weaver

5¾ in (14.5 cm). The forehead, crown and nape are yellow, the upper surfaces dark brown, with feathers margined with yellow. The lores, sides of the head and ear-coverts are blackish-brown (the blackish-brown area may be more extensive in some subspecies); the underparts are yellow, paler towards the abdomen. The bill is horn-coloured, the legs and feet flesh-coloured. The female, and male in non-breeding plumage, has brown upper surfaces with darker streaking; the underparts are yellow-buff, darker on the flanks.

Range Pakistan, India, Sri Lanka, Bangladesh, Burma, Thailand, South Vietnam, Malaysia, Sumatra, Java; south-west China.

Note There are some plumage differences among the 5 described subspecies.

Status Abundant in some parts of the range.

Avicultural rating Occasionally available, they are hardy when acclimatised, and superlative nest-builders.

Feeding Diet A, with millet sprays, seeding grasses, green food, some live food. Grit is essential.

Breeding It builds a suspended, elongated globular nest with a spout entrance at the base; usually in colonies. Clutch 2 (occasionally 3–4); incubation 12–13 days. Abundant live food is needed for rearing. The male is a persistent nest-builder during the breeding season and is capable of providing homes for several females. They are very aggressive to other species.

General management These pugnacious, active birds are hardy after acclimatisation. They do best in an outdoor shelter and flight, but will damage plant growth during nest-building activities.

Quelea quelea [PLATE 145]
Red-billed Quelea, Red-billed Weaver, Common Quelea

PLATE 145 *Quelea quelea*, Red-billed Quelea

$4\frac{3}{4}$ in (12 cm). The forehead, sides of head, chin and throat are black; the crown, nape, sides of neck, rump, tail coverts and underparts are buff-pink. The mantle, back, wings and tail are warm buff and brown with darker streaking, the feathers of the wings and tail margined with buff-yellow. The bill is red; the legs and feet are flesh-pink. The female (and male in non-breeding plumage) has grey-brown upper surfaces with darker streaking; the underparts are buff-white, paler on the chin and throat, with more yellow in the centre of the breast.

Range Senegal eastwards to Chad, Sudan, Ethiopia, Somaliland; south to Angola, Zaire, Zambia, Tanzania, South Africa.

Status Abundant in many parts of the range, but in vast flocks that move between feeding areas; it is a crop pest in many parts of the range and is destroyed whenever the opportunity occurs.

Avicultural rating These inexpensive and easily maintained seedeaters are hardy after acclimatisation.

Feeding Diet A, with millet sprays, seeding grasses, green food, some live food. Grit is essential.

Breeding Several breeding successes have been recorded; they are colony breeders. A flimsy nest of plant fibres is woven onto grass stems. Clutch 2–5; incubation 12–13 days. Abundant live food is needed for rearing; also give germinated seeds, millet sprays etc. They are monogamous, and generally not aggressive with other species in large aviaries.

General management Acclimatised birds are weather-resistant in an outdoor shelter and flight throughout the year. They usually mix well with other birds. They are very industrious nest-builders during the breeding season and, like many of their relatives, will seriously damage growing plants in their quarters.

Foudia madagascariensis [PLATE 146]
Madagascar Fody, Madagascar Weaver, Madagascan Red Fody

PLATE 146 *Foudia madagascariensis*, Madagascar Fody

5 in (13 cm). The lores and a small area around the eyes are black; the head, nape, rump, tail coverts and underparts are vermilion. The mantle and back are similar but laced with black; the wings are black, the feathers of the secondaries and flights with buff-brown margins. The bill is black, the legs and feet flesh-coloured. The female (and male in non-breeding plumage) has olive-brown upper surfaces with darker streaking; the underparts are buff-grey.

Range Madagascar, Mauritius, Reunion.

Status There are still good populations, although suitable habitat is contracting.

Avicultural rating They are occasionally available but are fairly expensive. These brilliantly-coloured aviary birds are easily managed.

Feeding Diet A, with millet sprays, seeding grasses, regular live food; green food may be eaten. Grit is essential.

Breeding Several breeding successes have been recorded. It is not a colony nester. It builds a globular nest of plant fibres usually low down among vegetation. Clutch 4; incubation 13 days. The nestlings are fed almost exclusively on live food; some germinated seed, millet sprays and green food should also be provided.

General management They are hardy after acclimatisation and can be over-wintered in an outdoor shelter and flight. They mix well with other species in large aviaries; they may be pugnacious (in breeding condition) in smaller accommodation. I have found them nervous and very subject to stress in a restricted space; they fare best in large planted aviaries.

Euplectes afer (III)
Napoleon Weaver, Golden Bishop, Yellow-crowned Bishop

$4\frac{3}{4}$ in (12 cm). The forehead, crown, nape, sides of the neck, upper breast, flanks, rump and upper tail coverts are yellow; the cheeks, chin, throat, a narrow stripe across the nape, the lower breast, underparts and abdomen are black. The wings and tail are grey-brown; there are some brown feathers on the mantle. The bill is black, the legs and feet flesh-coloured. The female (and male in non-breeding plumage) has light brown upper surfaces with darker streaking; the underparts are buff-white, darker on the flanks and breast. The bill is horn-coloured.

Range Senegal, Gambia, eastwards to Chad, Central African Republic, Sudan, Ethiopia, Uganda, northern Kenya; south to Gabon, Congo, South Africa.

Note Four subspecies are described.

Status Abundant in many parts of the range.

Avicultural rating Inexpensive and widely kept, they are hardy.

Feeding Diet A, with millet sprays, seeding grasses, green food, some live food. Grit is essential.

Breeding A few captive breeding successes have been recorded. A ratio of one male to three or four hens in a reed-planted aviary may produce results. The nest is a compact globe of plant fibres built among reeds or tall grasses, invariably in a marshy area. Clutch 2–4; incubation 12–13 days. They are very insectivorous when breeding; supply abundant small live food for rearing, plus germinated seeds, extra millet sprays etc. The cock builds several nests; when not thus occupied, he puffs out his feathers and engages in much aerial activity. These birds are very disruptive to other species of smaller or equal size.

General management Hardy after acclimatisation, they are good occupants of an outdoor shelter and flight; they mix with other birds, except when approaching breeding condition. Quail and doves are good companions. They are very colourful in nuptial plumage, but sparrow-like for a proportion of the year.

Euplectes orix (**III**)
Red Bishop

---

5 in (13 cm). The forehead, crown, sides of the head, breast and underparts are black; the rest of the plumage is orange except the brown wings and tail. The bill is black, the legs and feet flesh-coloured. The female, and male in non-breeding plumage, has brown upper surfaces with darker striations, and buff-brown underparts, paler towards the abdomen.

Range Senegal eastwards to Chad, Central African Republic, Sudan, northern Ethiopia, Uganda, Kenya; south to northern Angola, Mozambique, South Africa.

Note Four subspecies are described, two of particular interest to aviculturists. The nominate subspecies, *E. o. orix* (Angola, Mozambique, South Africa), is the Grenadier Weaver. *E. o. franciscana* (Senegal to Ethiopia) is the Orange Weaver.

Status Abundant in many parts of the range.

Avicultural rating Popular and widely kept, they are hardy and easily managed.

Feeding Diet A, with millet sprays, seeding grasses, green food and some live food. Grit is essential.

Breeding It builds a globular nest with a side entrance usually in a bush. Clutch 2–4; incubation 12–14 days. Abundant live food is needed for rearing. A ratio of one cock to three to four hens, or one cock to one hen, have both produced breeding success. They do best in a well-planted aviary with ample space for the cock's excitable behaviour during the breeding season.

General management They are weather-resistant after acclimatisation. They are not recommended for the close confinement of a cage or indoor compartment; loss of the brilliant nuptial plumage usually results and the birds ultimately moult annually to a dull rather than a fiery orange. They are disruptive with other species when in full colour and breeding condition. They have a wheezing, staccato song.

Euplectes macrourus (III)
Yellow-mantled Whydah, Yellow-backed Whydah

9½ in (24 cm). The mantle, back and lesser wing coverts are yellow; the rest of the plumage is black, the feathers of the secondaries and flights with light brown margins. The bill is black; the legs and feet are dark grey. The female, and male in non-breeding plumage (6½ in, 16.5 cm), has grey-brown upper surfaces with darker streaking and buff-brown underparts, paler on the chin and throat.

Range Senegal eastwards to Chad, Central African Republic, Uganda, Sudan, western Kenya; south to Angola, Mozambique.

Note Four subspecies are described; *E. m. macrocercus* (Uganda and western Kenya) is known as the Yellow-shouldered Whydah.

Status Abundant in many parts of the range.

Avicultural rating Frequently available, but not as popular and widely-kept as Viduine Whydahs; they are easily managed and hardy.

Feeding Diet A, with millet sprays, seeding grasses, green food, some live food. Grit is essential.

Breeding Occasional breeding successes have been recorded. Ideally they should be kept in a spacious planted aviary. They are probably polygamous but a 1 : 1 male-female pairing has also produced results. A flimsy globular nest of plant fibres and fresh grasses is made. Clutch 2–4; incubation 12–14 days. Provide germinated seed, live food etc for rearing.

General management Acclimatised birds are hardy and can be wintered in a suitable shelter and flight; they are generally aggressive when in nuptial plumage so choose companions with care, and it is best if they are segregated if breeding is hoped for.

Euplectes ardens (III)
Red-collared Whydah

15 in (38 cm). The entire plumage is black, except for a red band across the throat and upper breast; there are buff-brown margins to the feathers of the secondaries and flights. The bill, legs and feet are black. The female, and male in non-breeding plumage (6 in, 15 cm), has tawny upper surfaces with black streaking; the underparts are buff, with more yellow on the chin and throat, paler towards the abdomen.

Range Senegal to southern Sudan; to Angola, northern Namibia, eastern Cape Province, Natal.

Note Four subspecies are described, including *E. a. laticauda* (south-eastern Sudan and Ethiopia), the Red-naped Whydah.

Status Abundant in many parts of the range.

Avicultural rating See Yellow-mantled Whydah; this species is less often available.

Feeding/Breeding/General management Similar to Yellow-mantled Whydah. They are hardy and easily managed but aggressive, so choose companions with care and ensure they are *never* species with red in their plumage.

Euplectes progne
Long-tailed Whydah, Giant Whydah

24 in (60 cm). The plumage is mainly black; the shoulder of the wing is yellow-buff and orange, the secondaries and flights margined with buff-brown. The bill is blue-grey, the legs and feet flesh-brown. The female, and male in non-breeding plumage (8 in, 20 cm), has tawny-brown upper surfaces with darker streakings; her underparts are warm buff. The female is smaller (6 in, 15 cm) than the out-of-colour male.

Range Eastern Angola, Zambia, west and central Kenya, South Africa.

Note Four subspecies are described; *E. p. delamerei* (eastern Kenya), is Delamere's Whydah, which was once fairly regularly available.

Status Locally distributed; this is a familiar species within the range but it is not abundant.

Avicultural rating Now rarely available, it is hardy and easily managed. In-colour males are spectacular inhabitants of very large aviaries.

Feeding Diet A, with millet sprays, green food and some live food. They are also capable of eating larger seeds, including hemp.

Breeding Some breeding successes have been recorded. Spacious accommodation is important. They are polygamous, and one male to four to six females is probably the ideal ratio, if stocks are available. A large oval nest of plant fibres is built, usually low down among coarse grass. Clutch 2–4; incubation 13–14 days. Live food, germinated seeds etc are needed for rearing.

General management This is a hardy species which can over-winter in an unheated shelter and flight. Some birds in nuptial plumage during the late (northern hemisphere) winter can become rain-soaked and chilled in an unsheltered aviary; in Africa it is often possible to pick up waterlogged birds after storms. They are not especially aggressive, but the impressive size and frequent aerial activity makes it an unsuitable companion for smaller species.

Euplectes jacksoni
Jackson's Whydah

---

13 in (33 cm). The plumage is entirely black; there is some olive-brown on the wing shoulder. The bill is grey, the legs and feet black. The female, and male in non-breeding plumage (6½ in, 16.5 cm), has light brown upper surfaces with darker streaking; the underparts are buff, paler towards the abdomen. The female is smaller (5½ in, 14 cm) than the out-of-colour male.

Range Western and central Kenya, northern Tanzania.

Status Abundant in many suitable areas of habitat.

Avicultural rating Only occasionally available, they are hardy after acclimatisation, but frost-free winter quarters are recommended.

Feeding Diet A, with millet sprays, green food, some live food. Grit is essential.

Breeding This is a challenge. The males, with tails shaped like those of bantam cockerels, create rings in the grass to perform dancing displays when they leap and prance to attract the attention of the females. The nest is domed, made of grass or plant fibres, low down among vegetation. Clutch 2–4; incubation 13–14 days. Live food, germinated seed etc are necessary for rearing.

General management They are not suitable for close confinement and fare best in a suitable large shelter and flight, preferably sheltered and on well-drained ground. They are aggressive in the breeding season and best segregated.

Sturnidae (Starlings)

Despite the spread of the Common Starling (*Sturnus vulgaris*) in North America, this is essentially an Old World family; it comprises more than 100 species, which are primarily tropical and particularly well represented in the East Indies, Asia and Africa, with the Common Starling's various subspecies occupying a range through Europe and the Middle East to the Himalayas. The species is also found in the USA after 60 birds were brought from Europe and liberated in New York towards the end of the nineteenth century. Many tropical starlings have iridescent plumage and are popular avicultural subjects; they are usually adaptable and hardy, are fairly omnivorous in their diet, and several breed freely. Behaviour ranges from truculent to seriously aggressive: most are potential killers of smaller, weaker or less mobile companions when breeding, and pairs should always be segregated. Among the most specialised members of the family are two species of oxpecker (*Buphagus*) which, although they are specialist feeders on ticks etc, are not difficult to cater for in captivity. They have been exhibited in some zoos, when mock grazing animals, constructed of hide (or even hessian) stretched over a rudimentary wooden frame, were provided for them.

Aplonis panayensis
Philippine Glossy Starling, Red-eyed Starling

---

9 in (23 cm). The plumage is mainly glossy black with iridescent green and purple reflections. The eyes are ruby-red. The bill, legs and feet are black. The sexes are alike.

Range Western Burma, southern Thailand, South Vietnam, Malaysia, Andamans, Nicobars, Greater Sundas, Celebes, Philippines.

Note 13 subspecies are described.

Status Abundant in most parts of the range.

Avicultural rating Occasionally available, they are hardy and easily managed after acclimatisation.

Feeding Diet E and G, with berries as available; they appear to be slightly more frugivorous than African glossy starlings, but are clearly adaptable in dietary matters. Provide regular items of live food such as small locusts, crickets and a few mealworms.

Breeding In the wild they utilise a variety of nest sites, from tree holes, cavities in earth banks, to under eaves or in trees; the nest is built of plant fibres, moss etc, sometimes with less usual materials such as discarded paper or fabric. Clutch 3–4; incubation 14–16 days. Provide extra live food etc for rearing. The breeding pair should occupy separate accommodation; they are aggressive when nesting.

General management They are usually nervous after importation and quarantine, but settle quickly and prove as confident as others of the family. Eventually hardy, they can occupy a shelter and flight throughout the year but there should be frost-free (roosting) quarters in winter. They mix with other birds of comparable size, but watch for aggression if housed with weaker, less assertive species.

Lamprotornis purpureus
Purple Glossy Starling

9½ in (24 cm). The forehead, crown, sides of the head, chin and throat are metallic purple; the nape and rump are blue, the mantle, back and wings are golden-green. The underparts are an iridescent violet-blue. The eyes are orange. The bill, legs and feet are black. The sexes are alike.

Range Senegal, Gambia, Guinea, Ivory Coast, eastwards to Nigeria, northern Cameroons, southern Chad, Central African Republic, Sudan, Uganda and Kenya.

Status Abundant in many parts of the range.

Avicultural rating This popular and widely-kept species is frequently available.

Feeding Diets E and G, with some minced raw meat; also offer berries as available. Regular live food is important.

Breeding The difficulty of sexing is one of the biggest obstacles to more frequent captive-breedings with this species. They need a separate aviary with some natural or artificial cover; they will use a suitably-sized nest-box. The nest is bulky, made of twigs, plant fibres etc with a softer lining. In the wild this bird is a hole-nester, so ensure that the nesting receptacle has a small entrance (not open-fronted). Clutch 3; incubation 14–15 days. Provide extra live food (small locusts, crickets, mealworms, fly pupae etc), fruit for rearing; some pairs take bread and milk, honey water and sponge cake when the nestlings are very small.

General management They are easily managed after acclimatisation and usually extremely hardy. Choose companions with care (jay-thrushes, mynahs, starling etc); they are not suitable for association with smaller or weaker species, or for close confinement in a cage or indoor flight.

Lamprotornis iris
Emerald Starling

---

8 in (20 cm). The plumage is mainly an iridescent metallic-green; there are coppery-violet cheek patches, and a similar colour on the lower breast and abdomen. The bill, legs and feet are black. The sexes are alike.

Range Guinea, Sierra Leone, Liberia, Ivory Coast.

Status Much less abundant than related species.

Avicultural rating Only occasionally available, they are easily managed.

Feeding/Breeding/General management Similar to Purple Glossy Starling. One of the smallest and most beautiful of the glossy starlings, it can be as belligerent as its relatives. Acclimatised birds are fairly hardy, but do best with a frost-free shelter available in winter.

Lamprotornis purpureiceps
Purple-headed Glossy Starling

---

8 in (20 cm). The forehead, crown, nape, sides of the head, chin, throat and upper breast are an iridescent violet-blue; the mantle, back, lower breast and underparts are metallic green. There is some metallic blue on the sides of the breast, the wings and tail. The bill, legs and feet are black. The sexes are alike.

Range Ivory Coast eastwards to Nigeria, Congo, Central African Republic, southern Sudan, Uganda.

Status There are good populations in some parts of the range, but it is locally distributed.

Avicultural rating It is fairly frequently available, but there is much confusion with the Purple Glossy Starling (*P. purpureus*); the smaller size and more metallic green in the plumage of this species is a good guide to correct identification.

Feeding/Breeding/General management Similar to Purple Glossy Starling. Numerous breedings are recorded. Established birds are aggressive and not safe companions for smaller or weaker species; they are good with larger doves, pheasants, waterfowl etc.

Lamprotornis chalybaeus
Greater Blue-eared Glossy Starling, Green Glossy Starling

9½ in (24 cm). The plumage is mainly iridescent green, except for metallic-blue on the ear-coverts and violet-blue on the lower underparts. The bill, legs and feet are black. The sexes are alike.

Range Mauritania, Senegal, Guinea, Ivory Coast, eastwards to northern Nigeria/Cameroon, Chad, Sudan, Ethiopia, Somaliland, northern Uganda and Kenya.

Status Abundant in many parts of the range.

Avicultural rating This popular and widely-kept species is best known in aviculture as the Green Glossy Starling.

Feeding/Breeding/General management Similar to Purple Glossy Starling. Numerous breeding successes are recorded. Like their relatives, they are of uncertain temperament and companions should be chosen with care. An easily managed species, it is hardy after acclimatisation.

Lamprotornis chloropterus
Lesser Blue-eared Glossy Starling, Swainson's Glossy Starling

7½ in (19 cm). Similar to Greater Blue-eared Glossy Starling, but distinguished by its smaller size and lack of notches on the inner webs of the primaries. The sexes are alike.

Range Senegal eastwards to southern Chad, Central African Republic, south-eastern Sudan, Ethiopia, Uganda, Kenya, Tanzania, Zambia, Mozambique.

Status Abundant in many parts of the range.

Avicultural rating See Purple Glossy Starling.

Feeding/Breeding/General management Similar to Purple Glossy Starling.

Lamprotornis caudatus
Long-tailed Glossy Starling

20–24 in (51–60 cm). The forehead, crown, cheeks, chin and throat are bronze-green; the rest of the plumage is iridescent green, except for some violet-blue on the sides of the breast and flanks and a purple and blue tail. The bill, legs and feet are black. The sexes are similar but the female is smaller.

Range Senegal, Gambia, Guinea, eastwards to northern Ghana, Togo, northern Nigeria/Cameroon, Chad, Central African Republic, central and southern Sudan.

Status There is wide but local distribution, with good populations in some areas.

Avicultural rating Available from time to time, it needs spacious accommodation. It is easily managed and hardy.

Feeding/Breeding/General management Similar to Purple Glossy Starling. It is often in poor feather condition after import and quarantine, but usually recovers quickly with care; watch for foot problems caused by unclean conditions.

Cinnyricinclus leucogaster [PLATE 147]
Amethyst Starling, Violet Starling, Violet-backed Starling, White-bellied Amethyst Starling, Plum-coloured Starling

PLATE 147 *Cinnyricinclus leucogaster*, Amethyst Starling

7 in (18 cm). The head, the whole of the upper surfaces, the chin, throat and breast are an iridescent violet-blue; the lower breast to the under tail coverts are white. The plumage reflects plum, bronze, green etc according to the light. The bill, legs and feet are black. The female's upper surfaces are tawny-brown with darker streaking and mottling; her underparts are buff with some streaks on the throat and breast.

Range Senegal, Gambia, Gabon, Congo, Central African Republic, Sudan, Ethiopia, Uganda, Kenya, western Tanzania.

Status There are good populations in many areas but distribution is patchy.

Avicultural rating Occasionally available, and a much-prized avicultural subject, it is easily managed but is not as hardy as some of the *Lamprotornis* species.

Feeding Diets E and G, with occasional finely minced raw meat and a variety of live food such as small locusts, crickets, mealworms, fly pupae, wax moth larvae etc; they are more insectivorous than glossy starlings. They also enjoy berries in season.

Breeding Successes are infrequent. They will use a box or hollow log with a nest of plant fibres, leaves etc. Clutch 2–3; incubation 14–15 days. Provide abundant live food for rearing.

General management This species is difficult to establish; immature birds will often remain in poor plumage without a moult for a year or more. This static situation changes rapidly with little indication of what is the trigger; birds are often transformed in a matter of a few weeks after making no progress for a long period. They are generally peaceful with most other species and good occupants of a tropical house, or an outdoor shelter and flight, but do best in frost-free winter quarters.

Spreo superbus [PLATE 148]
Superb Starling, Spreo Starling, Superb Spreo Starling

---

8 in (20 cm). The head, sides of the face, chin and throat are black; the nape, mantle, sides of the neck and breast are an iridescent blue. The mantle, wings and back are metallic green, with several sooty-black spots on the wings. A narrow white band separates the iridescent blue of the breast from the chestnut of the underparts. The under tail coverts are white, the tail black with green and blue reflections. The bill, legs and feet are black. The sexes are alike.

Range Ethiopia and Somalia south to Uganda, Kenya, southern Tanzania.

PLATE 148 *Spreo superbus*, Superb Starling

Status Widespread and abundant in many parts of the range.

Avicultural rating This popular and widely-kept species is easily managed and hardy.

Feeding Diets E and G, with some minced raw meat and a variety of live food. It also enjoys berries when available.

Breeding Numerous breeding successes are recorded. They will use a box similar to that used by grass parrakeets; a substantial nest of twigs, plant stems etc is built. Clutch 2–4; incubation 13–15 days. Provide abundant live food for rearing including small locusts, crickets, fly pupae, mealworms etc. Insects appear to form the bulk of the rearing food until the young are fledged.

General management These intelligent, confident birds thrive in an outdoor shelter and flight after acclimatisation. They are generally good mixers, except when breeding, although they are best not associated with smaller species. They usually develop a close relationship with human owners and occasionally become mimics.

Spreo pulcher
Chestnut-bellied Starling

---

8 in (20 cm). The whole of the head is dull sooty-brown; the mantle, back, wings, tail, throat and

breast are a dull metallic green, with some blue reflections on the wings. The lower underparts are chestnut. The bill, legs and feet are black. The sexes are alike.

Range Mauritania, Senegal, Guinea, eastwards to northern Nigeria/Cameroon, Chad, central and eastern Sudan, northern Ethiopia and Somaliland.

Status Abundant in many parts of the range.

Avicultural rating Occasionally available, but not as popular as their more brightly-coloured close relatives.

Feeding/Breeding/General management Similar to Superb Starling. They are generally hardy and easily managed after acclimatisation, and have been bred in aviaries.

Spreo hildebrandti
Hildebrandt's Starling

---

8 in (20 cm). The whole of the head, nape, chin, throat and breast are an iridescent violet-blue; the rest of the upper parts including the wings and tail are similar, but with greenish reflections. The lower breast and underparts are chestnut. The bill, legs and feet are black. The sexes are alike.

Range Southern half of Kenya, northern Tanzania.

Status There is scattered distribution with good numbers in some areas.

Avicultural rating Only occasionally available, this is a good avicultural subject.

Feeding/Breeding/General management Similar to Superb Starling. Hand-reared young examples occasionally arrive in shipments from East Africa; they are delightful, confident birds often with an extraordinary vocabulary of whistling calls. Like others of the family they are very susceptible to infection by a small parasite (*Syngamus*) which causes a condition known as 'gapes'. This is usually caused by housing starlings in already contaminated quarters, although the problem may not have flared up with other species on the same ground; modern treatments usually effect a cure, and the affected area should have all soil, to a depth of 12 in (30 cm), removed and replaced. Flame-gunning may not eradicate parasite eggs.

Cosmopsarus regius
Royal Starling, Golden-breasted Starling

---

14 in (35.5 cm). The head and nape are metallic green, the cheeks, mantle, back and wings violet-blue. The chin, throat and breast are an iridescent violet; the rest of the underparts are yellow. The tail is bronze-green with blue reflections. The bill, legs and feet are black. The sexes are alike.

Range Southern Ethiopia and Somaliland, eastern Kenya and Tanzania.

Status Distribution is local, and they are reasonably abundant in a few areas.

Avicultural rating This is one of the most highly-prized of all exotic birds. It is expensive and rarely available.

Feeding Diets E and G, with finely minced raw meat and varied live food. It enjoys various berries, including elder, blackberry and cultivated types.

Breeding It is a hole-nester; the nest usually has a rudimentary lining of plant fibres, roots etc. Clutch 2–6; incubation 14–15 days. Provide adequate live food, bread and milk, soft fruits etc for rearing.

General management Immature Royal Starlings are difficult to distinguish from immature Ashy Starlings (*C. unicolor*), which are sometimes inadvertently offered for sale as the former. They are good tropical house birds; an outdoor shelter and flight is best during the summer months, then frost-free winter quarters. Pay particular attention to the ground area (especially in a previously-occupied tropical house or conservatory) to prevent infection by parasites which affect the trachea.

Sturnus malabaricus
Malabar Starling, Ashy-headed Starling, Chestnut-tailed Starling, Grey-headed Mynah

8 in (20 cm). The forehead is white, the head, nape, chin and throat grey, with some feathers elongated and with paler shafts. The rest of the upper surfaces are grey-buff, the wings and tail darker, the latter tipped with chestnut; the breast and underparts are rufous. The bill is blue at the base, with a yellow tip; the legs and feet are flesh-brown. The sexes are alike.

Range India, Assam, Burma, Thailand, Indo-china, south-west China.

Status Abundant in many parts of the range.

Avicultural rating Occasionally available, they are easily managed and hardy after acclimatisation.

Feeding Diets E and G, with a variety of live food; occasionally provide small quantities of minced raw meat. It also enjoys various wild and cultivated berries. Fruit and live food are favourite diet items.

Breeding Established true pairs are good breeders. They nest in tree holes and a box is often accepted in an aviary. Clutch 3–5; incubation 13–15 days. Provide live food etc for rearing.

General management They are generally hardy and can over-winter in an outdoor shelter and flight. Non-breeding stock can be mixed with birds of comparable size in spacious quarters, but, like their relatives, they need watching in company and are best not associated with smaller or weaker companions.

Sturnus pagodarum
Pagoda Starling, Pagoda Mynah, Brahminy Mynah

8 in (20 cm). The top of the head (including the crest) is black; the face, sides of the head, neck and underparts are warm buff, paler towards the abdomen and with some elongation of the neck, throat and upper breast feathers to form a hackle. The rest of the upper surfaces are grey, the flight feathers black; the tail is dark brown and the central feathers are tipped white. There is a small white patch at the back of the eye. The bill is blue at the base, with an orange tip; the legs and feet are flesh-brown. The sexes are alike.

Range Afghanistan, Pakistan, India, Sri Lanka.

Status Abundant in many parts of the range except the north-east.

Avicultural rating See Malabar Starling.

Feeding/Breeding/General management Similar to Malabar Mynah. Established true pairs are often prolific. They are intelligent, amusing birds; but should be housed only with similar-sized species.

Leucopsar rothschildi [PLATE 149] (I)
Rothschild's Starling, Rothschild's Mynah, Rothschild's Grackle, Bali Starling

10 in (25 cm). The plumage is almost entirely white except for black tips to the flight feathers and a black terminal band on the tail. An area of bare skin from the base of the bill through the eye is blue; there is a long backward sweeping crest. The bill is blue-grey, yellow towards the tip; the legs and feet are blue-grey. The sexes are similar but the female is usually smaller and with a less luxuriant crest.

Range Island of Bali.

Status Critically endangered in the wild.

PLATE 149 *Leucopsar rothschildi*, Rothschild's Starling

Avicultural rating There are good captive populations through breeding programmes, but they are still rare and expensive.

Feeding Diets E and G, with small amounts of minced raw meat and a variety of live food: locusts, crickets etc.

Breeding They are often prolific, but pairs vary in their willingness to breed. Provide separate accommodation; they will use nest-boxes, and build a nest of twigs, grass etc. Clutch 3–4; incubation 13–15 days.

General management This is usually straightforward; they are hardy but need frost-free winter quarters, although a temperate rather than tropical bird house environment seems to suit them better. They can be very aggressive, so choose companions with care.

Acridotheres tristis
Common Mynah

$9\frac{1}{2}$ in (24 cm). The head, nape, chin, throat and upper breast are sooty black; the rest of the upper surfaces are dark brown, with more black on the wings and tail and a white wing-patch and tip of the tail. The underparts are light brown with grey-white on the abdomen and under tail coverts. There is an area of bare skin on the ear coverts. The bill is orange, the legs and feet yellow. The sexes are alike.

Range Afghanistan, Pakistan, India, Sri Lanka, northern and central Burma eastwards to southwest China. Introduced to Australia, New Zealand, South Africa, Andamans, Nicobars, Hawaii.

Status Abundant in most parts of the range; numbers are increasing in several areas where they have been introduced.

Avicultural rating Usually inexpensive and frequently available, they are hardy when acclimatised. True pairs are willing breeders.

Feeding Diets E and G, with some minced raw meat and regular items of live food. They enjoy various wild and cultivated berries when available.

Breeding They are good breeders, but can be savage towards companions when nesting, and even birds of comparable or larger size are unsafe sharing the same aviary. Clutch 3–5; incubation 14–15 days. Provide abundant live food, eggfood etc for rearing. Young birds are best removed from the breeding quarters as soon as they are seen to be self-supporting.

General management They are good occupants of an outdoor shelter and flight when acclimatised; and usually prove completely weather-resistant. Single birds can be housed with comparable-sized companions, but pairs need separate quarters, especially as breeding condition is assumed. They have loud, cheerful calls, which are not at all discordant.

Acridotheres fuscus
Jungle Mynah

9 in (23 cm.) The head is black; the rest of plumage is a dark grey-brown, blackish on the

wings and tail. The throat and breast are dark grey becoming lighter towards the abdomen. There is a short tuft of erect feathers at the base of the bill. There is a white wing patch and the tip of the tail is white. The bill, legs and feet are yellow. The sexes are alike.

Range India, Burma, Thailand, northern and central Malaysia, Java, southern Celebes.

Note Six subspecies are described.

Status Abundant in many parts of the range, including cultivated areas.

Avicultural rating See Common Mynah.

Feeding/Breeding/General management Similar to Common Mynah. It has been bred on several occasions.

Gracula religiosa
Southern Grackle, Hill Mynah

9–15 in (23–38 cm). The plumage is mainly glossy black with purple reflections; there is a white patch on the primaries. There is an area, varying in size and shape, of bare skin on the sides of the head and nape which is yellow-orange. The bill is orange-yellow, the legs and feet yellow. The sexes are alike.

Range India, Sri Lanka, Burma, Thailand, Indochina, Malaysia, Sumatra, Borneo, Java, Bali, Andamans, Nicobars, Sundas, Palawan; southern China, Hainan.

Note Ten subspecies are described. In aviculture they are usually described under the following headings. Lesser Hill Mynah (*G. r. indica*), southwest India, Sri Lanka. Nepal/Assam Hill Mynah (*G. r. intermedia*), northern India, Burma, Thailand, Indochina. Greater/Javan Hill Mynah (*G. r. religiosa*), Malaysia, Sumatra, Java, Bali, Borneo, Bangka Island.

Status There are good populations in some parts of the range, but over-collecting for the pet trade often has an adverse effect if concentrated in specific areas.

Avicultural rating This species is very popular, but mainly as pet birds which, if obtained young, may make good mimics; few are maintained as aviary subjects.

Feeding Diets E and G, with some live food; berries are enjoyed when available. There are also various brands of 'complete' mynah-foods on the market, ranging from pellets to powder; some are useful diets but the birds should also have fruit and live food.

Breeding Several breeding successes have been recorded with various subspecies. A simple nest is usually made in a hole or cavity; they will use a box in aviaries. Clutch 2–3; incubation 13–16 days. Provide live food etc for rearing. Adults can prove very aggressive when nesting, and should not be housed with other birds.

General management They are fairly hardy after acclimatisation, but do best with frost-free (even slightly warm, 50°F, 10°C) winter quarters. Intelligent and aggressive, many become accomplished imitators of familiar sounds including the human voice; but it is essential that hand-reared young birds are obtained for training, as unschooled adults will remain intractable and do not make talkers. Beware of 'bargains' in pet shops.

Oriolidae (Old World Orioles)

The Oriolidae comprises more than 20 species; their distribution includes Europe, south-east Asia, Africa, Australia and the Pacific Region. They are mainly black and yellow or black and red, although some species in Australia and the Moluccas are drably-coloured. Excellent avicultural subjects, they are generally easy to manage after acclimatisation. Despite the similarity of the name, they are not related to New World Orioles. They are strong flyers and best kept in aviaries or a tropical house; fruit and live food form important diet items, but insectile mixtures are also consumed.

Oriolus chinensis [PLATE 150]
Black-naped Oriole

PLATE 150 *Oriolus chinensis*, Black-naped Oriole

10½ in (27 cm). The forehead and crown are orange-yellow; the area from the base of the bill, through the eye, to the nape is black. The upper and lower surfaces are yellow, with a slight green suffusion on the lower back and wings; the flight and tail feathers are black, with some black at the shoulder of the wing, the latter tipped with yellow. The bill is pink-brown, the legs and feet blue-green. The female is similar but her mantle and back are more olive-green.

Range India, Burma, Thailand, Indochina, Malaysia, south-west China, Hainan, Taiwan, Borneo, Sundas, Celebes, Philippines.

Note There is considerable variation of plumage patterns among the 23 described subspecies.

Status Abundant in many parts of the range.

Avicultural rating Only occasionally available, this is an excellent avicultural subject.

Feeding Diets E and G, plus berries as available and live food. Additional protein can be provided by the use of trout pellets (see p. 27), but for many birds fruit is the main item of diet.

Breeding A cup-shaped nest of plant fibres with a softer lining is built and is often suspended cradle-fashion between a fork at the extremities of a high branch. Clutch 4–5; incubation 14–15 days. Provide abundant live food in addition to normal diet items for rearing.

General management They are good inhabitants of a tropical house or a large conservatory; they can also occupy an outdoor shelter and flight during the summer months but need moderate warmth (50°F, 10°C) in winter, even when acclimatised. They usually mix well with birds of comparable size; segregate if breeding is likely. They have pleasant calls and song.

Paradisaeidae (Birds of Paradise)

The 43 species of birds of paradise are confined to New Guinea and some nearby small islands, northern Australia and the Moluccas. Between them the various members of this family represent the most colourful and spectacularly-plumaged group of birds in the world. They range in size from starling to crow size. The various kinds have feather adornments including elongated plumes, wire-like quills, capes and frontal shields of shimmering, iridescent feathers. The cock's display is a highly specialised ritual which varies from species to species. The exotic feathers were in great demand in Europe (to provide decorations for women's hats) from the middle of the nineteenth century; more than 50,000 skins a year were shipped out of New Guinea at the height of this craze. Although no longer obtainable by aviculturists they have been fairly widely kept in the past and several species proved reasonably easy to manage, as well as proving hardy and long-lived. Despite welcome conservation measures in New Guinea, the fact that all but the most remote areas are liable to exploitation of one kind or another may yet place a question mark against the long-term future of some species. Breeding successes in aviculture have been few and far between, but in some American zoos more consistent results are being achieved which augur well for any future captive-breeding programmes that might be initiated.

Paradisaea raggiana (II)
Count Raggi's Bird of Paradise, Empress of Germany's Bird of Paradise, Raggiana Bird of Paradise

20 in (51 cm). The forehead, lores and chin are velvety-black with iridescent green reflections. The rest of the head, including a collar, and the nape of the neck, is straw-yellow; the upper wing coverts are also yellow. The rest of the upper surfaces are deep vinous-brown, brighter on the wings and tail. The breast is blackish-maroon with the lower underparts showing less black. Plumes of bright red feathers grow from the sides of the breast and flanks. The bill is blue-grey; the legs and feet are brown. The female is essentially brown, darker on the face, throat, back and wings, more earthy on the nape and neck and with a distinctly rufous hue on her flanks and lower underparts.

Range Eastern and southern areas of New Guinea.

Status Still abundant in many areas; this is probably the best known member of the family in New Guinea.

Avicultural rating A good avicultural subject, but presently impossible to obtain from the wild.

Feeding Diets E and G, with plenty of live food including locusts and crickets. Small dead mice should also be provided three or four times a week. A small quantity of trout pellets will provide valuable extra protein; a dessertspoonful of ground pellets can be mixed with a cupful of Diet J (or moistened whole pellets may be mixed with a fruit and cooked rice supplement). Many berries are enjoyed and should form a regular part of the diet. Fresh blackcurrants, redcurrants, raspberries etc can be offered in season; at other times frozen berries are adequate. Wild berries including blackberries, elders and bilberries are also enjoyed.

Breeding A single pair of these birds may prove unwilling breeders even in spacious accommodation such as might be provided in a tropical house; it seems likely that additional males, housed separately but within sight and sound of the potential breeders, will provide an effective stimulus through display. The breeding cock is best removed from the hen's quarters should she start to lay, for the very ardour of his nuptial performance will prove disruptive and may lead to both the nest and its contents being destroyed. Clutch 1–2; incubation 17–19 days. A cup-shaped nest of plant fibres is made. The chicks are fed mainly on small live food, to begin with; crickets, small locusts, wax moth larvae and mealworms should be offered. Pinky mice are also a valuable rearing food. The chicks fledge at about three weeks but are fed by the hen for several weeks more. The young cocks are fairly easy to distinguish even when young, but it may take six years for them to assume full adult plumage.

General management This species will do well in temperate house environment and can also occupy outdoor accommodation in suitable areas during summer. Provide spacious aviaries with a simple and effective means of partitioning, so that the male can be excluded should nesting occur. Ideally, additional cock birds should be housed in separate adjoining or nearby flights to promote essential display behaviour. Double wiring may be necessary to prevent rival males injuring themselves through combative behaviour. Many species have been lost over the years through exposure in tropical houses to high temperatures or humidity; conditions which, in the medium to long-term, are quite unsuitable.

Corvidae (Crows, Jays)

This is a family of more than 100 species with worldwide distribution, except the polar regions,

New Zealand and some Polynesian islands. Their main characteristics are boldness combined with a high degree of intelligence. Many are gregarious and form large flocks outside the breeding season. They are mainly omnivorous; several are nest predators taking eggs and nestlings. Although black or black and white are the main colours for larger members of the family and some magpies, many smaller species, and particularly jays, are brightly-coloured. There are several species in aviculture; they are generally adaptable, easy to feed and, in varying degrees, hardy. Many breed well, but infanticide can be a problem if an unsuitable rearing diet is provided; abundant live food etc is vital if chicks are to fledge successfully.

Cyanocorax yncas [PLATE 151]
Green Jay

---

11 in (28 cm). The forehead, cheek patches and a small area above the eye are cobalt-blue, the crown and nape blue, white and yellow. The upper surfaces are a bright grass-green; the tail is green, blue and yellow. The sides of the head, chin, throat and upper breast are black; the rest of the underparts greenish-yellow and green. The bill, legs and feet are black. The sexes are alike.

Range Southern Texas, Mexico, Central America, north-west Colombia, northern Venezuela, eastern Ecuador, northern and central Peru, north-west Bolivia.

Note Considerable plumage differences occur among the 13 described subspecies.

Status There are good populations in some parts of the range.

Avicultural rating This species is well known in aviculture and easily managed, but is not completely weather-resistant.

PLATE 151 *Cyanocorax yncas*, Green Jay

Feeding Diets E and G, with some minced raw meat and a variety of live food including locusts, crickets, mealworms, beetles etc. It is slightly more insectivorous than its relatives. Nuts and acorns are enjoyed.

Breeding It builds a flimsy platform of twigs with a grass lining; provide open boxes. Clutch 3–5; incubation 17–18 days. Provide abundant live food *prior* to the expected hatching date: locusts, crickets, mealworms, beetles, pinky mice etc. Nestlings are likely to be ejected, or eaten by their parents if the supply of suitable rearing food proves inadequate.

General management They are fairly hardy after acclimatisation, but need some winter warmth (50°F, 10°C) and are not happy in cold, wet conditions. They are good occupants of a temperate bird house, but unsafe companions for smaller birds; they can be housed in spacious compartments but do not thrive in a restricted space. The green plumage is likely to fade if they are not kept in accommodation with adequate shade, so heavily-planted aviaries are best for these birds. They can be fairly long-lived (10–12 years) if properly housed and fed. They are active and noisy.

Cyanocorax chrysops [PLATE 152]
Plush-crested Jay

13 in (33 cm). The forehead, crown (including the plush-like crest), chin, throat and breast are black; small areas above and below the eye are blue. The nape is pale blue, with darker cobalt-blue on the mantle, back, wings and tail, darkest on the wings and tail; the underparts and tip of the tail are buff-white. The bill is black, the legs and feet grey. The sexes are alike.

Range North-east Bolivia, Paraguay, northern and north-eastern Argentina, western and south-western areas of Brazil.

Status There are good populations in some parts of the range.

PLATE 152 *Cyanocorax chrysops*, Plush-crested Jay

Avicultural rating Occasionally available, they are easily managed and often good breeders.

Feeding/Breeding/General management Similar to Green Jay. Established pairs, properly housed and fed, are often willing breeders. They are intelligent and amusing birds, but not weather-hardy in most areas. Infanticide may be a problem.

Urocissa erythrorhyncha
Red-billed Blue Magpie

26 in (66 cm). The head, chin, throat and breast are black, the nape blue and blue-white. The

upper surfaces are grey-blue, browner on the mantle and with white tips to most of the secondary and primary feathers; the underparts are a pale lilac-white to grey-white. The long, graduated tail (16–18 in, 41–46 cm) is mauve-blue tipped with black and white. The bill, legs and feet are red and orange. The sexes are alike.

Range Western Himalayas, Assam, north-east Burma, Indochina, northern, central and south-western areas of China.

Note Some plumage variations occur among the five described subspecies.

Status There are good populations in suitable areas of habitat.

Avicultural rating This is a popular avicultural subject but it needs space; it is hardy after acclimatisation, easy to manage and a good breeder.

Feeding Diet E and G, with minced raw meat, small dead mice and sectioned day-old chicks, live food including locusts, crickets etc. Dusting sectioned chicks or small rodents with a multivitamin preparation for these and related species is recommended.

Breeding Potential breeding pairs should be housed in spacious aviaries with some natural or artificial cover. They build a fairly shallow nest of twigs etc and will use an open box or even a shelf. Clutch 3–5; incubation 17–19 days. Provide abundant live food and step up the normal supply just prior to the date on which the eggs are due to hatch, which may help to forestall any murderous tendencies on the part of the parents; locusts, pinky mice etc are all important in the early stages together with collected natural live food.

General management They are fairly easy to establish, although they are often in poor plumage after import and quarantine; do not house them outside, even during the summer months, until their condition improves. They are generally hardy and can eventually be wintered in a suitable outdoor shelter and flight without heat; they are at their best in spacious, lofty aviaries. They are not suitable companions for many species other than large pigeons, pheasants, waterfowl; but eggs of all three will be vulnerable with Blue Pies present. Active and fairly noisy, these are intelligent and extremely rewarding birds to keep.

Cissa chinensis [PLATE 153]
Green Magpie, Hunting Cissa

---

14 in (35.5 cm). The plumage is mainly bright blue-green in colour with a prominent black streak from the base of the bill, through the eyes, to the nape; the flight feathers are bright chestnut, and there are black and white tips to the inner secondaries. The bill, legs and feet are orange-red. The sexes are alike.

Range Himalayas, Burma, Thailand, Cambodia, South Vietnam, southern China, Malaysia, Sumatra, north-west Borneo.

Status There are good populations, but it is not abundant.

Avicultural rating Occasionally available, it is fairly easy to maintain in good health, but loses its natural green colour unless kept in shady, well-planted accommodation.

Feeding/Breeding/General management Similar to the Red-billed Blue Magpie but more secretive in behaviour, and therefore accommodation requirements. An outdoor shelter and flight suits them, and they are hardy after acclimatisation; but good shrub cover where they can remain out of sight for long periods is ideal. In sparsely-furnished aviaries, their attractive green plumage will quickly become powder-blue; shady aviaries appear to help them retain their natural colour. They have been bred but are probably more difficult than most related species, and nestling mortality is high.

PLATE 153 *Cissa chinensis*, Green Magpie

Dendrocitta vagabunda
Rufous Treepie, Indian Treepie, Wandering Treepie

18 in (46 cm). The whole of the head, nape, chin, throat and breast are sooty-black; the mantle, back, rump and upper tail coverts are rufous-brown. The underparts are rufous-chestnut; there is a large area of grey-white on the wing. The long tail is grey, with a black terminal band. The bill, legs and feet are lead-grey. The sexes are alike.

Range Western Himalayas, India, Burma, Thailand, Indochina.

Status This is a familiar and abundant species in many parts of the range.

Avicultural rating Rarely available at present, but they are good avicultural subjects.

Feeding/Breeding/General management Similar to the Red-billed Blue Magpie. They are hardy after acclimatisation and will over-winter in a suitable shelter and flight; it is best if the accommodation is planted with shrubs etc, with some higher branches as furnishings for these fairly arboreal birds. Breeding successes are only occasional.

Pica pica
Magpie

18 in (46 cm). The whole of the head, nape, mantle, back and breast are glossy black with green and purple reflections; the wings are black with white on the scapulars and flights. The lower breast and underparts are white, the tail black. The bill, legs and feet are black. The sexes are alike.

Range Europe eastwards to Siberia, China, western Canada and the USA.

Status Abundant in many areas; increasing in some parts of the range.

Avicultural rating An occasional avicultural subject which is hardy and easily managed.

Feeding They can be given left-over insectivorous mixtures, but many feed on scalded dog meal or poultry crumbs. Some raw meat should be given from time to time. Live food is enjoyed and they also eat a variety of suitable household scraps.

Breeding True pairs breed fairly readily in suitable accommodation. Provide a box with a half-front; the nest is bulky and made of twigs, stout plant fibres etc. Clutch 4–6; incubation 21 days. Abundant animal matter is essential for rearing food; dead mice, day-old chicks, pieces of rabbit (complete with fur) should supplement the normal diet, supplemented with some raw minced meat. Use a multivitamin supplement to dust the food.

General management They are hardy and undemanding, but need to be in a spacious shelter and flight to be seen at their best; they are not suitable companions for other birds, but individuals are sometimes kept with other single members of the crow family. They are hardy but need a dry roosting area.

Note In the UK only aviary-bred and closed-ringed birds of this species may be offered for sale under the provisions of the Wildlife and Countryside Act, 1981 (Schedule 3, Part 1).

APPENDIX

PRINCIPAL AUTHORITIES RESPONSIBLE FOR CITES LICENSING

ARGENTINA
Direccion Nacional de Fauna Silvestre,
Paseo Colón 922-2 Piso Of. 201,
1063 *BUENOS AIRES*
Telephone 3620145; 3620274; 3620266

AUSTRALIA
Australian National Parks and Wildlife Service,
PO Box 636,
CANBERRA A.C.T. 2601
Telephone (062) 466211

BELGIUM
Ministère de l'agriculture,
Service de l'inspection vétérinaire,
Manhattan Center,
21, Avenue du Boulevard, 6° étage,
B-1000 *BRUXELLES*
Telephone 02/211.72.11

BOLIVIA
Ministerio de Asuntos Campesinos y Agropecuarios,
Centro de Desarrollo Forestal,
Jefatura Nacional de Vida Silvestre,
Parques Nacionales, Caza y Pesca,
Av. Camacho 1471-6° Piso,
Casilla de Correo No 1862,
LA PAZ
Telephone 371268; 367301

CANADA
The Administrator,
Convention on International Trade in Endangered Species,
Canadian Wildlife Service,
Department of the Environment,
OTTAWA Ontario KIA OE7
Telephone (819) 9971840

CHINA
The People's Republic of China,
Endangered Species of Wild Fauna and Flora,
Import and Export Administrative Office,
Ministry of Forestry,
Hepingli,
BEIJING
Telephone 464180

DENMARK
Fredningsstyrelsen,
Miljoministeriet,
Amaliegade 13,
DK-1256 *KOBENHAVN K*
Telephone (451) 119565

ECUADOR
Director Ejecutivo del Programa Nacional Forestal,
Ministerio de Agricultura y Ganaderia,
Casilla 2919,
QUITO

FRANCE
Direction de la protection de la nature,
Convention de Washington,
Ministère de l'Environment,
14, Bd du Général Leclerc,
F-92524 *NEUILLY-SUR-SEINE* Cedex
Telephone (1) 475 81212

WEST GERMANY
Bundesministerium fur Ernährung,
Landwirtschaft und Forsten,
Referat 623,
Postfach 140270,
D-5300 *BONN 1*
Telephone (0228) 529-3777, 3352, 3572

GHANA
Department of Game and Wildlife,
PO Box M 239,
Ministries Post Office,
ACCRA
Telephone 64654, 66476

GUYANA
The Permanent Secretary,
Ministry of Agriculture,
PO Box 1001,
GEORGETOWN
Telephone 67800, 53851-9

Appendix

INDIA
The Director of Wildlife Preservation,
Government of India,
Department of Environment,
Room 240,
Krishi Bhavan,
NEW DELHI 110001
Telephone 388071, 384556

INDONESIA
Directorate General of Forest Protection and Nature Conservation,
Departemen Kehutanan,
Direktorat Jenderal Perlindungan Hutan dan Pelestarian Alam,
Jalan Ir. H. Juanda No. 9,
BOGOR
Telephone (0251) 24013-24015-23067

ITALY
Ministero dell'Agricoltura e delle Foreste,
Direzione generale per l'Economia Montana e per le foreste,
Divisione II,
Via G. Carducci, 5,
I-00187 *ROMA*
Telephone 0039-6 463984; 4665

LIBERIA
Forestry Development Authority,
PO Box 3010,
MONROVIA
Telephone 262250/251/252

MALAYSIA
Director General,
Department of Wildlife and National Parks,
Kompleks Pejabat-pejabat Kerajaan,
Blok K-19, Jalan Duta,
KUALA LUMPUR 11-04
Telephone Kuala Lumpur 941466

NETHERLANDS
Hoofd van de Directie Natuur-en Landschaps-bescherming,
Ministerie van Landbouw en Visserij,
Prins Clauslaan 6,
Postbus 20401,
2500 EK *'S-GRAVENHAGE*
Telephone 070-793911

PERU
Dirección General Forestal y de Fauna,
Ministerio de Agricultura,
Jirón Natalio Sanchez 220, 3er piso,
Jesús María,
LIMA
Telephone 233978

SENEGAL
Direction des eaux et Forêts et Chasses,
Parc forestier de Hann,
B.P. 1831,
DAKAR
Telephone 217614, 210628, 217105

SOUTH AFRICA
Department of Environment Affairs,
Environmental Conservation Branch,
Private Bag X 447,
PRETORIA 0001
Telephone 299-2567

SURINAME
Head of Suriname Forest Service,
c/o Head Nature Conservation Division,
PO Box 436,
10, Cornelis Jongbawstraat,
PARAMARIBO
Telephone 75845, 71856

SWITZERLAND
Office vétérinaire fédéral,
Schwarzenburgstrasse 161,
CH-3097 *LIEBFELD*
Telephone (031) 598508-09

TANZANIA
The Director of Wildlife,
Wildlife Division,
Ministry of Lands, Natural Resources and Tourism,
PO Box 1994,
DAR ES SALAAM
Telephone Dar es Salaam 27811-14

THAILAND
Wildlife Conservation Division,
Royal Forest Department,
Paholyothin Road, Bangkhen,
BANGKOK 10900
Telephone 5791565, 5792776, 4794847

UNITED KINGDOM
Department of the Environment,
Tollgate House,
Houlton Street,
BRISTOL BS2 9DJ
Telephone 0272-218811

UNITED STATES OF AMERICA
Chief of the Federal Wildlife Permit Office,
Room 611, Broyhill Building,
1000 North Glebe Road,
ARLINGTON, VA 22201
Telephone (703) 2352418

BIBLIOGRAPHY

Ali, S. (1979), *Field Guide to the Birds of the Eastern Himalayas*, Oxford University Press.
Ali, S. (1979), *Indian Hill Birds*, Collins.
Ali, S. and Dillon Ripley, S. (1968–74), *Handbook of the Birds of India and Pakistan*, Oxford University Press.
Austin, O.L. and Singer, A. (1961), *Birds of the World*, Hamlyn.
Avon, D. and Tilford, T. (1975), *Birds of Britain and Europe*, Blandford Press.
Avon, D., Tilford, T. and Woolham, F. (1974), *Aviary Birds in Colour*, Blandford Press.
Bannerman, D.A. (1951–53), *Birds of West and Equatorial Africa* (Volumes *1–2*), Oliver and Boyd.
Bond, J. (1979), *Birds of the West Indies*, Collins.
Boosey, E.J. (1962), *Foreign Bird Keeping*, Iliffe.
Campbell, B. (1974), *The Dictionary of Birds in Colour*, Peerage Books.
Cayley, N.W. (1932), *What Bird is That?*, Angus and Robertson, Sydney.
Delacour, J. (1951), *The Pheasants of the World*, Country Life.
Delacour, J. and Scott, P. (1954–64), *The Waterfowl of the World* (Volumes *1–4*), Country Life.
De Schauensee, R.M. (1964), *The Birds of Colombia*, The Academy of Natural Sciences of Philadelphia, USA.
De Schauensee, R.M. (1970), *A Guide to the Birds of South America*, Oliver and Boyd.
Dunning, J.S. (1982), *South American Land Birds*, Harrowood Books, Pennsylvania, USA.
Forshaw, J.M. and Cooper, W.T. (1978), *Parrots of the World*, David and Charles.
Gerrits, H.A. (1961), *Pheasants—Including their Care in the Aviary*, Blandford Press.
Glenister, A.G. (1951), *The Birds of the Malay Peninsula*, Singapore and Penang.
Gooders, J. (1969), *Birds of the World* (Volumes *1–9*), IPC.
Gruson, E.S. (1978), *A Checklist of the Birds of the World*, Collins.
Harrison, C. (1978), *A Field Guide to the Nests, Eggs and Nestlings of North American Birds*, Collins.
Haverschmidt, F. (1968), *The Birds of Surinam*, Oliver and Boyd.
Hayward, J. (1979), *Lovebirds and their Colour Mutations*, Blandford Press.
Henry, G.M. (1971), *A Guide to the Birds of Ceylon*, Oxford University Press.
Howard, R. and Moore, A, (1984), *A Complete Checklist of the Birds of the World*, Macmillan.
King, B., Woodcock, M. and Dickinson, E.C. (1980), *A Field Guide to the Birds of South-east Asia*, Collins.

Lint, K.C. and Lint, A.M. (1981), *Diets for Birds in Captivity*, Blandford Press.
Low, Rosemary (1977), *Lories and Lorikeets*, Van Nostrand Reinhold Company.
Low, Rosemary (1980), *Parrots: Their Care and Breeding*, Blandford Press.
Mackworth-Praed, C.C. and Grant, C.H.B. (1980), *African Handbook of Birds* (Volumes *1* and *2*—Birds of Eastern and North-eastern Africa), Longman.
Martin, R.M. (1980), *Cage and Aviary Birds*, Collins.
Martin, R.M. (1983), *The Dictionary of Aviculture*, B.T. Batsford Ltd.
Meaden, F. (1979), *A Manual of European Bird Keeping*, Blandford Press.
Pearce, D.W. (1983), *Aviary Design and Construction*, Blandford Press.
Peterson, R.T. (1963), *The Birds*, Time-Life.
Pizzey, G. (1980), *A Field Guide to the Birds of Australia*, Collins.
Robbins, C.S., Bruun, B. and Zim, H.S. (1966), *A Guide to Field Identification: Birds of North America*, Golden Press, New York.
Robbins, G.E.S. (1981), *Quail: Their Breeding and Management*, World Pheasant Association.
Roots, C. (1970), *Softbilled Birds*, John Gifford Ltd.
Rutgers, A. (1977), *The Handbook of Foreign Birds* (Volumes *1* and *2*), Blandford Press.
Rutgers, A. and Norris, K.A. (1970–77), *Encyclopedia of Aviculture* (Volumes *1–3*), Blandford Press.
Serle, W., Morel, G.J. and Hartwig, W. (1977), *A Field Guide to the Birds of West Africa*, Collins.
Soothill, E. and Whitehead, P. (1978), *Wildfowl of the World*, Blandford Press.
Trollope, Jeffrey (1983), *The Care and Breeding of Seed-eating Birds*, Blandford Press.
Thompson, A.L. (1964), *A New Dictionary of Birds*, Neilson.
Whistler, H. (1935), *Popular Handbook of Indian Birds*, Gurney and Jackson.
Williams, J.G. and Arlott, N. (1980), *A Field Guide to the Birds of East Africa*.

Journals

Avicultural Magazine (1894–1981), Avicultural Society.
Cage and Aviary Birds, Business Press International.
Zoonooz, San Diego Zoological Society.

ACKNOWLEDGEMENTS

I make no apologies for including here the names of one or two people who are now no longer with us. For they made significant contributions to this book, either through conversation or correspondence—although neither they nor I were aware of it at the time. Among them are John Yealland, former Curator of Birds at London Zoo, and two eminent collectors, Cecil Webb and Wilfred Frost.

An aviculturist who willingly passed on to me as much of his considerable knowledge as I could assimilate—and, incidentally, had as much to do with shaping my attitudes to aviculture as anyone—was the much-missed Canon J.R. Lowe, truly a man who deserves a book to himself instead of a mere acknowledgement.

Jack Lowe was, by any standards, an impressive figure. Standing some $6\frac{1}{2}$ feet tall, with distinguished—even slightly autocratic—features and a fine head of silver-white hair, he was undoubtedly capable of putting the fear of his Heavenly 'Boss' into any congregation that incurred his wrath. Which is not to suggest his sermons were of the lengthy, hellfire and brimstone variety beloved by some churchmen. At least not often.

For the good Canon's bird-keeping activities were nothing if not all-embracing. So alongside aviaries occupied by European and exotic species, the odd Hummingbird buzzing around the bathroom, indoor flights housing Norwich canaries, mules and hybrids, there was a flock of prize Maran poultry ... and Jack Lowe's racing pigeons.

It is whispered still in some quarters that many a sermon was cut short if its deliverance happened to coincide with the impending return from some cross-country flight of the Canon's pigeons.

More direct help with *The Handbook of Aviculture* has been forthcoming from many people during the two-year period occupied by research and writing. I am especially grateful to Gerald Durrell OBE, Honorary Director of the Jersey Wildlife and Preservation Trust for agreeing to contribute the foreword and to David H.W. Morgan for the information regarding CITES.

My thanks are also due to the following individuals and organisations who provided information and assistance: Mr C.D. Jenkins, Managing Director, Consolidated Chemicals, Wrexham; Mr M.J. Ricketts, Special Diets Services Ltd., Witham, Essex; Mr G. Coombs, Director, E.W. Coombs Ltd., Strood, Kent; Clare and Ian Hinze, Birdquest International, Whitefield, Manchester; Sharron and Charles O'Gorman, Golden Valley Aviaries, Pontrilas, Herefordshire; Donald Bruning, Curator, Department of Ornithology, New York Zoological Society; Dr Karl-L. Schuchmann, Department of Ornithology, Zoological Research Institute and Alexander Koenig Museum, Bonn; Herr W.W. Brehm, Director, Vogelpark, Walsrode; Peter Olney, Curator of Birds, Zoological Society of London; Peter Lowe, Dartmoor Wildlife Park; Peggy, Jack and David Brown, Padstow Bird and Butterfly Gardens, Padstow, Cornwall; Bill Timmis, Curator, Harewood Bird Gardens, Leeds; Dr Janet Kear, Curator, The Wildfowl Trust, Martin Mere, Lancashire; Dave Coles, The Avicultural Society, Ascot, Berkshire; Herr Eberhard Mussler, Biotropic-Verlag GmbH, Baden-Baden; Andrew Storrar and Peter Wrather, Storrar and Wrather, Veterinary Surgeons, Chester; Herr Karl Claus, Claus GmbH, Limburgerhof; Malcolm Ellis, Wadebridge, Cornwall.

Special thanks are due to Alison Copland, the Blandford editor who had the misfortune to be assigned to the project and who has, throughout, displayed considerable qualities of tact, diplomacy and patience as she has pulled together the threads of the book—despite my best efforts at disruption!

I must add a word about three people who have not the slightest connection with aviculture but without whose help the *Handbook* might never have seen the light of day ... or at best, been much delayed.

Between them, Dr John Peaston, Consultant Physician at the Countess of Chester Hospital, Chester, and Dr Virginia Clough of the same establishment have not only donated time to ensure I stay in one piece but also sought (successfully, I regret to say) to curb some of my brighter ideas which they clearly regard as the products of an eccentric mind.

I fell into their hands following an illness in 1984. Thanks largely to the two of them—and members of their staffs—I made a speedy recovery. But a short period of partial immobility persuaded Virginia Clough it was safe to browbeat me into making all kinds of unwelcome changes in my way of life—such as taking more exercise and giving up smoking.

'Ah, that Dr Clough,' said a somewhat lugubrious porter, propelling me at indecent speed in a wheelchair while I was hospitalised, 'she's a bleedin' expert, y'know.' It took me a second or two to realise what a splendidly succinct and accurate description this was of the hospital's Consultant Haematologist.

To Cloughie, now saddled with my presence at her outpatient's clinic, I can only add—'serves you right—and my grateful thanks to you.'

Dr Ian Russell, the Woolham family's long-suffering GP, is another who views many of my less orthodox schemes with some scepticism—but is too much of a gentleman to admit it.

Finally, my thanks are due to my wife Meg and our sons, John, Mark, Andrew and Michael—all of whom have had to take second place to many of my feathered friends for far too long. I owe them a great debt.

Frank Woolham

Picture Credits

The publishers would like to thank the following for permission to reproduce illustrations.

Ardea London Ltd: all colour photographs except those listed below.
Frank Woolham: Plates 37 and 76.
Dennis Avon: Plate 103.
Anita Lawrence: all line drawings.

Index of Scientific Names

Acanthis 296–7
Acridotheres 351
Acryllium 118
Actophilornis 129
Aegintha 317
Aethopygia 261
Agapornis 172–5
Agelaius 291
Aidemosyne 325
Aix 85–6
Alectoris 102–3
Amadina 333
Amandava 314–15
Amazilia 198
Amazona 188
Anas 87–94
Anisognathus 279–80
Anitibyx 131
Anser 78–80
Anthreptes 257
Anthropoides 121–2
Aplonis 345
Ara 179–82
Arachnothera 261
Aratinga 183–4
Athene 194
Aulacorhynchus 217–18
Auripasser 337
Aythya 95

Balearica 123
Bambusicola 107
Barnardius 163
Bombycilla 235
Branta 80–2
Brotogeris 187
Bubo 194
Bubulcus 70
Bucephala 97–8

Cacatua 155–7
Callipepla 101
Callonetta 85
Caloenas 144
Calyptomena 222

Cardinalis 275
Carduelis 295–6
Cariama 128
Carpodacus 297
Centropus 192
Cereopsis 83
Chalcophaps 140
Charadrius 132
Chenonetta 88
Chiroxiphia 225
Chloebia 324
Chloephaga 83–4
Chlorophanes 287
Chloropsis 232–4
Chrysolophus 114
Chunga 128
Ciccaba 195
Cinnyricinclus 347
Cissa 357
Cissopsis 278
Coccothraustes 298–9
Coereba 290
Colibri 196
Colinus 102
Colius 200
Columba 136
Columbina 142–4
Copsychus 238–9
Coracias 208
Coryphospingus 271
Corythaixoides 190
Coscoroba 77
Cosmopsarus 349
Cossypha 238
Coturnix 104–5
Crossoptilon 112–13
Cyanerpes 287–8
Cyanocorax 355–6
Cyanoptila 253
Cyanorhamphus 166–7
Cygnus 75–7

Dacelo 206
Dacnis 286
Dendrocitta 358

Dendrocygna 74–5
Deroptyus 189
Dicaeum 256–7
Diglossa 288

Eclectus 160
Emberiza 265–6
Emblema 317–18
Eolophus 154
Eos 148–9
Eremopterix 227
Erithacus 237
Erythrura 322–3
Estrilda 309–13
Eubucco 212
Eudocimus 71
Eulabeornis 124
Eumyias 255
Euphonia 281
Euplectes 341–4
Eurypyga 127
Eurystomus 208
Euschistospiza 304
Excalfactoria 105

Ficedula 252–3
Forpus 186
Foudia 341
Francolinus 103
Fringilla 293

Gallicolumba 145–6
Gallus 109–10
Garrulax 244–5
Geopelia 141–2
Geotrygon 144
Goura 146
Gracula 352
Gubernatrix 272

Halcyon 205
Heterophasia 247
Hypargos 303
Hypergerus 250

Index of Scientific Names

Irena 234
Ispidina 204

Jacana 130

Lagonosticta 304–5
Lamprotornis 345–7
Larosterna 133
Laterallus 125
Leiothrix 245–6
Lepidopygia 325
Leucopsar 350
Lonchura 326–32
Lophophorus 108
Lophortyx 102
Lophospingus 267
Lophura 110–11
Loriculus 175–7
Lorius 152–3
Lybius 215

Macropygia 137
Malurus 251
Manacus 224
Mandingoa 302
Megalaima 214
Melanerpes 221
Melopsittacus 169
Mergus 98
Microhierax 100
Minla 247
Molothrus 291
Momotus 207
Myiopsitta 185

Nandayus 184
Nectarinia 258–60
Neochmia 318–19
Neophema 167–9
Netta 95
Niltava 254
Nothoprocta 69
Numida 119
Nymphicus 158

Ocyphaps 140
Oena 139
Opopsitta 159
Oriolus 353
Ortygospiza 316–17

Oxyura 99

Padda 332
Panurus 249
Paradisaea 354
Paradoxornis 249
Paroaria 273–4
Passerina 275–7
Pavo 116–17
Perdicula 106
Pericrocotus 228–9
Pharomacrus 202
Phoeniconaias 73
Phoeniconparrus 71
Phoenicopterus 71–3
Phoenicurus 240
Pica 358
Pionus 187
Pitta 226–7
Platycercus 164
Ploceus 338–9
Pluvianus 140
Poephila 319–21
Poicephalus 171–2
Polyplectron 115
Polytelis 162–3
Pomatorhinus 243
Poospiza 267–8
Popelairia 198
Porphyrio 126
Porzana 126
Probosciger 153
Prunella 237
Psephotus 165–6
Pseudeos 149
Psilopogon 213
Psittacula 177–8
Psittaculirostris 159
Psittacus 170–1
Psophia 123
Pterocles 135
Pterocnemia 68
Pteroglossus 218
Pycnonotus 229–31
Pyrenestes 302
Pyrrhula 298
Pyrrhura 185
Pytilia 299–301

Quelea 340

Ramphastos 219–20
Ramphocelus 279
Rhea 67
Rhodospingus 272
Rollulus 106
Rupicola 223–4

Selenidera 219
Semnornis 213
Serinus 292–4
Sicalis 268
Somateria 96
Spermophaga 303
Spiza 274
Sporophila 269
Sporopipes 337–8
Spreo 348–9
Streptopelia 136–7
Sturnus 350
Syrmaticus 113

Tadorna 84
Tangara 281–6
Tauraco 190–1
Tersina 289
Thalurania 198
Thinocorus 133
Tiaris 270
Tockus 210–11
Trachyphonus 216
Tragopan 107
Treron 147
Trichoglossus 150–1
Turdus 242–3
Turnix 120
Turtur 137–8
Tyto 193

Upupa 209
Uraeginthus 306–8
Urocissa 356

Vanellus 131–2
Vidua 334–6

Yuhina 248

Zoothera 241
Zosterops 262–4

INDEX OF COMMON NAMES

Accentors 237
Aracari, Lettered 218
Avadavats 314–15

Babblers 243, 250
Baldpate 89
Bananaquit 290
Barbets 212–16
Bib-finch 325
Birds of Paradise 354
Bishops 341–2
Bittern, Sun 127
Blackbirds 242, 291
Bluebill, Western 303
Bluebirds, Fairy 234
Bobwhite 102
Broadbill, Lesser Green 222
Budgerigar 169
Bufflehead 97
Bulbuls 229–31
Bullfinch 298
Buntings 265–6, 275–7
Bustard Quail, Indian 120
Button Quail 120

Canaries 292–4
Canvasback 95
Cardinals 267, 272–5
Cariama, Crested 128
Chaffinch 293
Chat, Robin 238
Chloropsis 232–4
Cissa, Hunting 357
Cockatiel 158
Cockatoos 153–7
Cock-of-the-rock 233–4
Colies 200
Colin, Virginian 102
Combassou, Senegal 334
Conures 183–5
Cordon Bleu 306–7
Coucal, Senegal 192
Cowbirds 291
Crakes 125–6
Cranes 121–3

Crocodile Bird 130
Cuckoo-dove, Barred 137
Cut-throats 333

Dacnis, Blue 286
Dhyal Bird 238
Dickcissel 274
Dollarbird 208
Doves 136–44
Ducks 85–6, 88, 93, 95, 99
Dunnock 237

Egret, Cattle 70
Eider, European 96
Euphonia, Orange-crowned 281

Falconets 100
Finches 267–72, 294, 297, 299, 301, 309, 314, 316–20, 322–5, 329, 331–3
Finch-lark, Chestnut-backed 227
Firefinches 303–5
Flamingos 71–3
Flowerpeckers 256–7
Flower-piercer, Masked 288
Flycatchers 252–5
Fodies 341
Francolin, Black 103
Fruitsuckers 232, 234

Galah 154
Gallinules 126
Garganey 94
Geese 78–81, 83–4, 94
Goldeneyes 97–8
Grackles 350, 352
Grassfinches 321
Grassquits 270
Greenbul, Stripe-cheeked 231
Greenfinches 295
Grenadiers 308
Grosbeaks 298–9
Guineafowl 118–19

Hawfinches 298–9

Hemipode, Yellow-legged 120
Heron, Buff-backed 70
Honeycreepers 287–8
Hoopoe 209
Hornbills 196, 198

Ibis, Scarlet 71
Indigo Birds 334

Jacanas 129–30
Java Rice Bird 332
Jays 355–6
Jay-thrushes 244–5
Junglefowl 109–10

Kakarikis 167
Kingfishers 204–5
Kookaburra 206

Lapwing, Crowned 132
Leafbirds 232, 234
Leiothrix, Red-billed 246
Linnet 297
Lories 148–9, 152–3
Lorikeets 150–1
Lovebirds 172–5

Macaws 179–82
Magpies 356–8
Manakins 224–5
Mannikins 327–8, 330–2
Marshbird, Yellow-headed 291
Merganser, Hooded 98
Mesia, Silver-eared 245
Minivets 228–9
Minla, Blue-winged 247
Moho 250
Motmot, Blue-crowned 207
Mousebirds 200
Munias 329–31
Mynahs 350–2

Nene 80
Nightingale, Pekin 246
Niltavas 254

[367]

Index of Common Names

Nonpareil, Pin-tailed 322
Nuns 330–1

Oriole, Black-naped 353
Owls 193–5

Paddy Bird 332
Parrakeets 162–9, 177–8, 185, 187
Parrotbill, Grey-headed 249
Parrotlets 186
Parrots 158–60, 162, 166, 170–2, 176, 187–9
Partridges 102–3, 106–7
Peacock Pheasants 115
Peafowl 116–17
Pheasants 108, 110–14
Pigeons 136, 140, 144–7
Pintails 87, 90, 93
Pionus, Blue-headed 187
Pittas 225–7
Plantain-eater, Grey 190
Plovers 130–2
Pochard, Red-crested 95
Pytilias 299–301

Quail 101–2, 104–6
Queleas 340
Quetzal, Resplendent 202

Rail, White-breasted 125
Redpolls 296
Redstart, White-capped 240

Reedhen, King 126
Rheas 67–8
Robins 238, 246
Rollers 208
Rosellas 164
Roulroul 106
Rubythroat, Siberian 237
Ruficauda 319

Sandgrouse 135
Seedcracker, Black-bellied 302
Seedeaters 269, 292, 294
Seedsnipe, Least 133
Seriemas 128
Serin, Yellow-rumped 292
Shamas 239
Shelduck 84
Shovelers 94
Sibias 247
Silverbills 326
Siskin 296
Siva, Blue-winged 247
Sparrow-lark, Chestnut-backed 227
Sparrows 237, 317, 332–3, 337
Spice Bird 329
Spiderhunter, Grey-breasted 261
Starlings 345–50
Sugarbirds 286–8
Sunbirds 257–61
Swamphen, Purple 126
Swans 75–7

Tanagers 278–86, 289
Teals 87, 90–4
Tern, Inca 133
Thrushes 238, 241–5
Tinamou, Chilean 69
Toucanets 217–19
Toucans 219–20
Touracos 190–1
Tragopan, Satyr 107
Treepies 358
Trumpeters 123
Twinspots 302–4

Warbler, Oriole 250
Waxbills 301, 305–13
Waxwings 235
Weavers 303, 337–42
Whistling Ducks 74–5
White-eyes 262–4
Whydahs 334–6, 343–4
Wigeon 88–9
Wood Partridges 106
Woodpecker, White 221
Wood Rails 124
Wrens 251
Wren-warbler, Superb 251

Yellowhammer 265
Yuhina, Black-chinned 248

Zosterops 262–4